Evolution, Systematics, and Fossil History of the Hamamelidae

Proceedings of an International Symposium held at University of Reading, UK, March 1988

The Systematics Association
Special Volume No. 40A

Evolution, Systematics, and Fossil History of the Hamamelidae

VOLUME 1
Introduction and 'Lower' Hamamelidae

Edited by

PETER R. CRANE

Department of Geology
Field Museum of Natural History
Chicago

and

STEPHEN BLACKMORE

Department of Botany
British Museum (Natural History)
London

Published for the SYSTEMATICS ASSOCIATION by the
CLARENDON PRESS · OXFORD
1989

Oxford University Press, Walton Street, Oxford OX2 6DP
Oxford New York Toronto
Delhi Bombay Calcutta Madras Karachi
Petaling Jaya Singapore Hong Kong Tokyo
Nairobi Dar es Salaam Cape Town
Melbourne Auckland
and associated companies in
Berlin Ibadan

Oxford is a trade mark of Oxford University Press

Published in the United States
by Oxford University Press, New York

British Library Cataloguing in Publication Data
Evolution, systematics, and fossil history of the hamamelidae.
Vol. 1: Introduction and 'lower' Hamamelidae.
Vol. 2: 'Higher' Hamamelidae.
1. Hamamelidales
I. Crane, Peter R. II. Blackmore, Stephen III. Series
583'394
ISBN 0–19–857711–7
ISBN 0–19–857726–5

Library of Congress Cataloging in Publication Data
Evolution, systematics, and fossil history of the Hamamelidae/edited by Peter R. Crane and Stephen Blackmore.
p. cm.—(The Systematics Association special volume; no. 40)
"Proceedings of an international symposium held at University of Reading, UK, March 1988."
Includes indexes.
Contents: v. 1. Introduction and 'lower' Hamamelidae—v. 2. 'Higher' Hamamelidae.
1. Dicotyledons—Classification—Congresses. 2. Dicotyledons—Evolution—Congresses. 3. Dicotyledons, Fossil—Congresses. I. Crane, Peter R. II. Blackmore, Stephen. III. Systematics Association. IV. Title: Hamamelidae. V. Series.
QK495.A12E96 1989
583'.394—dc20 89–16344 CIP
ISBN (v. 1) 0–19–857711–7
ISBN (v. 2) 0–19–857726–5

Typeset by Cotswold Typesetting Limited, Gloucester
Printed in Great Britain
by Bookcraft (Bath) Ltd
Midsomer Norton, Avon

Preface

The Hamamelidae have long been at the centre of controversies concerning the origin and evolution of angiosperms. Although it is now clear that this group is not among the most primitive flowering plants, comparative studies of extant taxa, and detailed analyses of the palaeobotanical evidence, have shown that several hamamelid families occupy critical positions in the phylogeny and evolution of the dicotyledons. Some Hamamelidae exhibit putatively primitive characters seen among the Magnoliidae, others have been implicated in the origin of the Rosidae, while some families have been allied to various clades of 'higher' dicotyledons. The Hamamelidae as a whole have been treated as monophyletic, paraphyletic and polyphyletic: even the spelling of Hamamelid(id)ae has generated controversy! The aim of the symposium held at the University of Reading, UK (22–25 March 1988) was to address some of the many outstanding phylogenetic and evolutionary questions centred on the Hamamelidae and in the process to attempt to clarify some of the major features of dicotyledonous evolution.

Although two other symposia in the past 15 years have focused on the Hamamelidae we felt that rapid progress in three areas required a renewed synthetic effort. Recent work has enormously increased both the quality and quantity of comparative data available for extant Hamamelidae, outstanding efforts in palaeobotany have demonstrated that many of the constituent families have extensive and informative fossil histories, and explicit techniques of phylogenetic analysis have considerably sharpened our understanding of several critical phylogenetic issues. One of the major objectives of the symposium was to present and synthesize recent progress in these three areas, and in particular to integrate data from extant and fossil plants.

The published symposium includes almost all of the presentations from Reading, together with several short review papers that were included to improve our coverage of the group. The scope of the symposium has required the publication of two volumes, but fortunately the necessary division is not completely arbitrary. Volume 1 provides an introduction to the Hamamelidae (Chapters 1, 2), a consideration of its phylogenetic position with respect to other dicotyledons (Chapters 3, 4), and syntheses of recent data on leaf architecture, sieve element plastids, karyology and Cretaceous fossil history (Chapters 5–8). Volume 1 also includes papers on relationships within the Trochodendrales and

Hamamelidales *sensu lato* (Chapter 9), the phylogenetic position of the Myrothamnaceae (Chapter 10), and the morphology and systematics of the Hamamelidaceae (Chapters 11–13), Buxaceae (Chapter 14), Didymelaceae (Chapter 15) and Daphniphyllaceae (Chapter 16). Volume 2 includes treatments of the 'higher' Hamamelidae (Fagales, Juglandales, Myricales, Urticales). Several contributors (see Chapters 2, 4, 5 in Volume 1, Chapters 3, 6, 11 in Volume 2) interpret these 'higher' hamamelids as more closely related to certain Rosidae and Dilleniidae than to 'lower' hamamelid taxa (Trochodendrales and Hamamelidales *sensu lato*).

The papers in both volumes adopt a range of approaches and methods of analysis, but all make important contributions to increasing and systematizing our knowledge of the Hamamelidae. We are grateful to all of the contributors for their cooperation in preparing the volumes and especially their tolerance of our editorial demands. This collection of papers contains new data and new syntheses and we hope that these will prove a stimulus to future research on the Hamamelidae and their significance in the broader context of dicotyledonous evolution.

The conference was held under the auspices of the Systematics Association, and we are also pleased to acknowledge additional financial support from the Royal Society, the British Council, British Petroleum plc and Shell UK Exploration and Production. We are also indebted to the Vice Chancellor and School of Plant Sciences, University of Reading for their hospitality at the Whiteknights Campus. We thank Professor W. G. Chaloner, Professor A. Cronquist, Mr L. L. Forman, Professor J. B. Harborne, Professor D. M. Moore, and Professor R. F. Thorne for chairing the sessions, and Professor K. Kubitzki for leading the final discussion. We are especially grateful to Professor W. G. Chaloner, Professor H. G. Dickinson, Professor J. B. Harborne, Professor V. H. Heywood, and Dr S. L. Jury for their enthusiastic support, and to R. Hampshire, C. E. Jarvis, R. W. Scotland, P. J. Stafford, and D. A. Sutton for their help with many aspects of the Symposium.

Chicago and London P. R. C.
November 1988 S. B.

Contents

Contents

VOLUME 2: 'HIGHER' HAMAMELIDAE

Contributors

H.-DIETMAR BEHNKE
Zellenlehre, Universität Heidelberg, Im Neuenheimer Feld 230, D-6900 Heidelberg, Federal Republic of Germany.

A. LINN BOGLE
Botany and Plant Pathology Department, University of New Hampshire, Durham, NH 03824, USA.

PETER R. CRANE
Department of Geology, Field Museum of Natural History, Roosevelt Road at Lake Shore Drive, Chicago, IL 60605, USA.

WILLIAM C. DICKISON
Department of Biology, University of North Carolina, Chapel Hill, NC 27599, USA.

MICHAEL J. DONOGHUE
Department of Ecology and Evolutionary Biology, University of Arizona, Tucson, AZ 85721, USA.

JAMES A. DOYLE
Department of Botany, University of California, Davis, CA 95616, USA.

F. EHRENDORFER
Institut für Botanik und Botanischer Garten der Universität Wien, Rennweg 14, A-1030 Wien, Austria.

PETER K. ENDRESS
Botanischer Garten und Institut für Systematische Botanik der Universität Zürich, Zollikerstrasse 107, CH-8008, Zürich, Switzerland.

DAVID K. FERGUSON
Department of Botany, University of Antwerp (RUCA), Groenenborgerlaan 171, B-2020 Antwerp, Belgium.

ELSE MARIE FRIIS
Department of Palaeobotany, Swedish Museum of Natural History, S-104-05, Stockholm, Sweden.

LARRY A. HUFFORD*
Department of Botany, University of Kansas, Lawrence, KS 66045, USA.

CHARLES E. JARVIS
Department of Botany, British Museum (Natural History), Cromwell Road, London, SW7 5BD, UK.

WILFRIED MORAWETZ
Institut für Botanik und Botanischer Garten der Universität Wien, Rennweg 14, A-1030 Wien, Austria.

M. R. A. SAMUEL
Institut für Botanik und Botanischer Garten der Universität Wien, Rennweg 14, A-1030 Wien, Austria.

DAVID A. SUTTON
Department of Botany, British Museum (Natural History), Cromwell Road, London, SW7 5BD, UK.

ROBERT F. THORNE
Rancho Santa Ana Botanic Garden, Claremont, CA 91711, USA.

JACK A. WOLFE
Branch of Paleontology and Stratigraphy, United States Geological Survey, Box 25046, M.S. 919, Denver Federal Center, Denver, CO 80225, USA.

*Present address: *Department of Ecology and Evolutionary Biology, University of Arizona, Tucson, AZ 85721, USA.*

1. The phylogenetic position of the Hamamelidae

F. EHRENDORFER

Institut für Botanik und Botanischer Garten der Universität Wien, A-1030 Wien, Austria

Abstract

Hamamelidae are regarded as ancient and partly relictual survivors from a broad 'transitional field' between the ancestors of Magnoliidae and Rosidae/Dilleniidae. Data from several fields of comparative analysis support the notion that Hamamelidae represent a coherent, but paraphyletic group linked to both Magnoliidae and Rosidae/Dilleniidae. Simple hermaphroditic flowers with undifferentiated perianth, two or more whorls of stamens, pollenkitt, and a chori- to coenocarpous ovary with numerous ovules are a basic, symplesiomorphic character syndrome among more or less entomophilous, lower Hamamelidae of open (sub)tropical habitats. The origin of unisexual flowers, relative increase of male flowers, inactivation of pollenkitt, loss of perianth, and a coenocarpous ovary with one or few ovules are apomorphic features, particularly characteristic among temperate (or xeromorphic), anemophilous higher Hamamelidae of open xeric or temperate habitats. The origin of petals from stamens resulting in dialypetaly, the multiplication of stamens (secondary polyandry) and/or the development of various structures for nectar production among Rosidae/Dilleniidae is interpreted as a second phase of adaptive radiation towards specialized zoophily, from the lower Hamamelidae.

Introduction

Growing interest in the Hamamelid(id)ae during the last two decades is reflected in a series of international symposia, beginning with that at the

Evolution, Systematics, and Fossil History of the Hamamelidae, Volume 1: Introduction and 'Lower' Hamamelidae (eds P. R. Crane and S. Blackmore), Systematics Association Special Volume No. 40A, pp. 1–7. Clarendon Press, Oxford, 1989.

University of Minnesota in 1972 ('What happened to the Amentiferae?', *Brittonia* **25,** 316–405, 1973), to the one at Colorado State University in 1984 ('Phylogeny of the Hamamelidae', *Annals of the Missouri Botanic Garden* **73,** 225–441, 1986) and to the meeting from which the present volume is derived. That the Hamamelidae are of crucial importance for problems of angiosperm evolution as a whole, is also obvious from relevant recent symposia (e.g. Ehrendorfer and Dahlgren 1983; Dilcher 1984); and one particularly positive aspect of this interest is increasing cooperation of 'neobotanists' and paleobotanists on various major issues of seed plant evolution (cf. Ehrendorfer 1989; this symposium).

Stern (1973) has commented in detail on the historical development of the 'amentiferous concept', but has missed the important point that R. Wettstein (1903–8) was the first to use 'Hamamelidales' (incl. Hamamelidaceae, Platanaceae, Cercidiphyllaceae, Eupteleaceae, Eucommiaceae, and Leitneriaceae), and to interpret them as one of the phylogenetic links between 'Monochlamydeae' and 'Dialypetaleae'. Since this time the group has been the subject of much discussion, and in considering present opinions on the phylogenetic position of the Hamamelidae in the sense of Cronquist (1981) and Takhtajan (1987), I will briefly comment on the following questions: 'Are the Hamamelidae mono-, para-, or polyphyletic?' 'Are the Hamamelidae a coherent group?' 'Is there a primitive symplesiomorphic floral character syndrome in lower Hamamelidae, and what is its ecological significance?' 'Does the floral diversification among Hamamelidae and linked Rosidae/Dilleniidae signify adaptive radiation?', and finally, 'What is the phylogenetic position of the Hamamelidae?'

Review

Among current proposals for the classification of the dicotyledons, the placement of the orders comprising the Hamamelidae at the beginning (implying direct relationship with gymnosperms, cf. Melchior 1964), has been generally abandoned. However, there are still unresolved conflicts about their taxonomic treatment which have led to two opposing interpretations: first, that the Hamamelidae are a monophyletic (or at least paraphyletic) group of common descent, and thus acceptable as a 'natural' taxon; or, second, that the Hamamelidae are clearly polyphyletic and represent an 'artificial' taxonomic assemblage to be assigned to the different groups of Rosidae/Dilleniidae to which they are most closely related. Proponents of the first interpretation are Takhtajan (1987: the latest version of his system) and Cronquist (1981, 1986), who derive Hamamelidae from Magnoliidae and interpret them phylogenetically as 'dead ends'. There is only limited divergence between these two authors in the taxonomic concept of Hamamelidae (e.g. in respect to inclusion or exclusion of Urticales, Leitneriales, Buxales, etc.), and both exclude, for example, the Euphorbiales. Strong advocates of the second interpretation include Thorne (1973, 1983) and Dahlgren (1983). The former maintains a main core of Hamamelidiflorae (incl. Trochodendrinae, Eucommiineae, Fagales, Casuarinales, etc.), but links Buxineae with Pittosporales in Rosiflorae, Juglandineae and Myricineae with Rutales in Rutiflorae, and Urticales together with Euphorbiales in Malviflorae. Dahlgren's classification of 1983 was similar, but his last taxonomic arrangement (made available by his wife, G. Dahlgren, personal communication) has a much more inclusive superorder Rosiflorae (with Trochodendrales, Cercidiphyllales, Hamamelidales, Balanopales, Fagales, Juglandales, Myricales, Casuarinales, Buxales, etc.), and transfers Leitneriaceae to Rutales (Rutiflorae), and Eucommiales to Corniflorae, whereas Urticales and Euphorbiales are maintained in Malviflorae.

Discussion

It is obvious that there are several reasons for these diverging interpretations and conflicting classifications of Hamamelidae. First, the orders of Hamamelidae mostly represent very ancient lines, often documented from the Cretaceous (Muller 1984), and with long independent evolutionary histories that are commonly affected by extinction. This explains their heterogeneous, fractionated, and often isolated appearance as taxonomic groups.

Second, with the exception of obviously advanced characters that are mostly concerned with reproduction and linked to anemophily, there are hardly any apomorphous characters which would allow a clear circumscription of the Hamamelidae. Instead, one finds a large number of plesiomorphic features primarily woody habit, primitive or little advanced wood anatomy, chloranthoid or cunonioid leaf teeth, S-type sieve-tube plastids, mostly accumulation of phenolic compounds (tannins, protoanthocyanidins, elagic acid, myricetin, etc.), general formation of pollenkitt during early pollen development (Hesse 1978), basically tricolpate to triporate pollen grains, etc. (cf. Mølgard 1985; Giannasi 1986; Zavada and Dilcher 1986; and other symposium literature cited above). Also the available karyological features (cf. Ehrendorfer 1976; Morawetz and Rainer 1987, and this symposium) are not really disparate for the subclass, and serological analyses rather tend to link the divergent orders (e.g. Juglandales and Fagales). All these parameters (partly already exposed to numerical and cladistic analyses) signal nothing that would clearly contradict a loose mono- or paraphyletic relationship: that is, relative phylogenetic coherence among Hamamelidae. Thus, the Cronquist–Takhtajan interpretation of the subclass is supported.

Third, there is growing conviction that Hamamelidae are not only linked to Magnoliidae (e.g. via Trochodendrales; cf. Endress 1986), but that the many similarities between Hamamelidae and several other members of the Rosidae/Dilleniidae complex also reflect 'real' and more or less direct phylogenetic ties. This more clearly favours the Thorne–Dahlgren interpretation of the Hamamelidae.

The apparent unity of the Hamamelidae, and the apparent connection of the group with both the Magnoliidae and the Rosidae/Dilleniidae can only be maintained at the same time if the hypothesis of the Hamamelidae as phylogenetic 'dead ends', and the hypothesis of their polyphyletic derivation from several independent groups of Rosidae/Dilleniidae, are both abandoned. A way out of this seeming dilemma was discussed more than ten years ago (Ehrendorfer 1977, 1983*a*,*b*), and was later convincingly elaborated (Walker and Walker 1984), but has not gained widespread recognition. These contributions, together with the ideas presented above, suggest that the Hamamelidae can be regarded as ancient and partly relictual survivors from a broad 'transitional field' between the ancestors of Magnoliidae (from which they have ultimately originated) and the Rosidae/Dilleniidae complex (into which the Hamamelidae ancestors have continued to evolve).

Floral evolution

Can we reconstruct a primitive flower model for basic Hamamelidae which could account for their postulated transient position between Magnoliidae and Rosidae/Dilleniidae? If one compares for example, *Trochodendron*, *Tetracentron*, lower Hamamelidaceae and Fagaceae, Buxaceae, *Rhoiptelea*, Ulmaceae, etc. approximately the following plesiomorphic floral character syndrome emerges (see Fig. 1.1: II). Flowers hermaphrodite, small and simple, cyclic, mostly 5-, 4- or 3-merous, with undifferentiated perianth, primarily apetalous, polyandrous or mostly with two(one) whorls of stamina, pollen more or less sticky (pollenkitt), carpels free to fused, many to five, three, or two, each still with several ovules. There is some fossil evidence for the occurrence of such flower types among possible ancestors of Hamamelidae in the Upper Cretaceous (Friis 1984). Such basic Hamamelidae would still have been entomophilous or perhaps with tendencies towards anemophily. It appears likely that they had rather small and anemochorous fruits (or seeds) and grew in marginal to dense forests or in open pioneer

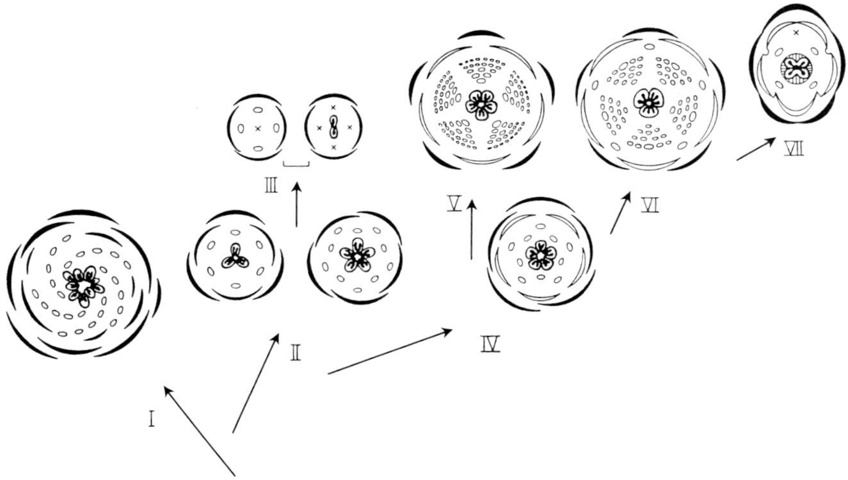

Fig. 1.1. Proposed relationships between several major dicotyledonous angiosperm groups and their flower types. I, Basic *Magnoliidae* (polymerous perigone, primary polyandry); II, Basic *Hamamelididae* (and some basic *Rosidae–Dilleniidae*) (oligomerous, hermaphrodite); III, Advanced *Hamamelididae* (strongly oligomerous, unisexual); IV, Basic *Rosidae– Dilleniidae* (calyx and corolla, androecium 1–2 cyclic); V, Some *Dilleniidae* (secondary polyandry: centrifugal); VI, Some *Rosidae* (secondary polyandry: centripetal), VII, Advanced *Asteridae* (synsepaly, sympetaly, dorsiventrality, oligomery, axial nectary).

woodlands or humid (sub)tropical and warm temperate areas (Tiffney 1984, 1986).

It is generally accepted that most of the Hamamelidae have developed highly specialized inflorescences and flowers during their long evolution, often as an adaptation to secondary anemophily (Fig. 1.1: III). The typical apomorphic character syndrome linked to these evolutionary lines is well known and includes the origin of unisexual and often extremely reduced flowers, relative increase of male flowers (commonly in aments), rapid inactivation of pollenkitt, loss of perianth, reduction of stamen, carpel and ovule numbers, and carpel fusion (often linked with hypogyny). Fruits are often much larger and zoochorous, a development frequently correlated with the penetration into more dense and shady forests, both in tropical and temperate regions (Tiffney 1984, 1986). With other advanced anemophilous and anemochorous members, 'higher Hamamelidae' have greatly expanded, both towards polar, mountainous and xeric areas.

It is very remarkable that the syndrome of full-fledged zoophily developed in the Rosidae/Dilleniidae complex (Fig. 1.1: IV, V, VI) is radically different from that of the Magnoliidae: their perianth now appears as a calyx, whereas new semaphylls (i.e. true petals) have developed from staminodia, resulting in typical dialypetaly; there is often secondary polyandry and/or new mechanisms of nectar production (cf. Ehrendorfer 1977; Walker and Walker 1984). All this is very difficult to explain, without an earlier complexity phase of reduced zoophily and flower morphology (Fig. 1.1: II). This second flush of zoophily in dicot evolution evidently has been linked with the development of dense, humid, (sub)tropical angiosperm-dominated forests and an enormous diversity of animal-pollinated flowers and flower-pollinating animals.

References

Cronquist, A. (1981). *An integrated system of classification of flowering plants.* Columbia University Press, New York.

Cronquist, A. (1986). Commentary on the status of the Hamamelidae. *Annals of the Missouri Botanical Garden* **73,** 227.

Dahlgren, R. (1983). General aspects of angiosperm evolution and macrosystematics. *Nordic Journal of Botany* **3,** 119–49.

Dilcher, D. (ed.) (1984). Dr. Jan Muller symposium on angiosperm evolution. *Annals of the Missouri Botanical Garden* **71,** 346–630.

Ehrendorfer, F. (1976). Evolutionary significance of chromosomal differentiation patterns in gymnosperms and primitive angiosperms. In *Origin and early evolution of angiosperms*, (ed. C. B. Beck), pp. 220–40. Columbia University Press, New York.

Ehrendorfer, F. (1977). New ideas about the early differentiation of Angiosperms. In *Flowering plants, evolution and classification of higher categories*, (ed. K. Kubitzki). *Plant Systematics and Evolution, Supplement* **1,** 227–34.

Ehrendorfer, F. (1983*a*). Summary statement. *Nordic Journal of Botany* **3,** 151–5.

Ehrendorfer, F. (1983*b*). Spermatophyta, Samenpflanzen. In *Lehrbuch der Botanik für Hochschulen ("Strasburger")*, **32**. Aufl., pp. 758–915. G. Fischer, Stuttgart.

Ehrendorfer, F. (1989). Woody plants—evolution and distribution since the Tertiary. Proceedings of a symposium organized by Deutsche Akademie der Naturforscher Leopoldina, 9–11 October 1986, Halle/S., DDR. *Plant Systematics and Evolution* **162,** 1–329, and separate book publication, Springer Verlag, New York.

Ehrendorfer, F. and Dahlgren, R. (eds.) (1983). New evidence of relationships and modern systems of classifications of the angiosperms. Proceedings of the XIII International Botanical Congress, Sydney, Australia. *Nordic Journal of Botany* **3,** 1–155.

Endress, P. K. (1986). Floral structure, systematics, and phylogeny in Trochodendrales. *Annals of the Missouri Botanical Garden* **73,** 297–324.

Friis, E. M. (1984). Preliminary report of Upper Cretaceous angiosperm reproductive organs from Sweden and their level of organization. *Annals of the Missouri Botanical Garden* **71,** 403–18.

Giannasi, D. E. (1986). Phytochemical aspects of phylogeny in Hamamelidae. *Annals of the Missouri Botanical Garden* **73,** 417–37.

Hesse, M., (1978). Entwicklungsgeschichte und Ultrastruktur von Pollenkit und Exine bei nahe verwandten entomophilen und anemophilen Angiospermensippen: Ranunculaceae, Hamamelidaceae, Platanaceae und Fagaceae. *Plant Systematics and Evolution* **130,** 13–42.

Melchoir, H. (ed.) (1964). *A. Engler's Syllabus der Pflanzenfamilien,* Vol. 2. Borntraeger, Berlin.

Mølgard, P. (1985). Caffeic acid as a taxonomic marker in dicotyledons. *Nordic Journal of Botany* **5,** 203–13.

Morawetz, F. and Rainer, H. (1987). Die Chromosomenzahlen der Hamamelidae. *Sitzungsberichte der Österreichischen Akademie des Wissenschaften Mathematisch-Naturwissenschaftliche Klasse,* **196,** 157–72.

Muller, J. (1984). Significance of fossil pollen for angiosperm history. *Annals of the Missouri Botanical Garden* **71,** 419–43.

Stern, W. L. (1973). Development of the amentiferous concept. *Brittonia* **25,** 316–33.

Takhtajan, A. (1987). *Systema Magnoliophytorum.* Nauka, Leningrad.

Tiffney, B. H. (1984). Seed size, dispersal syndromes, and the rise of the angiosperms: evidence and hypothesis. *Annals of the Missouri Botanical Garden* **71,** 551–76.

Tiffney, B. H. (1986). Fruit and seed dispersal and the evolution of the Hamamelidae. *Annals of the Missouri Botanical Garden* **73,** 394–416.

Thorne, R. F. (1973). The "Amentiferae" or Hamamelidae as an artificial group: a summary statement. *Brittonia* **25,** 395–405.

Thorne, R. F. (1983). Proposed new realignments in the angiosperms. *Nordic Journal of Botany* **3,** 85–117.

Walker, J. W. and Walker, A. G. (1984). Ultrastructure of Lower Cretaceous angiosperm pollen and the origin and early evolution of flowering plants. *Annals of the Missouri Botanical Garden* **71,** 464–521.

Wettstein, R. (1903–8). *Handbuch der Systematischen Botanik.* Deuticke, Leipzig.

Zavada, M. S., Dilcher, D. L. (1986). Comparative pollen morphology and its relationship to phylogeny of pollen in the Hamamelidae. *Annals of the Missouri Botanical Garden* **73,** 348–81.

2. 'Hamamelididae': a commentary

ROBERT F. THORNE

Rancho Santa Ana Botanic Garden, Claremont, CA 91711, USA

Abstract

The 'subclass' Hamamelididae is a polyphyletic assemblage of relatively unrelated orders brought together because most of their members possess the anemophilous syndrome of characteristics. Table 2.1 presents an alternative classification for those taxa conventionally included in the 'Hamamelididae'. A thorough cladistic study treating the whole anemophilous syndrome as a single characteristic equal to other independent features would help to place amentiferous taxa correctly. Changes from my last published classification pertinent to this group of taxa include submergence of Hamamelidanae into Rosanae, placing Pittosporales and Fagales between Hamamelidales and Rosales. Juglandaceae are placed as Juglandales in a separate superorder Juglandanae somewhere between Rosanae and Rutanae. Urticales, quite unrelated to the previous taxa, are, along with the closely related Malvales, Euphorbiales, and Rhamnales, treated in the Malvanae. Barbeyaceae surely do not belong to the Urticales, and for lack of adequate information about them are retained as a *taxon incertae sedis*. The monotypic family Simmondsiaceae, the jojoba, considered rather closely related to the Euphorbiaceae, is included in the Euphorbiales.

Introduction

This brief paper presents an alternative classification for those taxa conventionally treated as the 'Hamamelidae'. Because I treat the angiosperms as the class Angiospermae (Annonopsida) with two

Evolution, Systematics and Fossil History of the Hamamelidae, Volume 1: Introduction and 'Lower' Hamamelidae (eds P. R. Crane and S. Blackmore), Systematics Association Special Volume No. 40A, pp. 9–16. Clarendon Press, Oxford, 1989.

subclasses, Dicotyledoneae (Annonidae) and Monocotyledoneae (Liliidae), coordinate with the various classes and subclasses of gymnosperms, I could hardly accept the Hamamelididae as an acceptable subclass even if I considered it a natural grouping. Those taxa that are related within the grouping would have to be treated perhaps as a super-superorder or sub-subclass between my subclass Annonidae and superorder Rosanae (including my former Hamamelidiflorae or Hamamelidanae). Also, in my classification the 'subclasses' of Cronquist (1981) and Takhtajan (1987) are superfluous considering the more inclusive nature of my superorders and orders. In addition I regard all of those 'subclasses' as polyphyletic. I still insist that our classification of the Angiospermae should be as phylogenetic and monophyletic as our presently limited knowledge permits.

In *Brittonia* (Thorne 1974) I published a summary statement for an earlier symposium on the Amentiferae in which I detailed my reasons for regarding the group as highly artificial and the result of much convergent evolution involving the anemophilous syndrome of characteristics. Many of the old Amentiferae are still masquerading as Hamamelididae (Cronquist 1986) with only a few of the more obviously divergent groups having been removed to their proper phylogenetic niches.

Unfortunately, quantitative approaches to the classification of the Amentiferae have treated each of the various components of the anemophilous syndrome as separate characteristics carrying equal weight with those features not involved in the syndrome. Not surprisingly this huge syndrome thus can overweigh the other characteristics and show phyletic relationships where none really exist. Clearly, such syndromes of numerous, apparently unrelated features can be linked evolutionarily under the pressure of natural selection (Stebbins 1951) much as genes borne on the same chromosomes are linked. An objective cladistic study in which the syndromes are recognized and each syndrome treated as a single characteristic equal to other independent features would be more helpful in attempting to place the amentiferous taxa correctly.

Most of the arguments I presented in the *Brittonia* paper will not be repeated here, but I shall comment briefly on changes from my earlier classification of the amentiferous taxa that have become necessary because of our expanded knowledge of those taxa, including much information presented at the Reading symposium by the participating botanists. First, Table 2.1 presents my alternative classification of the taxa discussed at Reading, including four superorders and their included orders, suborders, families, subfamilies, and tribes. Classification within families is largely that accepted by the participants at Reading, especially Endress (1986), Bogle (1972, Vol 40A, Ch. 11),

Nixon (1982, Vol. 40B, Ch. 3), Crane (Vol. 40B, Ch.6), Stone (Vol. 40B, Ch. 7), Berg (1973, 1977*b*, 1978), Manchester (Vol. 40B, Ch. 12), and Friis (Vol. 40B, Ch. 16).

Rosanae

The biggest change from my last published classification (Thorne 1983) is the combining of the Hamamelidanae with the Rosanae (I have bowed to requests to replace the traditional superorder ending '-iflorae' with '-anae' since the former logically applies only to the flowering plants). For many years I have been troubled by the apparent interrelationships of orders placed in my Hamamelidanae and Rosanae, particularly the members of my Pittosporales and Hamamelidales. Additional evidence supplied by the participants in this symposium (Dickison Vol. 40A, Ch. 4; Nixon Vol. 40B, Ch. 3) links the Fagales perhaps more closely to the rosalian Cunonineae than to the Hamamelidales. Thus, I consider it expedient to combine the Hamamelidanae with the larger Rosanae, placing the Pittosporales and Fagales between the Hamamelidales and Rosales. I have discussed the classification of the Pittosporales elsewhere (Thorne 1975, 1976, 1977, 1981, 1983) so need say no more here about that fascinating, largely Old World group.

Juglandanae

Although I can see no natural relationship between the Juglandales and the Fagales and still think the Juglandales have a basic relationship to the Rutales (Thorne 1974, 1981), I have been much troubled by the reported (Petersen and Fairbrothers 1979) serological links between the Juglandaceae and Fagales. Until we know more about the true relationships of the Juglandales I think it best to treat the group as a separate superorder Juglandanae somewhere between the Rosanae and Rutanae. The placement of the Pistachieae and the 'Julianiales' in the rutalian Anacardiaceae (Young 1976, 1981; Petersen and Fairbrothers 1983) and of the Leitneriaceae in the Rutales (Petersen and Fairbrothers 1983) is now rather generally accepted.

Malvanae

The preceding taxa can with some reservations be accepted as having some relationship with each other. The Urticales, however, appear to

Table 2.1. Alternative classification of the members of 'Hamamelididae' discussed at the Reading symposium

- Rosanae (incl. Hamamelidanae)
  - Hamamelidales
    - Trochodendrineae
      - Trochodendraceae
        - Trochodendroideae
        - Tetracentroideae
      - Cercidiphyllaceae
      - Eupteleaceae
      - Eucommiaceae
    - Hamamelidineae
      - Hamamelidaceae
        - Disanthoideae
        - Hamamelidoideae
          - Hamamelideae
          - Eustigmateae
          - Corylopsideae
          - Distylieae
          - Fothergilleae
        - Rhodoleioideae
        - Exbucklandioideae (Symingtonioideae)
        - Liquidambaroideae
      - Platanaceae
  - Casuarinales
      - Casuarinaceae
  - Pittosporales
    - Buxineae
      - Buxaceae
        - Buxoideae
          - Buxeae
          - Sarcococceae
          - Pachysandreae
        - Styloceratoideae
      - Didymeleaceae
      - Daphniphyllaceae
      - Balanopaceae
    - (Pittosporineae)
    - Brunineae
      - (Bruniaceae, etc.)
      - Myrothamnaceae
  - Fagales
      - Fagaceae
        - Castaneoideae
        - Fagoideae
          - Fageae
          - Trigonobalaneae
          - Querceae
      - Nothofagaceae
      - Betulaceae

Table 2.1.—*continued*

- Betuleae
- Coryleae
- (Rosales, incl. Rosineae, Saxifragineae, and Cunoniineae but excluding Fabineae and Chrysobalanaceae)

- Juglandanae
  - Juglandales
    - Juglandineae
      - Rhoipteleaceae
      - Juglandaceae
        - Platycaryeae
        - Engelhardieae
        - Juglandeae
        - Hicoreae
    - Myricineae
      - Myricaceae (incl. *Canacomyrica*)

- Rutanae
  - Rutales
    - Rutineae
      - (Anacardiaceae, etc.)
      - Leitneriaceae
    - (Coriariineae, Sapindineae, and Fabineae)

- Malvanae
  - (Malvales)
  - Urticales
    - Ulmaceae
      - Ulmoideae
      - Celtidoideae
    - Moraceae
      - Moreae
      - Dorstenieae
      - Castilleae
      - Ficeae
    - Cecropiaceae
    - Urticaceae
      - Urticeae
      - Elatostemeae
      - Boehmerieae
      - Forsskaoleae
    - Cannabaceae
  - (Rhamnales, incl. Elaeagnaceae but excl. Vitaceae)
  - Euphorbiales
    - (Euphorbiaceae, etc.)
    - Simmondsiaceae

Taxon incertae sedis
- Barbeyaceae

me to be completely unrelated to the other amentiferous groups. Presumably the fruit and other superficial resemblances of *Eucommia ulmoides* Oliv. with the Urticales along with general possession of the anemophilous syndrome by members of the order have led to their placement in the 'Hamamelididae'. I have discussed elsewhere their rather obvious relationship to the Malvales and Euphorbiales (Thorne 1974, 1976). Bessey (1915) even included the Ulmaceae, Moraceae, and Urticaceae in his Malvales. I consider them, however, a very natural, well-defined group worthy of their own ordinal ranking. It would appear that urticalian experts, such as Berg (1977*a*), agree with the malvanian relationships of the Urticales.

Among some of the more salient similarities between the Urticales and Malvales are the often alternate, palmately-veined, oblique, broad, stipulate leaves; mucilage cells and receptacles and clustered crystals in stems and leaves; extruded nucleoli in the sieve-tube elements; alternating fibrous and non-fibrous layers in the phloem; often large and bent, folded, or straight embryos in the seeds; and $x = 10$ or 14 as basic chromosome numbers. Shared with the Euphorbiales in addition are the mostly unisexual, apetalous flowers; laticiferous elements and canals in the stem; similar petiolar structure; and unusual peptide alkaloids (also in Malvales and Rhamnales) (Mears 1973). The monotypic family Barbeyaceae of northeastern Africa and adjacent Arabia (Dickison and Sweitzer 1970) seem quite out of place in the Urticales. Since I can suggest no reliable placement for the family, I prefer at present to retain it as a *taxon incertae sedis*.

Simmondsiaceae

The monotypic family Simmondsiaceae, long misplaced in the Buxaceae, has been of special interest to us in California because the jojoba is an important new crop plant, possesses unique characteristics, and is often conspicuous in our southwestern desert plant communities. Because of serological, biochemical, leaf architectural, floral morphological, geographical, and other considerations, I now feel rather confident that the Simmondsiaceae are rather closely related to the Euphorbiaceae (Kakkar 1973; Scogin 1980; Thorne 1981, 1983, 1985).

References

Berg, C. C. (1973). Some remarks on the classification and differentiation of Moraceae. *Mededeelingen van het Botanisch Museum en Herbarium van de Rijks Universiteit te Utrecht* **386,** 1–10.

Berg, C. C. (1977*a*). Urticales, their differentiation and systematic position. *Plant Systematics and Evolution, Supplement* **1,** 349–74.
Berg, C. C. (1977*b*). Revisions of African Moraceae (excluding *Dorstenia, Ficus, Musanga,* and *Myrianthus*). *Bulletin du Jardin Botanique National de Belgique* **47,** 267–407.
Berg, C. C. (1978). Cecropiaceae, a new family of the Urticales. *Taxon* **27,** 39–44.
Bessey, C. E. (1915). The phylogenetic taxonomy of flowering plants. *Annals of the Missouri Botanical Garden* **2,** 109–64.
Bogle, A. L. (1970). Floral morphology and vascular anatomy of the Hamamelidaceae: the apetalous genera of Hamamelidoideae. *Journal of the Arnold Arboretum* **51,** 310–66.
Cronquist, A. (1981). *An integrated system of classification of flowering plants.* Columbia University Press, New York.
Cronquist, A. (1986). Commentary on the status of Hamamelidae. *Annals of the Missouri Botanical Garden* **73,** 227.
Dickison, W. C. and Sweitzer, E. M. (1970). The morphology and relationships of *Barbeya oleoides. American Journal of Botany* **57,** 468–76.
Endress, P. K. (1986). Floral structure, systematics, and phylogeny in Trochodendrales. *Annals of the Missouri Botanical Garden* **73,** 297–324.
Kakkar, L. (1973). Leaf vasculature in *Simmondsia californica. Indian Science Congress Association, Chandigarh, Proceedings, 60th Session* (Abstract) **3,** 293–4.
Mears, J. A. (1973). Chemical constituents and systematics of Amentiferae. *Brittonia* **25,** 385–94.
Nixon, K. C. (1982). In support of recognition of the family Nothofagaceae Kuprianova. *Publications of the Botanical Society of America, Miscellaneous Series* (Abstract), **162,** 102.
Petersen, F. P. and Fairbrothers, D. E. (1979). A serological investigation of selected amentiferous taxa. *Systematic Botany* **4,** 230–41.
Petersen, F. P. and Fairbrothers, D. E. (1983). A serotaxonomic appraisal of *Amphipterygium* and *Leitneria*—two amentiferous taxa of Rutiflorae (Rosidae). *Systematic Botany* **8,** 134–48.
Scogin, R. (1980). Serotaxonomy of *Simmondsia chinensis* (Simmondsiaceae). *Aliso* **9,** 555–9.
Stebbins, G. L. (1951). Natural selection and the differentiation of angiosperm families. *Evolution* **5,** 299–324.
Takhtajan, A. (1987). *Systema Magnoliophytorum.* Nauka, Leningrad.
Thorne, R. F. (1974). The "Amentiferae" or Hamamelidae as an artificial group: a summary statement. *Brittonia* **25,** 395–405.
Thorne, R. F. (1975). Angiosperm phylogeny and geography. *Annals of the Missouri Botanical Garden* **62,** 362–7.
Thorne, R. F. (1976). A phylogenetic classification of the Angiospermae. *Evolutionary Biology* **9,** 35–106.
Thorne, R. F. (1977). Some realignments in the Angiospermae. *Plant Systematics and Evolution, Supplement* **1,** 299–319.
Thorne, R. F. (1981). Phytochemistry and angiosperm phylogeny: a summary statement. In *Phytochemistry and angiosperm phylogeny* (ed. D. A. Young and D. S. Seigler), pp. 233–95. Praeger Scientific, New York.

Thorne, R. F. (1983). Proposed new realignments in the angiosperms. *Nordic Journal of Botany* **3,** 85–117.

Thorne, R. F. (1985). Phylogenetic relationships of the monotypic family Simmondsiaceae. *Jojoba Happenings* (Abstract) **13** (2), 8.

Young, D. A. (1976). Flavonoid chemistry and phylogenetic relationships of the Julianiaceae. *Systematic Botany* **1,** 149–62.

Young, D. A. (1981). The usefulness of flavonoids in angiosperm phylogeny: some selected examples. In *Phytochemistry and angiosperm phylogeny* (ed. D. A. Young and D. S. Seigler), pp. 205–32. Praeger Scientific, New York.

3. Phylogenetic analysis of angiosperms and the relationships of Hamamelidae

MICHAEL J. DONOGHUE

Department of Ecology and Evolutionary Biology, University of Arizona, Tucson, AZ 85721, USA

JAMES A. DOYLE

Department of Botany, University of California, Davis, CA 95616, USA

Abstract

In order to explore the basal radiation of angiosperms and the phylogenetic position of 'Hamamelidae', we conducted a numerical cladistic analysis of monosulcate taxa ('Magnoliidae', monocots) and six tricolpate groups (assumed to be appropriate placeholders for 'Rosidae', 'Amentiferae', and other 'higher dicots'). Based on outgroup comparison with Bennettitales plus Gnetales and with Mesozoic seed ferns, angiosperms are best rooted either within Magnoliales or with Magnoliales forming a basal clade. The remaining angiosperms are united by columellar exine structure. Within this group there are three main clades: i) Laurales, including *Austrobaileya*, Calycanthaceae, Trimeniaceae, and Chloranthaceae; ii) Winteraceae plus Illiciales; and iii) a group united by palmate leaf venation and differentiated stamen filaments, which contains the tricolpate groups (except Illiciales) and a 'paleoherb' clade consisting of *Lactoris*, Aristolochiaceae, Piperaceae plus Saururaceae, Nymphaeaceae plus Cabombaceae, and monocots. Within the palmate clade, Trochodendrales plus Hamamelidales and Ranunculidae plus *Nelumbo* may form a clade united by tricolpate pollen and loss of ethereal oil cells, or Ranunculidae and *Nelumbo* may be linked with the paleoherbs based on anomocytic stomata and herbaceous habit. Under both arrangements the chloranthoid teeth of Chloranthaceae are not homologous with those in Hamamelidae and Ranunculidae, but there are almost equally parsimonious

Evolution, Systematics, and Fossil History of the Hamamelidae, Volume 1: Introduction and 'Lower' Hamamelidae (eds P. R. Crane and S. Blackmore), Systematics Association Special Volume No. 40A, pp. 17–45. Clarendon Press, Oxford, 1989.

trees in which they may be homologous. Studies of Early Cretaceous cordate leaves and tricolpate pollen and of living paleoherb groups could provide valuable tests of these results. In future cladistic studies of Hamamelidae and other 'higher' dicots, Ranunculidae and paleoherbs should be considered closer outgroups than Magnoliales or Chloranthaceae.

Introduction

The position of the predominantly wind-pollinated groups assigned to the 'Hamamelidae' has long been a controversial issue in angiosperm phylogeny. Although the older view that 'Amentiferae' (i.e. 'higher hamamelids') are primitive angiosperms is now generally dismissed, there is still debate on whether Hamamelidae are a natural group or independent derivatives of diverse entomophilous ancestors, and whether their reduced flowers and anemophily are 'dead end' specializations or typical of an intermediate phase in the evolution of 'higher dicots' from monosulcate 'Magnoliidae' (cf. Ehrendorfer Vol. 40A, Ch. 1).

In our view, progress in unraveling these questions depends critically on understanding phylogenetic relationships at the base of the angiosperms. Although it is generally agreed that magnoliids are most similar to the first angiosperms (a view supported by the fossil record: Doyle 1969, 1978; Wolfe *et al.* 1975; Hickey and Doyle 1977), there remain major uncertainties on relationships among magnoliids, the course of character evolution, and ecological trends. Although most workers consider Winteraceae or Magnoliaceae most primitive, some suggest that Chloranthaceae or monocots are basal (e.g. Burger 1977, 1981), and there is little agreement on exactly how higher dicots are related to magnoliids. This is due in part to uncertainty on the relationships of angiosperms as a whole, and hence the polarity of character evolution within the group. However, this situation has improved as a result of recent cladistic analyses of seed plants (Crane 1985; Doyle and Donoghue 1986), which provide much more robust conclusions on the relationships of angiosperms to other groups.

In this paper, building on our seed plant study, we present a preliminary numerical cladistic analysis designed to resolve relationships at the base of the angiosperms, and consider its bearing on the phylogenetic position of Hamamelidae. We cannot address all aspects of this question, since this would require an analysis of many hamamelid groups and all of their potential relatives, which would be impractical at the present time. Instead, we will concentrate on analyzing relationships of presumably primitive hamamelids in the context of monosulcate groups and several tricolpate groups that are thought to be basal members of other higher dicot lines. Although this analysis is preliminary and incomplete, we hope that it will serve to point the way to a more complete solution. The present description of our analysis is necessarily abbreviated; detailed justification of the inclusion of taxa and the choice and coding of characters will be presented elsewhere.

Background

One popular view of hamamelid relationships is that of Takhtajan (1966, 1980, 1987) and Cronquist (1968), who considered Hamamelidae to be a natural group derived directly from magnoliids, parallel to other major groups (rosids, dilleniids, caryophyllids, monocots). They interpreted Hamamelidae as products of an early trend toward increasing anemophily, with Trochodendrales and Hamamelidales (which have presumably primitive features such as tricolpate pollen and, in Trochodendrales, vesselless wood) occupying an intermediate position between magnoliids and Amentiferae (most of which have triporate pollen). This contrasts with the schemes of Thorne (1968, 1976) and Dahlgren (1975, 1983), in which Hamamelidales and Fagales are thought to be derived from magnoliids but are also related to groups classified by Takhtajan and Cronquist as rosids, and other hamamelid groups are derived from unrelated triaperturate ancestors (e.g. Thorne placed Juglandaceae in Rutales, Urticales in Malviflorae). All of these authors left open the critical question of which magnoliid groups are most closely related to hamamelids.

One of the first attempts to link hamamelids with specific magnoliids was that of Hickey and Wolfe (1975), based on leaf architecture. Among their major contributions was the recognition of several marginal tooth types, defined on form, venation, and glandularity. Most interesting for our purposes is the chloranthoid tooth, with two lateral veins joining the medial vein and a glandular papilla, which they noted in several tricolpate groups—Illiciales, Ranunculales, and a few 'lower' Hamamelidae (*Trochodendron, Tetracentron, Cercidiphyllum*)—and in the magnoliid family Chloranthaceae. The Chloranthaceae, which contrast with most magnoliids in having extremely simple flowers, have been considered reduced Laurales, members of the Piperales (with which they share minute flowers, spicate inflorescences, and orthotropous ovules), or transitional between the two. The presence of chloranthoid teeth led Hickey and Wolfe to associate Chloranthaceae with Illiciales, Ranunculidae, and Hamamelidae. They argued that lower hamamelids are in turn linked with Rosidae, based primarily on similarities between Early

Cretaceous palmately lobed platanoid leaves (believed to be hamamelids) and pinnately lobed and compound *Sapindopsis* leaves (believed to be primitive rosids).

Pollen morphology has been another source of recent ideas on magnoliid–hamamelid relationships. Walker and Walker (1984) inferred a general trend within magnoliids from large, boat-shaped monosulcate pollen with a complete tectum and granular infratectal structure, to small, round monosulcate pollen with an open (semitectate), reticulate tectum, and columellar infratectal structure, which in turn gave rise to tricolpate pollen. Looking in more detail, they proposed two links between monosulcate and tricolpate groups. One was between the basically tricolpate Illiciales (*Illicium*, Schisandraceae) and Winteraceae (with monoulcerate tetrads), both of which have a coarse reticulum of peaked muri supported by short columellae. They also included Aristolochiaceae in this group, based on the reticulate monosulcate pollen of *Saruma*. The other proposed link was between Hamamelidae and Chloranthaceae plus Piperales (which they dissociated from Laurales), based on the spinulose muri and sculptured apertures of Chloranthaceae and some lower hamamelids (*Cercidiphyllum*, *Euptelea*). They noted that these conclusions are generally consistent with the chloranthoid tooth character, although their scheme would imply that chloranthoid teeth arose independently in Illiciales. They also argued that lower hamamelids are transitional to rosids and dilleniids, implying that the showy flowers of the latter groups are secondarily elaborated from reduced, apetalous flowers.

These indications of a closer connection between Hamamelidae and Rosidae were acknowledged by Cronquist (1981), who also noted supporting chemical evidence (cf. Kubitzki 1973). They are also reflected in Ehrendorfer's concept (Vol. 40A, Ch. 1) that Hamamelidae form a 'transitional field' between Magnoliidae and higher dicots. However, we caution against viewing the Hamamelidae as a single unit in this regard; most of the evidence for a transitional status concerns tricolpate Trochodendrales and Hamamelidales and does not necessarily apply to triporate Juglandales, Betulales, and Urticales, which may instead be nested within rosids or dilleniids.

Recently new evidence has appeared on the magnoliid group most widely discussed in connection with Hamamelidae, the Chloranthaceae. In a study emphasizing reproductive characters, Endress (1987) found that Chloranthaceae share the greatest number of similarities with Trimeniaceae, then with *Amborella*, then with other Laurales, and the fewest with Piperales. These observations suggest that Chloranthaceae are nested within Laurales and cast doubt on a direct relationship between Chloranthaceae and either Hamamelidae or Piperales, since this would require reversal of several lauralean advances (e.g. opposite leaves, unilacunar nodes, uniovulate carpels). However, this argument does not rule out the homology of chloranthoid teeth or the possibility that hamamelids and Laurales are sister groups: Endress (1987) noted chloranthoid teeth in Trimeniaceae, and we have seen them in *Amborella* (based on a photograph kindly provided by J. A. Wolfe).

We believe that the best way to resolve such diverse opinions on relationships and apparent conflicts among character sets is through parsimony analysis, attempting to determine which phylogenetic hypothesis best accounts for the totality of character information. There have been several previous cladistic analyses of basal relationships in angiosperms (Young 1981; Dahlgren and Bremer 1985; Lammers *et al.* 1986). However, all these studies suffer from problems concerning the inclusion of taxa, aspects of character analysis (e.g. polarity assessment, treatment of multistate characters), and computer-assisted exploration of the data (for a detailed critique of Young's analysis, see Riggins and Farris 1983). As a result, phylogenetic conclusions were limited, and the resulting cladograms show little congruence. We have endeavoured to overcome these problems in the present analysis.

Methods

Taxa

Considering the wide range of ideas on relationships of Hamamelidae and the tremendous diversity of angiosperms as a whole, a definitive analysis of hamamelid relationships would have to include a prohibitively large number of taxa. In our experience with existing algorithms, it is difficult to be confident of finding most parsimonious trees when there are more than about 25–30 taxa, especially when character conflict is common, as it is in the present data set. This problem is compounded when one wishes to explore alternative trees. Choice of taxa therefore involved compromises. Our main priority was to include all taxa needed to resolve relationships at the base of the angiosperms and between magnoliids and the most primitive hamamelids, but our desire to keep the number of taxa within manageable limits led us to collapse many large groups whose monophyly and internal relationships should be tested in future analyses.

Our primary analysis involved the 27 taxa listed in Table 3.1. These include all dicot families with monosulcate and related pollen types (other than triaperturate), or Magnoliidae in the sense of Takhtajan (1987), except for the highly specialized and questionably related *Ceratophyllum*, Rafflesianae, and Nepenthanae. However, we performed a separate cladistic analysis of Laurales (to be presented elsewhere),

Table 3.1. List of taxa and abbreviations.

Taxon	Abbreviation
Hypothetical ancestor (=ancestral states based on outgroup comparison)	ANC
Magnoliaceae	MAG
Winteraceae	WIN
Degeneria (=Degeneriaceae)	DEG
Eupomatia (=Eupomatiaceae)	EUP
Himantandraceae (=*Galbulimima*)	HIM
Annonaceae	ANN
Canellaceae	CAN
Myristicaceae	MYR
Austrobaileya (=Austrobaileyaceae)	AUS
Amborella (=Amborellaceae)	AMB
Trimeniaceae	TRI
Chloranthaceae	CHL
Calycanthaceae (including *Idiospermum*)	CAL
Monimiaceae *sensu lato*, Gomortegaceae, Hernandiaceae, Lauraceae	MON
Lactoris (=Lactoridaceae)	LAC
Saururaceae	SAU
Piperaceae	PIP
Aristolochiaceae	ARI
Nymphaeaceae (including *Barclaya*)	NYM
Cabombaceae	CAB
Nelumbo (=Nelumbonaceae)	NEL
Illicium (=Illiciaceae)	ILL
Schisandraceae	SCH
Ranunculidae (Takhtajan 1987)	RAN
Trochodendrales (=*Trochodendron, Tetracentron*)	TRO
Hamamelidales (=*Cercidiphyllum, Euptelea, Platanus*, Hamamelidaceae)	HAM
Liliopsida (=monocotyledons)	LIL

which indicated that *Hortonia*, Monimiaceae *sensu stricto*, Atherospermataceae, Siparunaceae, *Gomortega*, Hernandiaceae, and Lauraceae form a very robust clade that could be reduced to a single taxon (MON in the figures, or 'core Laurales'), united by stamen filaments, nectaries, and inaperturate pollen, and that Calycanthaceae and *Idiospermum* can also be lumped. The remaining groups are monocots (reduced to a single taxon) and six taxa with tricolpate pollen: *Illicium*, Schisandraceae, *Nelumbo*, Ranunculidae, Trochodendrales, and Hamamelidales (*Cercidiphyllum, Euptelea, Platanus,* Hamamelidaceae). We did not include any exclusively fossil groups; many Cretaceous pollen and leaf types, such as *Afropollis, Eucalyptophyllum, Vitiphyllum,* and *Proteaephyllum reniforme*, may represent extinct families of primitive angiosperms, but these are known only as isolated organs, and with so much missing character information their positions would be highly unstable. However, in several cases fossils did influence our scoring of taxa, and relationships between our results and the early angiosperm fossil record are discussed below.

At first sight, our concentration on primitive angiosperms may seem inappropriate, since primitive angiosperms alone constitute a paraphyletic group, whereas in cladistic analysis the study group should be monophyletic. However, we emphasize that we regard this study not as an analysis of primitive angiosperms alone but rather of angiosperms as a whole, with certain taxa serving as 'placeholders' for much larger clades, on the assumption that they possess the basic states of these clades. In other words, 'higher' groups are implicitly included under the assumption that they are nested within (derived from) groups in the analysis, such that they are unlikely to exert a significant effect on estimates of relationships near the base of the angiosperms. Specifically, we assume that Hamamelidales are a placeholder for most remaining dicots (Rosidae, Dilleniidae, Asteridae, and possibly Caryophyllidae); this view is supported by some character data (e.g. secondary chemistry, floral symmetry, pollen: Kubitzki 1973; Ehrendorfer Vol. 40A, Ch. 1) and preliminary cladistic analyses (Donoghue 1987). Obviously, these are rather bold assumptions, which we fully intend to test by expanding the analysis to include additional groups. Nevertheless, the taxa included in this study illustrate much of the range of diversity in relevant characters and character combinations, and we therefore expect that the picture of relationships presented here will be fairly stable to the addition of taxa.

Characters

In amassing data, we tried to extract as many potentially informative characters as possible from the valuable compendium of Cronquist (1981) and the basic literature on the groups in question. For some taxa, we found surprisingly little information on rather basic characters, such as phyllotaxy (spiral *vs.* distichous) in Piperales. We excluded some characters because there was too much missing information or too much variation within taxa to be certain about scoring. Some of these may eventually prove useful as new data accumulate and relationships within taxa are resolved (e.g. ptyxis: Cullen 1978). We excluded autapomorphies and universal characters, since these provide no information on relationships among terminal taxa.

There are 54 characters in the resulting data set (see Appendix for a list of characters and the data matrix). Forty-two of these are binary characters and 12 are multistate characters; there are a minimum of 69 character state changes (steps) in the absence of homoplasy. In the case of multistate characters, the states were treated as unordered. This is a conservative approach; more information on the ordering of states may allow better resolution in the future.

One frequently encountered problem concerned assessing the basic character states of terminal taxa when there was internal variation. In some cases we were able to score taxa based on previous cladistic work (e.g. we used our analysis of Laurales to reconstruct basal states of the core Laurales, and in several cases we referred to a preliminary cladistic analysis of Ranunculidae by Loconte and Estes, unpublished). In other cases we relied on less explicit ideas on relations within groups (e.g. in Annonaceae we assumed that taxa with monosulcate monad pollen are basal; in Aristolochiaceae, *Saruma*). In a few instances fossils were brought into the decision; for example, we scored Trochodendrales as having palmate venation, based on leaves that apparently represent primitive members of this group (Wolfe Vol. 40A, Ch. 5). When these approaches still left us uncertain, we scored taxa as unknown for the character. This was especially common in the case of monocots, where hypotheses on which groups are basal are so varied (cf. Dahlgren and Rasmussen 1983). We also scored some characters as unknown when we encountered major difficulties in the initial assessment of homology. For example, we were reluctant to score monocots for many leaf characters because of uncertainty over homology of the leaf as a whole.

Rooting was accomplished by a strict and conservative application of outgroup comparison, based on the results of our analysis of seed plants (Doyle and Donoghue 1986). According to this study (and Crane 1985), seed plants are a monophyletic group, originally of a 'seed fern' type, with coniferopsids and Mesozoic seed ferns derived from a common ancestor with platyspermic seeds and saccate pollen, like the Carboniferous seed fern *Callistophyton*. Angiosperms, Bennettitales, *Pentoxylon*, and Gnetales form a strongly supported group within the platyspermic clade, designated anthophytes because they all have flower-like strobili. Anthophytes are most parsimoniously united with *Caytonia*, based on reflexed cupules and seed structure, and in turn with other Mesozoic seed ferns. Within anthophytes, angiosperms are the sister group of the remaining groups. Thus, the first outgroup in the present analysis consisted of Bennettitales, *Pentoxylon*, and Gnetales, and the second outgroup was *Caytonia*. When one of the states of a character occurred in both outgroups, or when it occurred in one outgroup and the condition in the other was unknown, this state was assumed to be ancestral; when conditions in the two outgroups differed, the ancestral state was considered equivocal (Maddison *et al.* 1984). In this way we were able to polarize 30 characters; in effect, the resulting trees are rooted by these characters alone. The list of ancestral states, with equivocal characters scored as unknown, was entered in the analysis as a hypothetical ancestor; this procedure makes it possible to find globally most parsimonious ingroup cladograms (Maddison *et al.* 1984; Donoghue and Maddison 1986).

Analyses

All cladistic analyses were conducted using Wagner parsimony algorithms, which attempt to find the branching diagram that minimizes the total number of character state changes, allowing both forward and reverse changes. The data set was too large to make use of 'branch and bound' algorithms that guarantee finding all most parsimonious trees. In order to search for most parsimonious trees, we used PAUP (version 2.4, D. L. Swofford), run on an IBM PC AT, employing various combinations of options including local and global branch swapping, several methods of stepwise OTU addition (ADD-SEQ), and retention of multiple equally parsimonious trees (MUL-PARS). An analogous data set, in which we used the 'X-coding' method of Doyle and Donoghue (1986) to express partial ordering of multistate characters, was analysed with the METRO program in PHYLIP (version 3.0, J. Felsenstein); this was often useful in finding unanticipated, slightly less parsimonious topologies. Trees resulting from these analyses were input to MacClade (version 2.1, W. P. and D. R. Maddison) in order to calculate the parsimony of a wide variety of alternative topologies and to explore implications of topologies for character evolution; this actually led to the discovery of several additional most parsimonious trees. The cladograms shown here were generated using MacClade; where lines are shaded, this indicates the most parsimonious 'mapping' (optimization) of the states of a particular character on the cladogram. There may be several equally parsimonious ways to map character states on a tree (e.g. a state may arise twice or arise once and be reversed); in such cases (shown by horizontal hatching in MacClade), we will say that a state 'may be' homologous or 'may unite' particular taxa on the tree.

Results and discussion

Altogether, we have identified over 30 most parsimonious trees of 178 steps. This may not be an exhaustive inventory of most parsimonious trees, but all those found have the same general structure, and we suspect that any additional trees will be fairly similar. The overall consistency index is 0.39, which is somewhat below average for this number of taxa (Sanderson and Donoghue unpublished). This agrees with the traditional impression that homoplasy is widespread in primitive angiosperms.

To provide a point of reference, we begin by discussing one most parsimonious cladogram in some detail (Fig. 3.1), summarizing relationships of major groups and character support, with emphasis on the position of hamamelids and characters of special relevance to this question. Some differences among most parsimonious trees will be mentioned during description of this tree, while others deserving more attention will be discussed separately. A consensus tree is shown in Fig. 3.2.

One of the basal branches in Fig. 3.1 consists of seven families of the

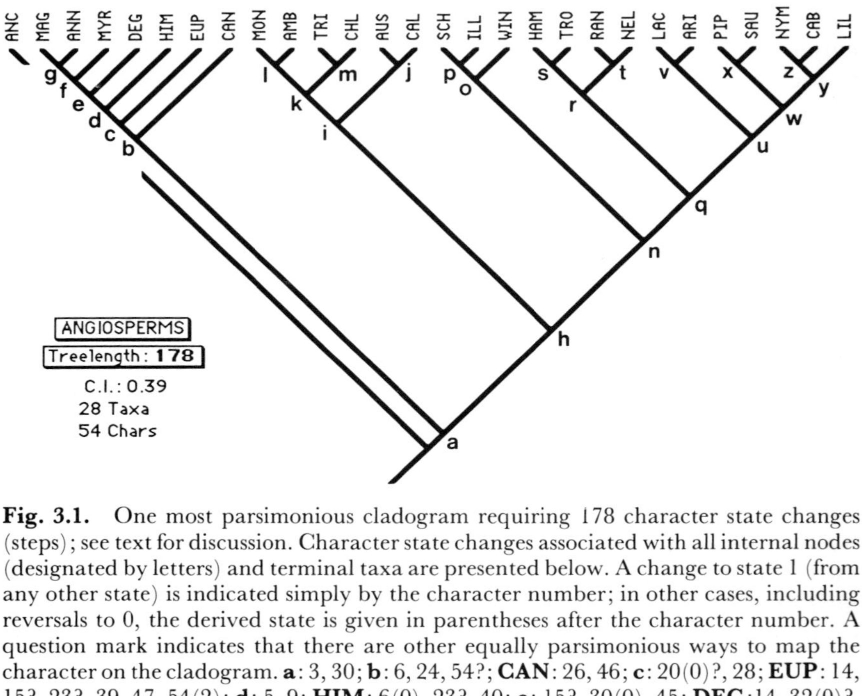

Fig. 3.1. One most parsimonious cladogram requiring 178 character state changes (steps); see text for discussion. Character state changes associated with all internal nodes (designated by letters) and terminal taxa are presented below. A change to state 1 (from any other state) is indicated simply by the character number; in other cases, including reversals to 0, the derived state is given in parentheses after the character number. A question mark indicates that there are other equally parsimonious ways to map the character on the cladogram. **a**: 3, 30; **b**: 6, 24, 54?; **CAN**: 26, 46; **c**: 20(0)?, 28; **EUP**: 14, 15?, 23?, 39, 47, 54(2); **d**: 5, 9; **HIM**: 6(0), 23?, 40; **e**: 15?, 30(0), 45; **DEG**:14, 32(0)?, 38?; **f**: 8, 28(0), 42(0); **MYR**: 6(0), 21(0), 22(2), 26, 35, 38?, 40(2), 50; **g**: 12, 32(0)?; **MAG**: 14, 52(0); **ANN**: 4, 47, 54(0); **h**: 33, 37; **i**: 13, 15(2), 35; **j**: 17?, 28, 54(2)?; **AUS**: 45; **CAL**: 4, 6, 20(0)?, 32(0), 39, 42(3), 51(2); **k**: 19, 21(0), 36, 40, 44?, 52(0)?; **l**: 7, 13(2), 33(0), 42(2), 54?; **MON**: 12?, 27; **AMB**: 3(0), 15, 17?, 37(0), 41; **m**: 38; **TRI**: 20(0)?, 46; **CHL**: 21(2), 22(3), 34, 41: **n**: 44?, 52(0)?; **o**: 8, 20(0)?, 34?, 46, 54?; **WIN**: 3(0), 29; **p**: 13(2), 19?, 31, 48; **SCH**: —; **ILL**: 40(2), 42(0)?, 44(0); **q**: 16?, 17(2), 27, 42(0)?,; **r**: 10(0), 19?, 31, 34?, 35; **s**: 21(2), 22(2); **HAM**: —; **TRO**: 3(0); **t**: 12, 20(2)?, 50; **RAN**: 16(0)?, 24, 46; **NEL**: 1, 2?, 3(2), 11, 18, 19(0)?, 32(0), 40, 42(3)?, 51(2), 53?; **u**: 20(2)?, 22, 24, 25; **v**: 4, 15?; **LAC**: 13, 22(2), 29, 37(0)?, 50; **ARI**: 34?, 39; **w**: 2, 54?; **x**: 14?, 15?, 20(3), 21(2), 22(3), 32(2), 41, 49, 51?; **PIP**: 40(2), 42(2); **SAU**: 35?; **y**: 1, 10(0), 30(0), 53?; **LIL**: 3(2)?, 14?, 37(0); **z**: 11?, 32(0), 33(0), 43, 46?, 51?; **NYM**: 3(0)?, 8, 22(0), 24(0), 25(0); **CAB**: 16(0), 18, 42(3).

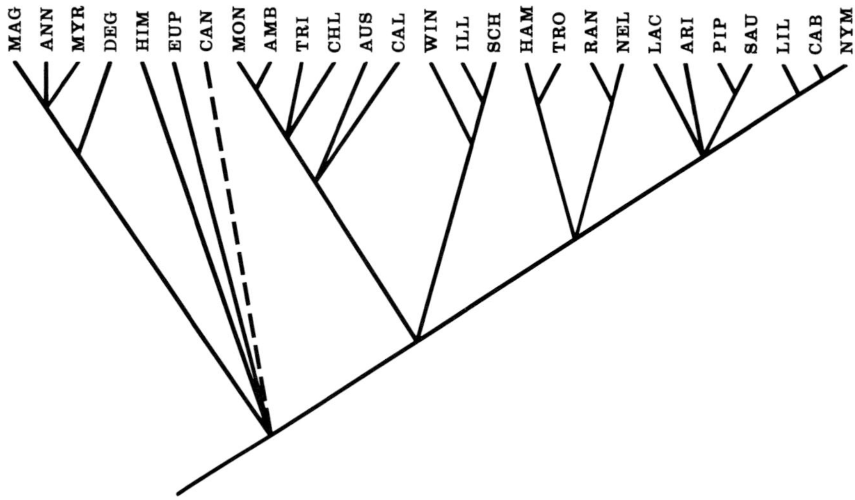

Fig. 3.2 A consensus of most parsimonious trees. This is a strict consensus tree, except that the alternative position of Canellaceae as sister group of Winteraceae and Illiciales is not indicated.

Magnoliales of Cronquist (1981), excluding Winteraceae, Austrobaileyaceae, and Lactoridaceae (which occupy near-basal positions in clades nested higher in the tree). These groups, which retain primitively granular monosulcate pollen (cf. Walker and Walker 1984), are united by trimerous perianth and PI-type sieve-tube plastids. For convenience, we will use the term 'Magnoliales' to refer to these seven taxa, even though they do not correspond exactly to Magnoliales in the sense of other authors and are not monophyletic in all trees. As discussed below, association of the basal members of this group (Canellaceae, *Eupomatia*, and Himantandraceae) is not secure. However, *Degeneria*, Myristicaceae, Annonaceae, and Magnoliaceae, or the 'core Magnoliales', form a more robust group, united by distichous phyllotaxy, boat-shaped pollen, and a sarcotesta or possibly derived aril. Himantandraceae may be associated with the core group on the basis of stratified phloem and septate pith.

The remaining angiosperms appear to form a monophyletic group, which we will term the 'columellate' clade because it is united primarily by two pollen characters, columellar infratectal structure and endexine. Exceptions, such as occurrences of granular structure in higher Laurales (associated with reduced, inaperturate pollen) and Nymphaeales, are most parsimoniously interpreted as reversals. Within this group there are three main clades, whose exact interrelationships are uncertain.

Thus, in Fig. 3.1 we show the second and third clades as united by non-multiplicative testa and non-ruminate endosperm, but it is not certain that these are synapomorphies: the testa may have become non-multiplicative in the common ancestor of columellates and then reversed in *Austrobaileya*–Calycanthaceae, and it is unclear whether ruminate endosperm is basic in angiosperms. Other characters potentially unite the first and third clades (sculptured pollen apertures, lack of a palisade exotesta), but these are also equivocal as synapomorphies. As all arrangements of the three clades are equally parsimonious, they are best regarded as forming an unresolved trichotomy, the resolution of which must await additional character data.

The first columellate clade corresponds to Laurales in a broad sense, with the notable inclusion of *Austrobaileya* and Chloranthaceae. This group is united by opposite leaves, unilacunar nodes, and sculptured pollen apertures (the spiral phyllotaxy of Lauraceae is most parsimoniously interpreted as a reversal). Within Laurales, Calycanthaceae and *Austrobaileya* are basal, and the remaining groups form a clade united by uniovulate carpels, chloranthoid teeth (apparently lost or transformed into monimioid teeth in advanced groups), and supratectal spinules. Chloranthaceae are linked with Trimeniaceae based on their unicarpellate flowers, which agrees with the view of Endress (1987); they are completely separated from Piperales where they are often placed (seven extra steps are needed to associate these two groups). *Amborella* is linked with the core Laurales on U-shaped sclereids, granular exine, and true drupes (in contrast to the one-seeded berries of other groups: Endress 1987).

The second columellate clade consists of Winteraceae and Illiciales (*Illicium*, Schisandraceae), united by coarsely reticulate, semitectate exine sculpture, as proposed by Walker and Walker (1984), plus branched sclereids and a palisade exotesta. Within this 'winteroid' clade, Illiciales are united by tricolpate pollen and unilacunar nodes.

The third columellate clade is united by palmately veined leaves and stamens with well-differentiated anthers and filaments. Two other possible synapomorphies are a shift to dehiscent fruits and the presence of stipules; we defined stipules adnate to the petiole as the derived state and scored other types (as in Trochodendrales and Hamamelidales) as unknown, because of their sporadic occurrence within terminal taxa and uncertain homology. In Fig. 3.1, this 'palmate' clade splits into two major lines, but other arrangements are equally parsimonious (see below). One line contains all the tricolpate groups except Illiciales. The other line, here called the 'paleoherbs', is characterized by anomocytic stomata, two perianth cycles, and trimery in both the perianth and the androecium (except for loss of one or both perianth cycles in *Lactoris* and Piperales and secondary multiplication of parts in Nymphaeaceae).

Arrangements within this clade are incompletely resolved. The *Lactoris*–Aristolochiaceae link is supported only by simple vessel perforations; it is not seen in all trees. In Fig. 3.1, Piperales are united with Nymphaeales (excluding *Nelumbo*) and monocots based on the loss of normal cambial activity, but in equally parsimonious trees Aristolochiaceae, *Lactoris*, and Piperales are united on distichous phyllotaxy. However, the link between Piperaceae and Saururaceae is very strongly supported (tetracytic stomata, spicate inflorescences, minute pollen, orthotropous ovules, sclerotic tegmen). Nymphaeales are united with monocots (as widely assumed) based on early replacement of the radicle by adventitious roots; another possible homology is lateral fusion of the cotyledons (Haines and Lye 1975), but we scored the monocotyledonous condition as unknown, to allow but not to assume homology.

The group of prime interest for our present purposes is the tricolpate clade, which contains the hamamelids. Besides tricolpate pollen, these taxa are united by loss of oil cells, sculptured pollen apertures, and possibly reticulate exine. Within this group, Ranunculidae are united with *Nelumbo* on the presence of benzylisoquinoline alkaloids, anomocytic stomata, nuclear endosperm, and possibly herbaceousness (for which ranunculids were scored unknown). Association of *Nelumbo* with Nymphaeales adds at least one step, despite similarities in vegetative morphology, pollen size, and seedling structure. Trochodendrales and Hamamelidales are united on spicate–racemose inflorescences and one perianth cycle. However, we emphasize that this does not mean that these two orders, or Hamamelidae as a whole, form a monophyletic group, since we assumed at the outset that Hamamelidales may be a placeholder for more derived groups. Our results are therefore consistent with the idea that lower Hamamelidae are relicts of an intermediate floral reduction phase in the evolution of Rosidae and other higher dicots, but they imply that this reduction was independent of that in Chloranthaceae and Piperales (contrary to Walker and Walker 1984).

Perhaps our most intriguing result is the inference that the first tricolpate dicots (except Illiciales) had palmate leaf venation. This was proposed for lower Hamamelidae by Hickey and Wolfe (1975), but they did not consider that the palmate venation of hamamelids might be homologous with that of herbaceous monosulcate groups, which they assumed originated independently three times. Since not all members of these groups are palmate, some explanation of our scoring may be in order. We scored Trochodendrales as palmate, although *Trochodendron* is pinnate, based on fossil relatives of *Trochodendron* with palmate venation (Wolfe Vol. 40A, Ch. 5), but we scored Hamamelidales as unknown because there are so many different conditions within them: palmate in *Cercidiphyllum*, pinnate but basally crowded in *Euptelea*, palmately lobed in *Platanus* and some Hamamelidaceae (*Liquidambar*, *Exbucklandia*) but pinnate in others. We scored ranunculids as palmate based on Menispermaceae and the assumption that the ternate organization of other groups is more readily derived from palmate than from pinnate. If our results are correct, palmate venation should not be considered a synapomorphy of those hamamelid groups with this trait; instead, it is the plesiomorphic state. Many rosids and dilleniids have pinnate venation; if these groups are nested within hamamelids, our results imply that this condition is secondary. For Rosidae, one may envision a scenario involving dissection into a platanoid leaf and then a pinnately compound *Sapindopsis* type, followed by reversion to simple leaves in some groups. Clearly, further analyses, with the addition of many more groups, are necessary to test these possibilities.

Alternative basal relationships

Some of the most interesting variations among most parsimonious trees concern the arrangement of groups at the base of the tree—in particular, different positions of the root, which in turn reflect uncertainties in polarizing characters based on outgroup comparison. In Fig. 3.1, the seven taxa of Magnoliales form a basal clade. However, in other most parsimonious trees Himantandraceae are the sister group of all other angiosperms, or Canellaceae, *Eupomatia*, and Himantandraceae are variously associated with the columellate clade. In some (Figs 3.3A, B), Canellaceae are linked with Winteraceae and Illiciales, based on presence of a palisade exotesta (cf. Corner 1976). In other words, angiosperms may be rooted within the Magnoliales, which would then be paraphyletic.

These different rooting arrangements have many different implications for early angiosperm evolution. For example, in Fig. 3.1 (with Magnoliales forming a basal clade), PI sieve-tube plastids and trimerous perianth unite the Magnoliales and originate only once or twice elsewhere, but in Fig. 3.3A (with Magnoliales paraphyletic), both features arise in a more scattered fashion. In addition, the various schemes often provide different estimates of primitive states for characters that we were unable to polarize using outgroup comparison. In Fig. 3.1, inner staminodes, stratified phloem, and septate pith arise within Magnoliales, but in Fig. 3.3B, these traits are or may be primitive in angiosperms. For these characters, Fig. 3.1 agrees better with conventional views. However, this tree implies that the boat-shaped pollen of core Magnoliales is a reversal, not primitive as widely assumed; it is primitive only if Magnoliales are paraphyletic. These results suggest that revision of conventional assumptions on one or more of these characters is in order.

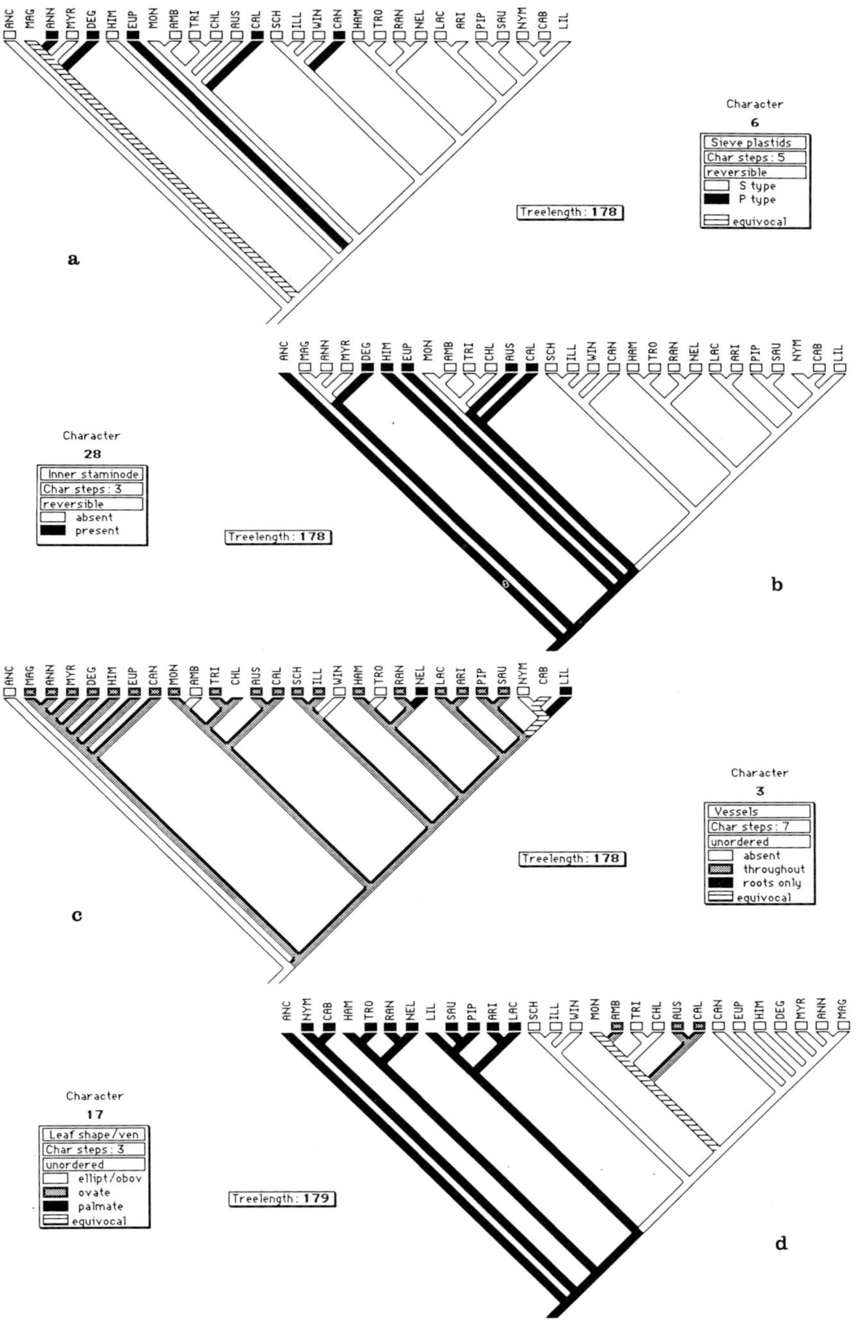

Fig. 3.3 Alternative arrangements at the base of the angiosperms, with implications for character evolution shown by shading. Boxes below names of taxa indicate character states (blanks = state scored as unknown). (A) 'Magnoliales' paraphyletic, showing the sieve-tube plastid character; (B) same tree, showing the inner staminode character; (C) Magnoliales monophyletic (Fig. 3.1), showing the vessel character; (D) Nymphaeales basal, showing the leaf shape/venation character.

All of our most parsimonious trees conflict with traditional views on vessel evolution (Fig. 3.3C). We coded vessels as primitively absent, based on outgroup comparison: although we assumed that Gnetales are related to angiosperms, vessels are presumably not homologous in the two groups, since Gnetales are nested among vesselless Bennettitales and *Pentoxylon* (Doyle and Donoghue 1986). Nevertheless, it is most parsimonious to assume that vessels originated at the very base of the tree and were subsequently lost in Winteraceae, *Amborella*, Trochodendrales, and Nymphaeales (another supposed vesselless group, *Sarcandra* in the Chloranthaceae, actually has vessels in the roots: Carlquist 1987). This conclusion was also reached by Young (1981), but we believe that it is now on more solid ground. Many will object to this result, on the grounds that there is no situation in which loss of vessels would be advantageous for a land plant. Three additional steps (eight origins) are required to assume that evolution of vessels was irreversible; this is not an overwhelming parsimony debt, but it does imply that the functional arguments for such an assumption should be scrutinized. We suggest that it is not inconceivable on functional grounds that vessels were lost in primitive angiosperms. First, what is implied is the loss of very primitive vessels, not specialized ones. In early vesseliferous angiosperms, other cell types and organs were presumably less dependent upon the presence of vessels (i.e. vessels were less 'burdened': Riedl 1978), and their loss might therefore be less deleterious than in advanced forms. Furthermore, Carlquist (1975) has argued that vessels are less vital in environments such as wet tropical uplands, where most vesselless angiosperms occur today. He interprets these as the only habitats where vesselless forms could survive, but one might also argue that these are the habitats where loss of vessels would have the least negative effect. Finally, there may be situations where vesselless wood is superior: for example, where freezing results in air embolisms. Here it may be relevant that one vesselless group, Trochodendrales, was abundant at high latitudes during the Late Cretaceous and early Tertiary (Wolfe Vol. 40A, Ch. 5).

We also experimented with several alternative positions of the root. For example, we found trees with Nymphaeales basal, then tricolpates, then other paleoherbs (including monocots) that are only one step longer than our best trees (Fig. 3.3D). This topology implies that the first angiosperms had palmate venation, anomocytic stomata, distinct filaments, and endexine; these traits are usually considered derived, but we were unable to polarize them on strict outgroup comparison. Trees rooted among the paleoherbs, with Nymphaeales plus monocots forming a basal clade, are only slightly worse (181 steps). However, trees rooted next to Chloranthaceae (cf. Burger 1977), which would be consistent with the prevalence of chloranthoid pollen, leaves, and flowers in the Early Cretaceous (Upchurch 1984; Walker and Walker 1984; Friis *et al.* 1986), are much worse (185 steps). Because Chloranthaceae share several features with Gnetales (opposite leaves, two-trace nodes, orthotropous ovules), these trees would be improved if angiosperms were linked directly with Gnetales (cf. Crane 1985). However, this arrangement is less parsimonious than one with the angiosperms basal in the anthophytes (Doyle and Donoghue 1986).

Alternative relationships within the columellate clade

In addition to the three equally parsimonious arrangements of Laurales, winteroids, and palmates, there are several equally parsimonious arrangements within the palmate clade. Most importantly, the ranunculid clade can be dissociated from the hamamelids and united directly with the paleoherbs on the basis of herbaceous habit and anomocytic stomata (Fig. 3.4A). This topology implies that tricolpate pollen originated three times (or originated twice and was lost at the base of the paleoherbs), and that oil cells were lost below the hamamelid branch and regained in Aristolochiaceae, *Lactoris*, and Piperales. Both might be seen as reasons to consider this arrangement less plausible than Fig. 3.1.

This observation led us to ask how much worse it would be to assume that tricolpate pollen evolved only once. The most parsimonious arrangement that allows this possibility unites the hamamelid–ranunculid clade with Illiciales, within the winteroid clade (Fig. 3.4B). Here, the tricolpate groups are united by both tricolpate pollen and chloranthoid teeth, and are linked with Winteraceae on reticulate exine sculpture. This topology adds only two steps overall. However, there are morphological reasons to suspect an independent origin of tricolpate pollen in Illiciales: in *Illicium* the colpi are unusually long, and in Schisandraceae they are fused at one pole and have a peculiar median ridge (Walker 1976*a*).

Along similar lines, we explored the possibility that chloranthoid teeth are homologous in all groups in which they occur. With our shortest trees, it is most parsimonious to assume that chloranthoid teeth arose independently at least three times: within Laurales, in Illiciales, and in the tricolpate (ranunculid–hamamelid) clade (Fig. 3.4C). However, on these trees only one extra step is required to assume that teeth arose once in the common ancestor of the whole columellate clade and were lost (four times) in derived groups lacking teeth. Considering alternative topologies, we found that uniting tricolpates directly with Chloranthaceae, as suggested by Hickey and Wolfe (1975), adds at least four steps, due to reversals of lauralean advances. Because chloranthoid teeth also occur in Trimeniaceae and *Amborella*, they can be homologous

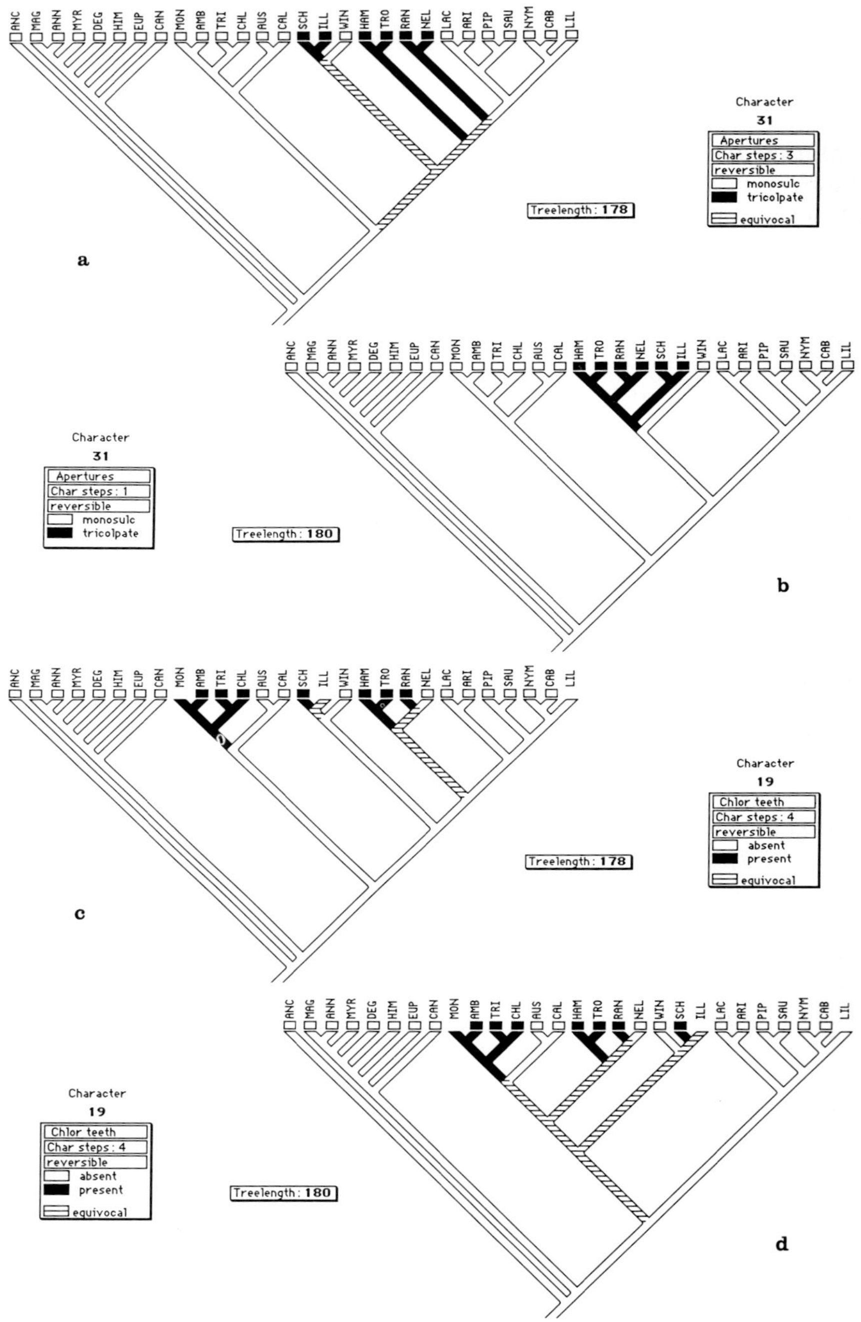

Fig. 3.4 Alternative relationships within the 'columellate' clade, with implications for character evolution (cf. Fig. 3.3). (A) Ranunculidae linked with paleoherbs, showing the pollen aperture character; (B) tricolpate clade (hamamelids and ranunculids) linked with Illiciales, showing the pollen aperture character; (C) cladogram in Fig. 3.1, showing the chloranthoid tooth character; (D) winteroids, tricolpates, and Laurales forming a clade, showing the chloranthoid tooth character.

in tricolpates and Laurales if these two clades are considered sister groups (a relationship supported by sculptured pollen apertures). Some such trees are just one step longer than our best trees (179 steps), but in these teeth must arise independently in Illiciales. Trees in which winteroids, Laurales, and tricolpates all form a clade, which do permit a single origin (plus three losses) of chloranthoid teeth, are only two steps longer than our best trees (Fig. 3.4D). These observations suggest that it would be premature to rule out a single origin of chloranthoid teeth, or a relationship of the tricolpate groups to Laurales rather than to paleoherbs.

Relations to fossil evidence

Although we did not include any fossil groups in our analysis and only occasionally consulted fossils in assessing basic states within terminal taxa, our results show interesting relations to the Cretaceous angiosperm record, and we see several instances in which fossils might help resolve problems.

Probable members of all of our major angiosperm clades can be recognized in the Early Cretaceous, all relatively primitive in morphology. In the Aptian, Magnoliales appear to be represented by large, granular, monosulcate pollen (Ward *et al.* 1988); winteroids by ulcerate tetrads (Walker *et al.* 1983) and zonasulculates with similar sculpture (*Afropollis*, *'Retimonocolpites' mawhoubensis*: Doyle *et al.* 1988); Laurales similar to Chloranthaceae by *Clavatipollenites* (Walker and Walker 1984); and monocots by *Acaciaephyllum* leaves and *Liliacidites* pollen (Doyle 1973; Walker and Walker 1984). Various Aptian leaves may represent Magnoliales, winteroids, or Laurales (Wolfe *et al.* 1975; Hickey and Doyle 1977; Upchurch 1984). The tricolpate clade is first clearly represented by early Aptian pollen and mid-Albian platanoid and *Sapindopsis* leaves (Doyle *et al.* 1977; Hickey and Doyle 1977; Crane *et al.* 1986). Most interesting for our purposes, though, are Albian actinodromous leaves with cordate or peltate bases, and Aptian reniform leaves with crowded secondaries, which have been variously compared with Nymphaeaceae, *Nelumbo*, Menispermaceae, and *Cercidiphyllum* (cf. Hickey and Doyle 1977). Even if these leaves remain unassignable to modern families, our results suggest that they may belong to a single natural group, the tricolpate–paleoherb clade. Further study of their morphology (especially stomata, which are mostly laterocytic in lower Hamamelidae but anomocytic in Ranunculidae and paleoherbs), or their association with flowers or pollen, might provide character combinations that would help resolve relations among hamamelids, ranunculids, and paleoherbs.

Fossils may also provide critical evidence by clarifying basic pollen features in the tricolpate clade. In Fig. 3.1, sculptured apertures unite the four tricolpate groups but arise independently in Laurales; this character is more consistent with the almost equally parsimonious alternative relationship between tricolpates and Laurales (Fig. 3.4D). Another trait that might seem to favour lauralean affinities is supratectal spinules, seen in both Chloranthaceae and some lower Hamamelidae (Walker 1976*b*); however, this may not be basic in either Laurales or tricolpates, since it is absent in *Austrobaileya*, Calycanthaceae, Trochodendrales (Endress 1986*b*), and *Nelumbo*, and sporadic in Hamamelidales and Ranunculidae. The oldest known tricolpates, from the early Aptian of Gabon (Doyle *et al.* 1977; Doyle and Ward in preparation), suggest that both of these similarities to Laurales are misleading. These fossils have narrow colpi (usually irregularly spaced and slightly oblique in orientation) that lack pronounced sculpture, and smooth muri; both sculptured colpi and supratectal spinules appear only later in the Albian radiation of tricolpates. If these grains are related to modern tricolpates (as suggested by transitional forms in the Aptian–Albian), and if they illustrate the primitive condition in tricolpates (as the irregular colpi might imply), they suggest that the tricolpate clade was actually derived from ancestors with smooth muri and non-sculptured apertures, like most living paleoherbs, and that sculptured colpi arose within the group.

Another connection between our results and fossil evidence concerns the ecological radiation of early angiosperms. Association of the hamamelid clade with Ranunculidae and paleoherbs (as in Fig. 3.1) raises the possibility that its early members were more herbaceous than most modern Hamamelidae. This might fit sedimentological evidence that Aptian–Albian angiosperms (including tricolpate platanoids and *Sapindopsis*) tended to occupy disturbed habitats, such as stream margins (Hickey and Doyle 1977). Hickey and Doyle saw this as evidence that such ecology was primitive in angiosperms (Stebbins 1974), but it may apply principally to the palmate clade. This habit might help explain the rapid spread of tricolpates and paleoherbs in the Aptian–Albian and the fact that most modern angiosperms belong to these clades. The most important exception to the early dominance of palmates is the Chloranthaceae (judging from the abundance of *Clavatipollenites*); perhaps significantly, modern Chloranthaceae differ from most other Laurales in being weedy and semiherbaceous. The spread of tricolpates (and Chloranthaceae) from tropical habitats (Brenner 1976; Doyle *et al.* 1982) might help explain the trend for reversion to wind pollination within these groups (cf. Walker and Walker 1984).

Conclusions

Despite the preliminary nature of this study, some relationships at the base of the angiosperms already seem rather well supported (Fig. 3.2), and we can identify hypotheses and characters that deserve more attention in the future. Thus it is uncertain, based on our conservative outgroup analysis, whether Magnoliales form a monophyletic basal line or a paraphyletic grade (as implied by many conventional views). However, Winteraceae (plus Illiciales), which are often included in Magnoliales, seem well nested within a clade with basically columellar exine structure that includes the bulk of the angiosperms. Our results confirm the unity of the Laurales, including Chloranthaceae, which seem to be quite unrelated to Piperales. An unexpected but apparently robust grouping is the paleoherb clade, including not only monocots and Nymphaeales but also Piperales, *Lactoris*, and Aristolochiaceae. Most surprising is the conclusion that Hamamelidae and other tricolpate taxa (except Illiciales) may be related to the paleoherbs, based on palmate leaf venation and differentiated stamens. However, trees linking tricolpates and Laurales (though not specifically Chloranthaceae), which would permit the homology of chloranthoid teeth, or linking tricolpates, Illiciales, and Winteraceae, which would permit a single origin of tricolpate pollen, are nearly as parsimonious and should be considered viable alternatives.

Although we have not considered relationships among hamamelids, our results have definite implications for future studies of this issue, and of the relationships of 'higher dicots' in general. Since 'lower hamamelids' are very likely paraphyletic, they must be split up into component monophyletic groups to allow the possibility that other groups are nested among them, and possibly derived groups should be added to the analysis, perhaps initially in the form of reconstructed primitive rosids, dilleniids, and caryophyllids. To address the origin of 'higher hamamelids', still more rosid and dilleniid groups would have to be added. For practical reasons, these additions would require reducing the number of monosulcate groups. One approach would be to use our most parsimonious results to polarize characters in hamamelids and their presumed relatives. The most probable first outgroup of these groups is either Ranunculidae (including *Nelumbo*) or Ranunculidae plus paleoherbs; in either case ranunculids would play a key role in determining polarities. Beyond this, the paleoherbs should be consulted, especially the most plesiomorphic groups—*Lactoris*, Aristolochiaceae, and Saururaceae. Next most important are either Winteraceae–Illiciales or Laurales. It is probably unnecessary to consider Magnoliales or Chloranthaceae, since these are 'screened off' from having a substantial effect on polarity by more closely related taxa. A more conservative approach, in view of the almost equally parsimonious relationship between tricolpates and Laurales or Winteraceae, and the possibility that 'higher' groups might (contrary to our assumption) alter the best position of tricolpates, would be to retain more monosulcate groups in the analysis to allow a better evaluation of the global parsimony of ingroup cladograms (Maddison *et al.* 1984). Here our results are helpful in indicating how the number of taxa can be reduced by condensing especially robust clades, such as core Magnoliales, 'higher' Laurales, Illiciales, Piperales, and Nymphaeales.

Acknowledgements

We thank Michael Sanderson for assistance in assembling and analysing the data matrix, and William Burger, Sherwin Carlquist, Peter Endress, Shirley Tucker, Gary Upchurch, James Walker, and Jack Wolfe for sharing their knowledge of characters. We are also grateful to Peter Crane and Stephen Blackmore for inviting us to participate in the symposium.

Appendix

Characters and character states

Polarized characters are indicated by (P), unpolarized by (U). All multistate characters are treated as unordered. Where not otherwise specified, data are from Cronquist (1981), supplemented as follows: anatomical characters (2–11, 13–14): Metcalfe and Chalk (1957, 1979), Metcalfe (1987); pollen characters (29–37): Erdtman (1952), Walker (1976*a,b*), Walker and Walker (1981, 1983) for Myristicaceae, Endress and Honegger (1980) for *Austrobaileya*; seed characters (43–49): Corner (1976), Endress (1980) for *Austrobaileya*, Endress and Sampson (1983) for Trimeniaceae, Endress (1987) for Chloranthaceae.

1. (P) 0 = radicle persistent; 1 = replaced by adventitious roots. Nymphaeales, *Nelumbo*: Corner (1976), Dahlgren *et al.* (1985); Piperales: Holm (1926), Vogel (1980).

2. (P) 0 = woody; 1 = herbaceous. Interfascicular cambium not producing normal vascular tissue scored as unknown (Esau 1953; Balfour 1958).

3. (P) 0 = vessels absent; 1 = present throughout; 2 = in roots only. Chloranthaceae: Carlquist (1987).

4. (P) 0 = perforations (or end-wall pitting) scalariform; 1 = simple. Taxa in which the most primitive condition is mixed scored as unknown.

5. (U) 0 = secondary phloem not stratified; 1 = stratified.

6. (P) 0 = sieve-tube plastids S type; 1 = PI type. Behnke (1981), Behnke and Barthlott (1983). PII type (monocots) scored as unknown to allow derivation from either state.

7. (U) 0 = U-shaped sclereids in pericycle absent; 1 = present.

8. (U) 0 = branched sclereids in leaves absent; 1 = present.

9. (U) 0 = pith without sclerenchymatous septa; 1 = with septa. Winteraceae: Metcalfe (1987) describes only scattered, sometimes grouped sclereids.

10. (U) 0 = oil cells absent; 1 = present.

11. (P) 0 = laticifers absent; 1 = present.

12. (P) 0 = benzylisoquinoline alkaloids absent; 1 = present. Hegnauer (1964–1973), Rezende *et al.* (1975), Gershenzon and Mabry (1983), Dahlgren and Bremer (1985). Taxa in which these alkaloids are generally lacking but have been reported are scored as unknown.

13. (U) 0 = nodes tri- or multilacunar; 1 = unilacunar, one-trace or with arc of several traces; 2 = unilacunar, two-trace.

14. (U) 0 = nodes uni- or trilacunar; 1 = multilacunar. Piperales: Holm (1926), Balfour (1958), S. Tucker (personal communication); Nymphaeaceae: Weidlich (1976); Cabombaceae: Moseley *et al.* (1984).

15. (P) 0 = phyllotaxy helical; 1 = distichous on at least some branches; 2 = opposite. Beck *et al.* (1982), Halle *et al.* (1978); *Lactoris*: S. Carlquist (personal communication); Saururaceae: S. Tucker (personal communication); Nymphaeaceae: Weidlich (1976).

16. (P) 0 = adnate/axillary stipules absent; 1 = present. Other stipule types scored as unknown (see text).

17. (U) 0 = leaves elliptical or obovate, secondary veins at constant angle or lower angle at base; 1 = ovate, basal secondaries crowded, at higher angle; 2 = actino- or acrodromous (palmate). Hickey and Wolfe (1975); JAD, personal observations.

18. (P) 0 = leaf base not peltate; 1 = peltate.

19. (P) 0 = chloranthoid teeth absent; 1 = present. See text. *Illicium* scored as unknown because teeth are present in *I. anisatum* but not in other species (Hickey and Wolfe, 1975).

20. (U) 0 = stomata mostly paracytic (> 90% of those on leaf); 1 = laterocytic or variable; 2 = anomocytic; 3 = tetracytic. Upchurch (1984; personal communication).

21. (U) 0 = cymose inflorescences; 1 = solitary flowers (occasionally 2–3); 2 = spikes or racemes. Trochodendrales, Hamamelidales: Endress (1986*b*, 1987).

22. (U) 0 = perianth with more than two whorls (spiral); 1 = two whorls; 2 = one whorl; 3 = absent. Himantandraceae, *Eupomatia* scored as unknown because of uncertain homology (Endress 1977); Trochodendrales: Endress (1986*b*).

23. (P) 0 = calyptra absent; 1 = present. Magnoliaceae, Winteraceae scored as unknown because of possible but uncertain homology (Endress 1977).

24. (P) 0 = perianth not trimerous; 1 = at least calyx trimerous. Endress (1986*a*), Kubitzki (1987). Taxa with some pentamerous members assumed to be basically trimerous.

25. (P) 0 = stamens not in threes; 1 = in multiples of three.

26. (P) 0 = stamens free; 1 = connate into a tube. Chloranthaceae: based on Albian flower of Friis *et al.* (1986); Schisandraceae: spiral mass questionably homologous with tube.

27. (U) 0 = stamens laminar; 1 = well-differentiated filament. Intermediate and excessively variable conditions scored as unknown. Aristolochiaceae based on *Saruma* (Dahlgren *et al.* 1985).

28. (U) 0 = inner staminodes absent; 1 = present.

29. (P) 0 = pollen shed as single grains; 1 = permanent tetrads.
30. (P) 0 = pollen boat-shaped; 1 = globose.
31. (P) 0 = monosulcate or derived inaperturate, zonate, etc.; 1 = tricolpate.
32. (P) 0 = pollen size large (>50 um); 1 = medium (ancestral state); 2 = small (<20 um).
33. (P) 0 = infratectal structure granular (including atectate); 1 = columellar.
34. (P) 0 = tectum continuous or finely perforate (foveolate); 1 = semitectate–reticulate.
35. (P) 0 = aperture membrane not conspicuously sculptured; 1 = sculptured.
36. (P) 0 = supratectal spinules absent; 1 = present.
37. (P) 0 = endexine absent in extra-apertural areas; 1 = present. Winteraceae: Roland (1971).
38. (P) 0 = carpels several-many; 1 = one (rarely 2).
39. (P) 0 = hypanthium absent; 1 = present.
40. (U) 0 = ovules several; 1 = one apical; 2 = one basal.
41. (U) 0 = ovules anatropous; 1 = orthotropous. *Amborella*: Endress (1987; personal communication).
42. (U) 0 = fruit dehiscent; 1 = berry; 2 = drupe (with endocarp); 3 = dry indehiscent. Laurales, Piperales: Endress (1987).
43. (P) 0 = seed not operculate; 1 = operculate.
44. (P) 0 = testa multiplicative; 1 = non-multiplicative. Where not characterized by Corner (1976), we adopted a limit of seven cells thick.
45. (U) 0 = sarcotesta absent; 1 = present. True arils scored as unknown to allow possible homology.
46. (U) 0 = palisade exotesta absent; 1 = present.
47. (U) 0 = fibrous mesotesta absent; 1 = present.
48. (U) 0 = sclerotic mesotesta absent; 1 = present.
49. (U) 0 = sclerotic (endo)tegmen absent; 1 = present.
50. (U) 0 = endosperm cellular; 1 = nuclear.
51. (U) 0 = endosperm only; 1 = perisperm plus endosperm; 2 = neither.
52. (U) 0 = endosperm not ruminate; 1 = ruminate. Himantandraceae: Endress (personal communication).
53. (P) 0 = cotyledons free; 1 = laterally fused. Haines and Lye (1975).
54. (P) 0 = chromosome number (n) 6–8; 1 = 12–19; 2 = 10–11. Raven (1975), Ehrendorfer (1976). Higher numbers scored as unknown because of ambiguous derivation.

Data matrix

```
                1         2         3         4         5
ANC 0000?0????00??00?00???0000??00010000000??00?????????00
MAG 00101?011101011??000010??000000000000000000001000000001
WIN 00000001?10?00000000?0?000001101110010000?0101000?0001
DEG 001011001100011000000100100010000000000100010010000000101
EUP 00100100010?011000001?1?00010101000000100100001000?0102
HIM 001?100011000000000001?1?000101010000?001010?0?000?0101
ANN 00111101110100100000?00100000000?00000000?00?010000100
CAN 0010010001 0?00000001?001?1?00101?0000000010?01000?0?01
MYR 0010100111000010000002010 1?00001??1?01020000??00010101
AUS 0010000001001 02?10011000000101011010100001001 00?0?010?
AMB 000000100?0?201010110000000001010 0?100?1120100000?0001
TRI 00100000?1001020001000000000010110111 1010101000?0000
CHL 0??0?000?10?102?0011230???000101111 1?101110?0000000000
CAL 001101?001001020100010000001010010?0101003000000002?02
MON 00100?1001012020?0??0000001?010100?110?1020?0000000001
LAC 001100000010?10112002?20110?01101?00?00000001000001000?
SAU 0110?00001000111200323 0?101001021010100010010000101001
PIP 0?1??000010?01?12003230?10100102?0???002120100001?1001
ARI ??110??0010?001?2002?101101001011 1?0?01000010000000?00
NYM 1100?0010010000120021000 00??00000000100000110 1000?1011
CAB 11?0?000001??0?02102110110100000000?100003110100001 0?1
NEL 1120?0000011???12102100000100110111010010301000001 2?10
ILL 00100001010020000 0?0100000000111 1100?0020000010 1000001
SCH 0010?00101002000001010000?000111110010000101010 1000001
RAN 0?10?0?000010?002012?001?0?001111?1?10000?0?01000100?0
TRO 000000??00000?0?2011220000100111111010000001 0000?0000?
HAM 001000?000000?0??011220000100 1111?1?10000001000?0?000?
LIL 1120??0000?001???????1011010000 11?0000000?0???000?00??
```

References

Balfour, E. (1958). The development of the vascular systems in *Macropiper excelsum* Forst. II. The mature stem. *Phytomorphology* **8,** 224–33.

Beck, C. B., Schmid, R., and Rothwell, G. W. (1982). Stelar morphology and the primary vascular system of seed plants. *Botanical Review* **48,** 691–815.

Behnke, H. D. (1981). Sieve-element characters. *Nordic Journal of Botany* **1,** 381–400.

Behnke, H. D., and Barthlott, W. (1983). New evidence from the ultrastructural and micromorphological fields in angiosperm classification. *Nordic Journal of Botany* **3,** 43–66.

Brenner, G. J. (1976). Middle Cretaceous floral provinces and early migrations of angiosperms. In *Origin and early evolution of angiosperms* (ed. C. B. Beck), pp. 23–47. Columbia University Press, New York.

Burger, W. C. (1977). The Piperales and the monocots. Alternative hypotheses for the origin of monocotyledonous flowers. *Botanical Review* **43,** 345–93.

Burger, W. C. (1981). Heresy revived: the monocot theory of angiosperm origin. *Evolutionary Theory* **5,** 189–225.

Carlquist, S. (1975). *Ecological strategies of xylem evolution.* University of California Press, Berkeley.

Carlquist, S. (1987). Presence of vessels in wood of *Sarcandra* (Chloranthaceae); comments on vessel origins in angiosperms. *American Journal of Botany* **74,** 1765–71.

Corner, E. J. H. (1976). *The seeds of the dicotyledons.* Cambridge University Press.

Crane, P. R. (1985). Phylogenetic analysis of seed plants and the origin of angiosperms. *Annals of the Missouri Botanical Garden* **72,** 716–93.

Crane, P. R., Friis, E. M., and Pedersen, K. R. (1986). Lower Cretaceous angiosperm flowers: fossil evidence on early radiation of dicotyledons. *Science* **232,** 852–4.

Cronquist, A. (1968). *The evolution and classification of flowering plants.* Houghton Mifflin, Boston.

Cronquist, A. (1981). *An integrated system of classification of flowering plants.* Columbia University Press, New York.

Cullen, J. (1978). A preliminary survey of ptyxis (vernation) in the angiosperms. *Notes from the Royal Botanic Garden, Edinburgh* **37,** 161–214.

Dahlgren, R. (1975). A system of classification of the angiosperms to be used to demonstrate the distribution of characters. *Botaniska Notiser* **128,** 119–47.

Dahlgren, R. (1983). General aspects of angiosperm evolution and macrosystematics. *Nordic Journal of Botany* **3,** 119–49.

Dahlgren, R., and Bremer, K. (1985). Major clades of the angiosperms. *Cladistics* **1,** 349–68.

Dahlgren, R., and Rasmussen, F. N. (1983). Monocotyledon evolution: characters and phylogenetic estimation. *Evolutionary Biology* **16,** 255–395.

Dahlgren, R. M. T., Clifford, H. T., and Yeo, P. F. (1985). *The families of the monocotyledons. Structure, evolution, and taxonomy.* Springer-Verlag, Berlin.

Donoghue, M. J. (1987). Phylogenetic analysis and the angiosperm system. *XIV International Botanical Congress (Berlin) Abstracts*, 281.

Donoghue, M. J., and Maddison, W. P. (1986). Polarity assessment in phylogenetic systematics: a response to Meacham. *Taxon* **35,** 534–8.

Doyle, J. A. (1969). Cretaceous angiosperm pollen of the Atlantic Coastal Plain and its evolutionary significance. *Journal of the Arnold Arboretum* **50,** 1–35.

Doyle, J. A. (1973). Fossil evidence on early evolution of the monocotyledons. *Quarterly Review of Biology* **48,** 399–413.

Doyle, J. A. (1978). Origin of angiosperms. *Annual Review of Ecology and Systematics* **9,** 365–92.

Doyle, J. A., and Donoghue, M. J. (1986). Seed plant phylogeny and the origin of angiosperms: an experimental cladistic approach. *Botanical Review* **52,** 321–431.

Doyle, J. A., Biens, P., Doerenkamp, A., and Jardiné, S. (1977). Angiosperm pollen from the pre-Albian Cretaceous of Equatorial Africa. *Bulletin des Centres de Recherches Exploration-Production Elf-Aquitaine* **1,** 451–73.

Doyle, J. A., Jardiné, S., and Doerenkamp, A. (1982). *Afropollis*, a new genus of early angiosperm pollen, with notes on the Cretaceous palynostratigraphy and paleoenvironments of Northern Gondwana. *Bulletin des Centres de Recherches Exploration-Production Elf-Aquitaine* **6,** 39–117.

Doyle, J. A., Ward, J. V., and Hotton, C. L. (1988). Possible relationships of Early Cretaceous ulcerate tetrads, zonasulculates, and Winteraceae. *VII International Palynological Congress (Brisbane) Abstracts*, 42.

Ehrendorfer, F. (1976). Evolutionary significance of chromosomal differentiation patterns in gymnosperms and primitive angiosperms. In *Origin and early evolution of angiosperms* (ed. C. B. Beck), pp. 220–40. Columbia University Press, New York.

Endress, P. K. (1977). Über Blütenbau und Verwandtschaft der Eupomatiaceae und Himantandraceae (Magnoliales). *Berichte der Deutschen Botanischen Gesellschaft* **90,** 83–103.

Endress, P. K. (1980). The reproductive structures and systematic position of the Austrobaileyaceae. *Botanische Jahrbücher für Systematik* **101,** 393–433.

Endress, P. K. (1986*a*). Reproductive structures and phylogenetic significance of extant primitive angiosperms. *Plant Systematics and Evolution* **152,** 1–28.

Endress, P. K. (1986*b*). Floral structure, systematics, and phylogeny in Trochodendrales. *Annals of the Missouri Botanical Garden* **73,** 297–324.

Endress, P. K. (1987). The Chloranthaceae: reproductive structures and phylogenetic position. *Botanische Jahrbücher für Systematik* **109,** 153–226.

Endress, P. K., and Honegger, R. (1980). The pollen of Austrobaileyaceae and its phylogenetic significance. *Grana* **19,** 177–82.

Endress, P. K., and Sampson, F. B. (1983). Floral structure and relationships of the Trimeniaceae (Laurales). *Journal of the Arnold Arboretum* **64,** 447–73.

Erdtman, G. (1952). *Pollen morphology and plant taxonomy—angiosperms*. Almqvist & Wiksell, Stockholm.

Esau, K. (1953). *Plant anatomy*. Wiley, New York.

Friis, E. M., Crane, P. R., and Pedersen, K. R. (1986). Floral evidence for Cretaceous chloranthoid angiosperms. *Nature* **320,** 163–4.

Gershenzon, J., and Mabry, T. J. (1983). Secondary metabolites and the higher classification of angiosperms. *Nordic Journal of Botany* **3,** 5–34.

Haines, R. W., and Lye, K. A. (1975). Seedlings of Nymphaeaceae. *Botanical Journal of the Linnean Society* **70,** 255–65.

Halle, F., Oldeman, R. A. A., and Tomlinson, P. B. (1978). *Tropical trees and forests. An architectural analysis*. Springer-Verlag, Berlin.

Hegnauer, R. (1964, 1966, 1969, 1973). *Chemotaxonomie der Pflanzen*, Band 3–6. Birkhäuser, Basel.

Hickey, L. J., and Doyle, J. A. (1977). Early Cretaceous fossil evidence for angiosperm evolution. *Botanical Review* **43,** 1–104.

Hickey, L. J., and Wolfe, J. A. (1975). The bases of angiosperm phylogeny: vegetative morphology. *Annals of the Missouri Botanical Garden* **62,** 538–89.

Holm, T. (1926). *Saururus cernuus* L., a morphological study. *American Journal of Science*, V **12,** 162–8.

Kubitzki, K. (1973). Probleme der Grossystematik der Blütenpflanzen. *Berichte der Deutschen Botanischen Gesellschaft* **85,** 259–77.

Kubitzki, K. (1987). Origin and significance of trimerous flowers. *Taxon* **36,** 21–8.

Lammers, T. G., Stuessy, T. F., and Silva O., M. (1986). Systematic relationships of the Lactoridaceae, endemic family of the Juan Fernandez Islands, Chile. *Plant Systematics and Evolution* **152,** 243–66.

Maddison, W. P., Donoghue, M. J., and Maddison, D. R. (1984). Outgroup analysis and parsimony. *Systematic Zoology* **33,** 83–103.

Metcalfe, C. R. (1987). *Anatomy of the dicotyledons* (2nd edn), Vol. III. Clarendon Press, Oxford.

Metcalfe, C. R., and Chalk, L. (1957). *Anatomy of the dicotyledons.* Clarendon Press, Oxford.

Metcalfe, C. R., and Chalk, L. (1979). *Anatomy of the dicotyledons* (2nd edn), Vol. I. Clarendon Press, Oxford.

Moseley, M. F., Mehta, I. J., Williamson, P. S., and Kosakai, H. (1984). Morphological studies of the Nymphaeaceae (sensu lato). XIII. Contributions to the vegetative and floral structure of *Cabomba. American Journal of Botany* **71,** 902–24.

Raven, P. H. (1975). The bases of angiosperm phylogeny: cytology. *Annals of the Missouri Botanical Garden* **62,** 724–64.

Rezende, C. M. A. da M., Gottlieb, O. R., and Marx, M. C. (1975). Benzyltetrahydroisoquinoline-derived alkaloids as systematic markers. *Biochemical Systematics and Evolution* **3,** 63–70.

Riedl, R. (1978). *Order in living organisms.* Wiley, London.

Riggins, R., and Farris, J. S. (1983). Cladistics and the roots of angiosperms. *Systematic Botany* **8,** 96–101.

Roland, F. (1971). The detailed structure and ultrastructure of an acalymmate tetrad. *Grana* **11,** 41–4.

Stebbins, G. L. (1974). *Flowering plants: evolution above the species level.* Harvard University Press, Cambridge, Mass.

Takhtajan, A. L. (1966). *Sistema i filogeniya tsvetkovykh rasteniy.* Nauka, Moscow.

Takhtajan, A. L. (1980). Outline of the classification of flowering plants (Magnoliophyta). *Botanical Review* **46,** 225–359.

Takhtajan, A. L. (1987). *Sistema magnoliofitov.* Nauka, Leningrad.

Thorne, R. F. (1968). Synopsis of a putatively phylogenetic classification of the flowering plants. *Aliso* **6,** 57–66.

Thorne, R. F. (1976). A phylogenetic classification of the Angiospermae. *Evolutionary Biology* **9,** 35–106.

Upchurch, G. R. (1984). Cuticle evolution in Early Cretaceous angiosperms from the Potomac Group of Virginia and Maryland. *Annals of the Missouri Botanical Garden* **71,** 522–50.

Vogel, E. F. de (1980). *Seedlings of dicotyledons.* Centre for Agricultural Publishing and Documentation, Wageningen.

Walker, J. W. (1976*a*). Comparative pollen morphology and phylogeny of the ranalean complex. In *Origin and early evolution of angiosperms* (ed. C. B. Beck), pp. 241–99. Columbia University Press, New York.

Walker, J. W. (1976*b*). Evolutionary significance of the exine in the pollen of primitive angiosperms. In *The evolutionary significance of the exine* (ed. I. K. Ferguson and J. Muller), pp. 1112–37. Academic Press, London.

Walker, J. W., and Walker, A. G. (1981). Comparative pollen morphology of the Madagascan genera of Myristicaceae (*Mauloutchia, Brochoneura,* and *Haematodendron*). *Grana* **20,** 1–17.

Walker, J. W., and Walker, A. G. (1983). Comparative pollen morphology of the American myristicaceous genera *Otoba, Iryanthera,* and *Osteophloeum. American Journal of Botany* **70,** 315–26.

Walker, J. W., and Walker, A. G. (1984). Ultrastructure of Lower Cretaceous angiosperm pollen and the origin and early evolution of flowering plants. *Annals of the Missouri Botanical Garden* **71,** 464–521.

Walker, J. W., Brenner, G. J., and Walker, A. G. (1983). Winteraceous pollen in the Lower Cretaceous of Israel: early evidence of a magnolialean angiosperm family. *Science* **220,** 1273–5.

Ward, J. V., Doyle, J. A., and Hotton, C. L. (1988). Probable granular magnoliid angiosperm pollen from the Early Cretaceous. *American Journal of Botany* **75**(6, Part 2), 118–9.

Weidlich, W. H. (1976). The organization of the vascular system in the stems of the Nymphaeaceae. I. *Nymphaea* subgenera *Castalia* and *Hydrocallis. American Journal of Botany* **63,** 499–509.

Wolfe, J. A., Doyle, J. A., and Page, V. M. (1975). The bases of angiosperm phylogeny: paleobotany. *Annals of the Missouri Botanical Garden* **62,** 801–24.

Young, D. A. (1981). Are the angiosperms primitively vesselless? *Systematic Botany* **6,** 313–30.

4. Comparisons of primitive Rosidae and Hamamelidae

WILLIAM C. DICKISON

Department of Biology, University of North Carolina, Chapel Hill, NC 27599, USA

Abstract

Comparisons of primitive Rosidae and Hamamelidae indicate the two groups are related. Shared features include: primitive wood structure; trilacunar, three-trace nodes; stipulate leaves with often toothed margins and pinnate (craspedodromous) venation; a similar floral groundplan with transitions from apocarpous to syncarpous and superior to inferior gynoecia; trends toward both the increase and decrease in numbers of floral parts; strong tendencies for both unisexuality and floral reduction, including the frequent loss of petals; formation of capitate inflorescences; frequent shifts from entomophily to anemophily; small, reticulate, triaperturate pollen; follicular and capsular fruit with lignified endocarps; and winged, albuminous seeds. Dispersal ranges from abiotic to more derived vectors. The two groups are held apart primarily by the possession of pinnately compound leaves with pinnate venation in the rosids versus basically simple leaves with palmate venation in the hamamelids. Like various Hamamelidales, many primitive rosids possess floral specializations that are presumably associated with wind pollination, indicating that these were early adaptations in the rosalean line. It is suggested that the Rosales and Hamamelidales arose from insect-pollinated ancestral stock having comparatively small, apocarpous, nectariferous flowers with a stable, intermediate number of parts and clearly differentiated sepals and petals. Systems of classification that regard the Rosidae and Hamamelidae to constitute natural groups independently derived from the Magnoliidae apparently do not represent accurate reflections of evolutionary history.

Evolution, Systematics, and Fossil History of the Hamamelidae, Volume 1: Introduction and 'Lower' Hamamelidae (eds P. R. Crane and S. Blackmore), Systematics Association Special Volume No. 40A, pp. 47–73. Clarendon Press, Oxford, 1989.

Introduction

Opinions have varied regarding the origin and relationships of the dicotyledonous plants currently recognized as the Rosidae and Hamamelidae. As summarized by Bogle (1970), one view envisions the hamamelids as directly derived from, and closely related to, magnolialean or proto-magnolialean stock (Hallier 1903, 1912; Tippo 1938; Tang 1943; Cronquist 1968, 1981; Takhtajan 1969, 1987). According to this scenario the rosids represent a distinct phyletic line with an independent origin from basal Magnoliidae. An opposing view links the hamamelids to rosalean or proto-rosalean ancestry, with resulting close affinities existing between certain Hamamelidae (Hamamelidaceae) and extant Rosaceae, Saxifragaceae, or more especially the Cunoniaceae and near relatives (Bentham and Hooker 1865; Baillon 1871; Eichler 1878; Bessey 1915; Hutchinson 1926, 1959; Harms 1930; Endress 1967; Airy Shaw in Willis 1973).

The similarities between rosoid families, such as Cunoniaceae, and the Hamamelidae, although long recognized (see Engler 1897), have been re-emphasized in some modern systems of classification. Dahlgren (1977), for example, placed the Cunoniales near Fagaceae and Hamamelidaceae in his Hamamelidaneae. In Dahlgren's 1980 system the Trochodendrales, Hamamelidales, and Fagales are included in the superorder Rosiflorae, along with Cunoniales, Saxifragales, Rosales, and others. Doyle and Donoghue (1987) discussed the occurrence within the Hamamelidae of both triporate and tricolporate pollen types, and proposed that this distinction is perhaps indicative of the fact that the Hamamelidae are polyphyletic, with the triporate Amentiferae actually being rosids. Based upon parsimony models, Crepet and Nixon (1987) further suggested that attempts to differentiate two distinct groups within the hamamelid–rosid complex have led to unsatisfactory phylogenetic patterns, and proposed a probable monophyletic origin for the entire assemblage.

The prevailing interpretation of the Rosidae is that of a loosely knit constellation of families that cannot be easily delineated (Cronquist 1968, 1981). Distinctions between more primitive members of the assemblage and those of other subclasses are not easily made. Generally speaking, the Rosales, when narrowly circumscribed, includes families with bisexual, pentamerous flowers having numerous parts, separate or variously united carpels, free styles, and endosperm absent or sparse. The Hamamelidales, on the other hand, appear as a group of families with mostly small, reduced, often unisexual flowers, having separate or more commonly united carpels, and seeds with abundant endosperm.

A major focus of this symposium is the question of how the

Hamamelidae may be related to other dicotyledonous subclasses. In order to assess these relationships, and thus clarify the significance of the group in the radiation of non-magnoliid dicotyledons, it is necessary to better understand the major patterns of evolution in putatively closely related groups. This contribution presents a brief character analysis and comparison of selected features among the primitive taxa of both Rosidae and Hamamelidae, with a goal toward re-evaluating relationships between, and major patterns of evolution within, families of these subclasses.

The primitive Rosidae

There is wide consensus among contemporary phylogenists that the woody families Cunoniaceae, Davidsoniaceae, Brunelliaceae, and Eucryphiaceae form a coherent unit, phylogenetically positioned at the base of the large rosalean complex (Takhtajan 1980, 1987; Cronquist 1981). Smith (1985) noted that these families are considered central to a broad rosoid concept, that in its entirety is regarded as a subclass (Takhtajan 1980, 1987; Cronquist 1981) or as a series of superorders (Thorne 1976; Dahlgren 1980). Although Cuatrecasas (1970, 1985) suggested that relationships between the Brunelliaceae and Cunoniaceae are more distant than currently implied in the systematic literature, the above four families closely resemble one another and as a matter of convenience will be subsequently referred to as the primitive Rosidae.

The predominantly southern hemisphere Cunoniaceae are of particular interest as archaic rosalean types, representing a complex from which a number of other groups, including some hamamelids, may have been derived. Wood anatomy, in particular, strongly reinforces the basal position of the Cunoniaceae within the broad rosalean alliance. *Acsmithia*, *Spiraeanthemum* and the isolated saxifragalean genus *Tetracarpaea* are of major morphological significance because they possess a very unspecialized xylem structure, containing vessel elements with only scalariform perforation plates, and nearly total apocarpy. These primitive aspects are together infrequent in the Rosales and identifies them as basal taxa.

The family Cunoniaceae, although long recognized, is unusually diverse morphologically and anatomically and is, therefore, difficult to define on the basis of one or a few uniquely derived character states. The features of opposite or whorled, mostly pinnately compound leaves accompanied by typically large, paired interpetiolar stipules are uncommon in other dicotyledons and are often cited as distinguishing family characters. However, some cunoniaceous taxa possess unfused stipules and it is apparent that there is no single feature that can totally characterize the group or convincingly demonstrate that the family is monophyletic.

The Brunelliaceae, containing the single genus *Brunellia* with about 62 species, are exclusively neotropical and are considered to have close, sister group affinities with the Cunoniaceae. Strong similarities exist between the two families in vegetative morphology and in the similar floral diagram, ovular type (particularly *Ascmithia* and *Spiraeanthemum*), and fruit anatomy (Engler 1928). *Brunellia* is uniquely characterized by a linear sutural stigma and a highly specialized carpel maturation and mature fruit (Cuatrecasas 1970).

The Davidsoniaceae, a monotypic family, is represented by *Davidsonia pruriens* of northeastern Australia. The species is allied with Cunoniaceae in all modern systems of classification, but differs most notably by the alternate leaf arrangement, presence of rigid, stinging hairs, having stamens scarcely exserted, and in having exalbuminous seeds (Bange 1952; Smith 1985). The epitropous ovules of *Davidsonia* also occur in *Acsmithia, Spiraeanthemum*, and *Brunellia*.

The Eucryphiaceae are another monogeneric, southern hemisphere family characterized by opposite, simple or pinnately compound leaves and interpetiolar stipules (Dickison 1978). Usually, solitary tetramerous flowers are produced in axillary positions and each possess from 4–12 multiovulate, united carpels and numerous stamens. The capsular fruits, which open to form boat-shaped valves that are attached basally to the central column, liberate winged, albuminous seeds.

Reproductive morphology

The Hamamelidae are widely interpreted as an ancient complex of closely related plants that have undergone major structural changes connected with wind pollination (Takhtajan 1969). Despite the fact that many taxa of the group are very specialized and entirely wind pollinated, some genera show both anemophily and entomophily and still others are exclusively insect pollinated (Faegri and van der Pijl 1979; Endress 1977). Endress (1977) argued persuasively that largely unspecialized, insect-pollinated flowers are primitive in the Hamamelidales. This opinion was supported by evidence from floral morphology, floral biology, chromosome number, and pollen morphology. Such primitive flowers are bisexual, pentamerous, diplo- or haplostemonous, nectariferous, with an intermediate number of ovules per carpel, a well-developed double perianth, imbricate calyx, and they produce reticulate, tricolpate pollen. A spike or system of spikes is considered the basic inflorescence type in the Hamamelidaceae from which capitate and dichasial systems evolved.

Extensive adaptive radiation in floral biology has resulted in the formation of diverse floral and inflorescence morphologies in the Hamamelidales–Fagales group. As summarized by Endress (1977), advancement has resulted in bird-pollinated inflorescences forming large pseudanthia (*Rhodoleia*); fly-pollinated flowers with ribbon-like petals and shiny perianth surfaces (*Disanthus, Hamamelis*); and bee-pollinated plants with bottlebrush-like inflorescences containing flowers with secondary (centrifugal) polyandry, showy stamen filaments, no corolla, extremely reduced calyx, and no nectaries in response to pollen feeding (*Fothergilla*). Different inflorescence types are encountered among the large number of wind-pollinated genera. One type resembles the bottlebrush form, but are andromonoecious, with flowers having less numerous stamens and inconspicuous filaments (*Parrotia, Sycopsis*). Ball-like inflorescences with unisexual flowers are found in Liquidambaroideae and Platanaceae. *Distylium, Matudaea*, and the Myrothamnaceae have inconspicuous, polygamous, or bisexual flowers (polyandrous in *Matudaea*) without a perianth in much condensed inflorescences. Unisexual catkins with an increased number of small flowers have evolved in *Sinowilsonia* (Hamamelidaceae) and higher Amentiferae.

The reproductive morphology of the primitive rosids also shows a broad spectrum of character states (Dickison 1975*b*; Hoogland 1960, 1979). Flowers of the Cunoniaceae, for example, are unisexual or bisexual, heterocyclic, and mostly pentamerous or tetramerous. Calyx members are valvate or less commonly imbricate. Petals are present or absent. The androecium is often diplostemonous, although haplostemonous and polyandrous androecia occur. The gynoecium is three- to five-carpellate and apocarpous, three- to five-carpellate and syncarpous, or, more commonly, bicarpellate with opposite concrescent carpels that are fused in the region of the ovary. A few genera have variously inferior ovaries. Styles are distinct and often diverging. Kennedy and Prakash (1981) reaffirmed the interesting circinate arrangement of styles in young buds of *Callicoma*, a feature common throughout the Cunoniaceae. Carpels are generally pluriovulate although biovulate and uniovulate carpels occur. The bitegmic ovules are, with exceptions, anatropous and placentation is axile or pendulous. A toral, lobed disk is usually present.

Flowers are either infrequently solitary, or borne in paniculate, racemose or raceme-like inflorescences (Fig. 4.1G, L), and dense, spherical heads (Fig. 4.1J) In Brunelliaceae the inflorescence is dichasial. Inflorescences are either axillary or terminal in position, with both conditions sometimes present in different species of the same genus. Paniculate inflorescences may be few to many flowered. Simple or compound racemes or raceme-like inflorescences (pseudoracemes) occur in a number of genera and are distinguished by having pedicellate

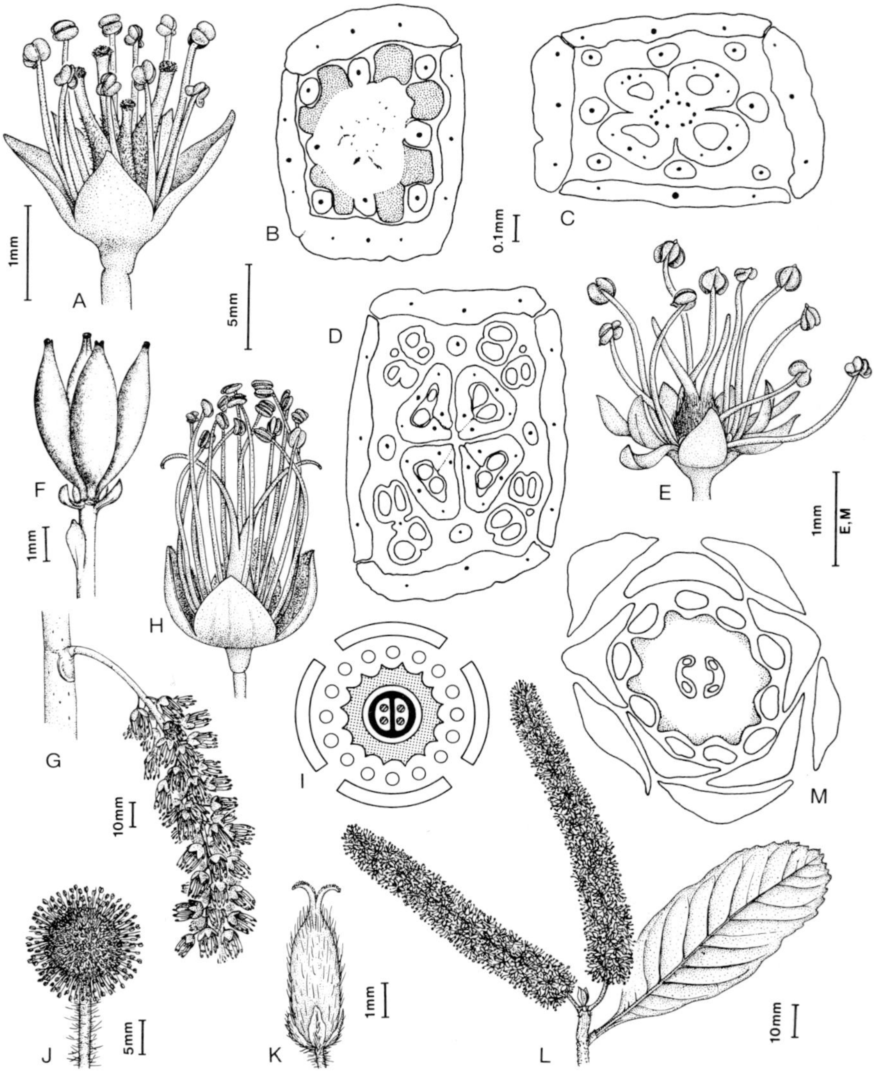

Fig. 4.1 Floral morphology of primitive Rosidae. A, *Ascmithia undulata* (Dickison 262). Flower from side. B–D, *Ascmithia reticulata* (Hoogland and Craven 10991). Diagrams of TS of flower from a level just below the gynoecium (B) to just above the level of placentation (D). Blackened areas represent vascular tissue. Stippled region in B represents disk tissue. E, *Caldcluvia paniculosa* (Dickison 296). Flower from side. F, *Tetracarpaea tasmannica* (Hoogland 11745). Immature follicles. Note persistent sepals and bract. (Used with permission of the *American Journal of Botany*). G, *Geissois pruinosa* (Dickison s.n.). Inflorescence. H, *Geissois pruinosa* (Dickison s.n.). Flower from side. I. *Geissois pruinosa* (Dickison s.n.). Floral diagram. Stippled region represents disk. J. *Callicoma serratifolia* (Hoogland 11962). Inflorescence. K, *Vesselowskya rubifolia* (Schodde 3210). Immature fruit. L, *Cunonia balansae* (Dickison 230). Inflorescence arrangement and structure. M, *Cunonia vieillardii* (McKee 2700). Flower in TS. Stippled region represents nectariferous tissue.

flowers of a single inflorescence develop simultaneously and not in spirals. In both *Weinmannia* and *Cunonia* some species with this type of inflorescence have sessile flowers, in which case the inflorescence is best called spicate.

Contrary to widely held belief, valvate, or open, aestivation may well be ancestral within the Rosidae. A valvate calyx is found in the majority of Cunoniaceae, as well as Brunelliaceae, Davidsoniaceae, and Rosaceae. Valvate sepals range between membranous and various grades of thick or thick-membranous, and are positioned so that the flattened margins of adjacent members meet (Fig. 4.1C, D).

The superior, usually four-carpellate, essentially apocarpous gynoecium containing five (or four) trace carpels is primitive in the Rosales (Sterling 1969; Eyde 1970; Dickison 1975*b*) (Figs 4.1D, 2A, B). Cuatrecasas (1970, 1985) thought apocarpy to be primitive in *Brunellia* but probably a derived state in *Spiraeanthemum*. I have previously rejected this notion (Dickison 1975*b*). Carpels of *Acsmithia* and *Spiraeanthemum* are laterally connate at the base and each carpel is weakly fused with surrounding tissue of the disk (Fig. 4.1B, C). The carpels of *Brunellia* are slightly inferior and free from one another, although each is partially united to the surrounding disk at lower levels (Eyde 1970). Among other rosalean genera the fusion of carpels to each other varies from being fully united in the ovarian region to only partially fused.

It has generally been assumed that the primitive rosalean carpel was multiovulate (Cronquist 1968). However, Sterling (1969) and Robertson and Rohrer (1987), following careful comparative studies, have reasoned that the biovulate state is more likely basic in the Rosaceae, Significantly, all species of *Spiraeanthemum* and many taxa of *Acsmithia* are also biovulate. A similar condition exists without exception in *Brunellia*. If the biovulate carpel is plesiomorphic within the primitive rosids, as it clearly seems to be, then trends exist toward both reduction to a single ovule per carpel and, independently, increases in ovular number. Ovules of *Acsmithia*, *Spiraeanthemum*, and *Brunellia* are epitropous in position, although the importance, if any, of this feature is unclear.

The apocarpous gynoecium of *Tetracarpaea* contains four or five carpels that, although free for most of their length, show fusion among their prominently stipitate bases. Individual carpels are erect, and have short, terminal, and lobed stigmas (Fig. 4.1F). The narrow, parietal placentae bear single, double, or triple rows containing numerous anatropous ovules. Ovules are bitegmic and appear to develop late in comparison with carpel size. Ovules immature at anthesis is a characteristic of many hamamelids. Carpels are vascularized by three bundles, a dorsal and two in the ventral position. The three bundles extend undivided the entire length of the carpel, terminating at the stigmatic region (Hils *et al.*, 1988).

I am in agreement with the conclusions of Sterling (1953, 1969) that the five-trace (or derivatives thereof), biovulate carpel represents an ancestral feature of rosalean plants and is an important character in relating these taxa (Fig. 4.1C).The more prevalent but derived bicarpellate, syncarpous cunoniaceous gynoecium (Figs 4.1E, 4.2D) is composed of pluriovulate, three-trace carpels. Higher numbers of carpels have evolved in *Eucryphia* and other Rosaceae.

An overall general similarity between the bicarpellate gynoecia of some Cunoniaceae and that of some Hamamelidaceae has been noted for some time. The gynoecium of the Hamamelidaceae is consistently syncarpous, generally bicarpellate, biloculate, with a pair of free and usually spreading styles. Ovary position ranges from superior to inferior. Pendulous ovules are borne on axile placentas. Ovular number has apparently both increased and decreased from an intermediate number of between three and six. In the entomophilous Hamamelidaceae the stigma surface is terminal, whereas in anemophilous genera the stigma extends along the entire ventral side of the style. A trend toward fusion of ventral bundles of adjacent carpels occurs as is also the case in some bicarpellate Cunoniaceae. Bogle (1986) interpreted the bicarpellate gynoecium of the Liquidambaroideae to be the most primitive within the family. The carpels are apocarpous above with a synovarious, half-inferior ovary. Each carpel is vascularized by a dorsal and two ventral bundles with the numerous ovules supplied by branches originating from the ventral strands. However, the observation by Bogle (1970) that carpels of subfamily Hamamelidoideae are five-trace structures is particularly noteworthy in view of the multitrace pattern in primitive rosids. As in the rosids, this vascular condition is correlated with a low ovular number. The relationship of such multitrace carpels with the classical three-trace megasporophyll with numerous ovules remains unresolved.

The Fagaceae have a three- to six-carpellate gynoecium. Ovaries range from inferior, half-inferior, to superior, with constantly two ovules per carpel. Carpels resemble other wind pollinated hamamelids in having styles with long stigmas along the upper surface. A satisfactory

---

Fig. 4.2 Floral morphology of primitive Rosidae (SEM). A, *Brunellia comocladifolia* subsp. *domingensis* (Dickison 305). Five-carpellate gynoecium (×11). B, *Acsmithia densiflora* (Dickison 242). Apocarpous gynoecium (×19). C, *Schizomeria ovata* (Dickison 297). Oblique view of flower. Note forked petal apices (×9). D, *Cunonia pterophylla* (Dickison 238). Oblique view of flower (×13). E, *Caldcluvia paniculosa* (Dickison 296). Anther from side showing longitudinal dehiscence and extended connective (Ae) (×29). F, *Ceratopetalum apetalum* (Hoogland 5177). Flower showing obdiplostemonous androecium, prominent disk, and half-inferior ovary (×7). G, *Gillbeea papuana* (Hoogland and Pullen 5873). Forked petal showing apical glands (arrows) (×30).

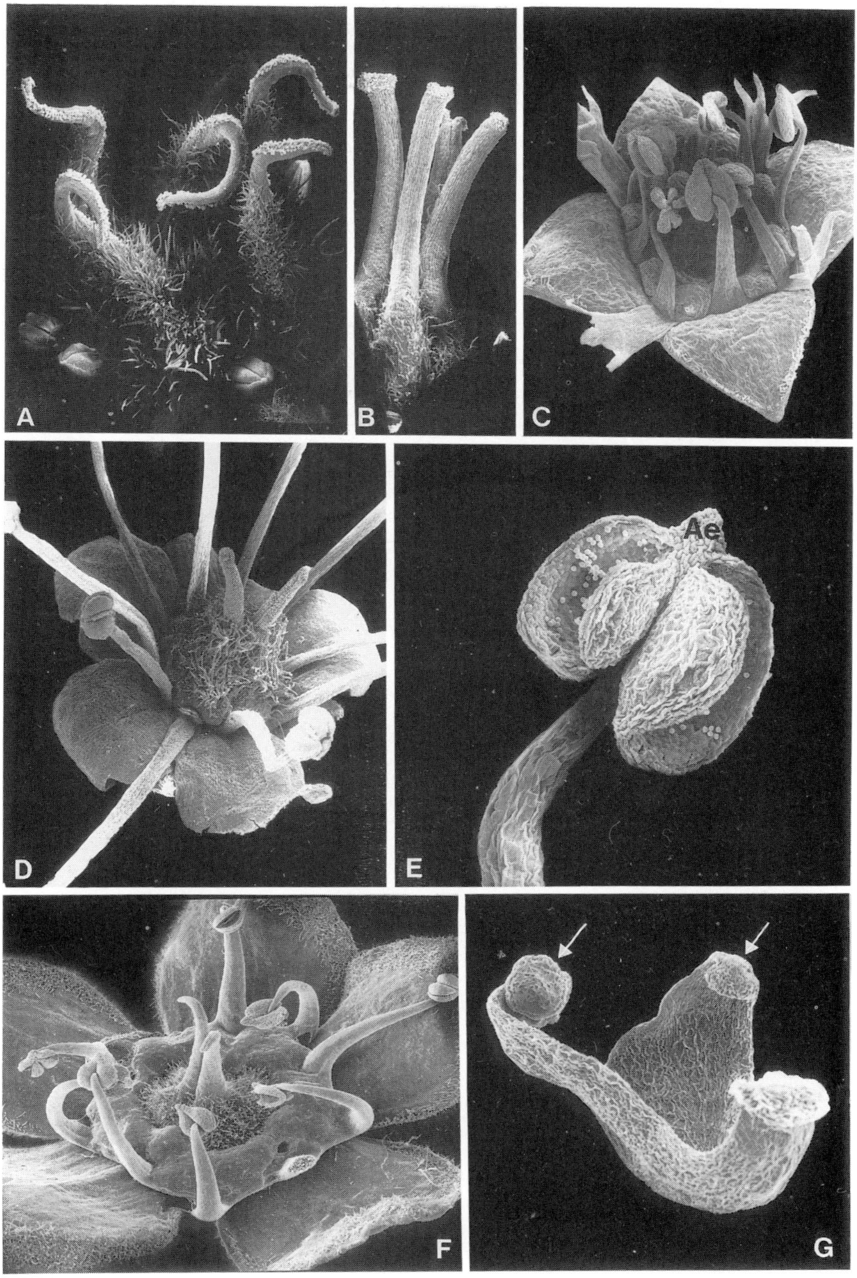
A
B
C
D
E
Ae
F
G

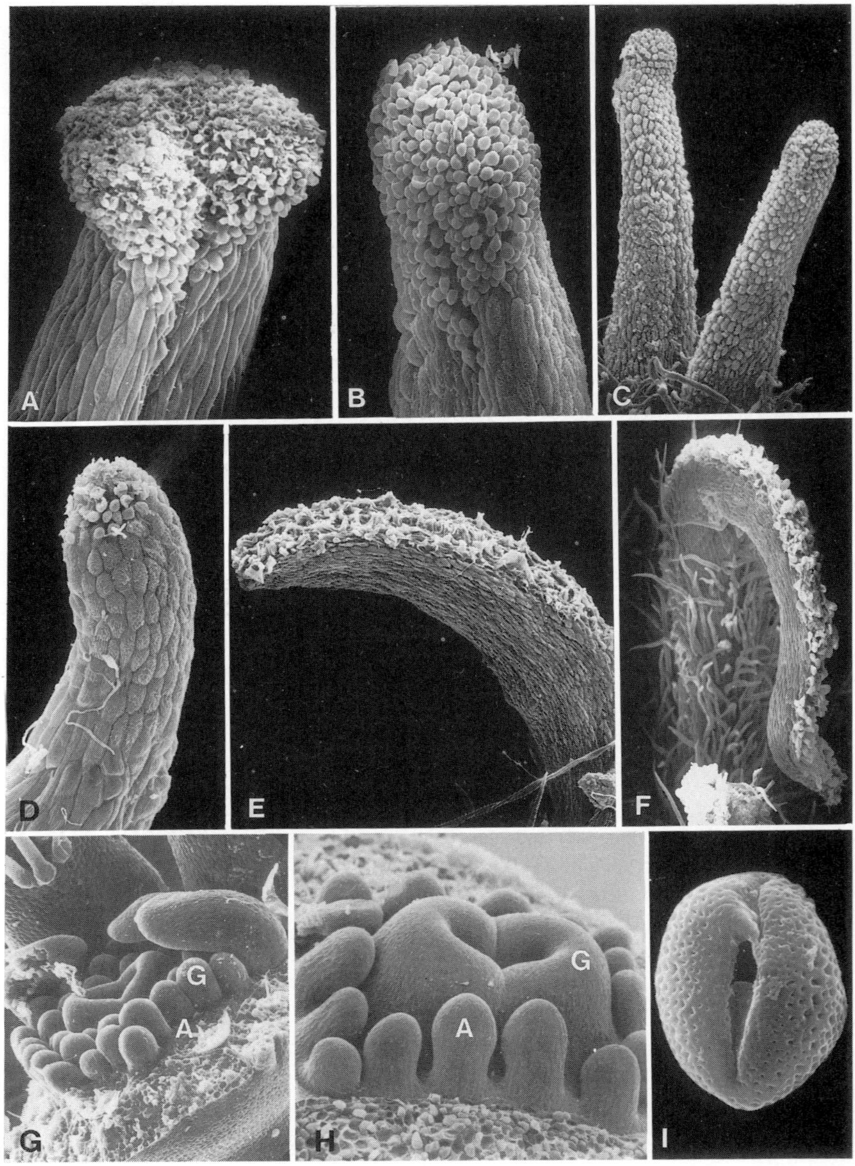

Fig. 4.3 Floral morphology of primitive Rosidae (SEM). A, *Acsmithia undulata* (Dickison 262). Stigma (×27). B, *Cunonia balansae* (Dickison 230). Stigma (×27). C, *Cunonia pterophylla* (Dickison 238). Styles and stigma (×33). D, *Caldcluvia paniculosa* (Dickison 297). Stigma (×30). E, *Vesselowskya rubifolia* (Schodde 3210). Extended stigmatic surface (×13). F, *Brunellia comocladifolia* subsp. *domingensis* (Dickison 305). Stigma (×30). G, *Bauera rubioides* (Hoogland 12226). Young flower with perianth partially removed showing numerous separate stamen primordia (A) and bicarpellate

interpretation of the cupule in Fagaceae remains a major unsolved problem.

Stigma morphology in the primitive rosids is more variable than generally described. The papillate stigmatic surface of the Cunoniaceae is mostly restricted to the terminal position of the style, with stigmas ranging from relatively undifferentiated (Fig. 4.3D) to capitate or subcapitate in *Acsmithia* (Fig. 4.3A) and *Spiraeanthemum*. In some species of *Cunonia* the papillose surface extends downward (Fig. 4.3B) and may surround the length of the style (Fig. 4.3C). *Vesselowskya rubifolia* is unique in the family with a conspicuously elongate, decurrent stigma on the ventral surface of the style (Figs 4.1K, 4.3E). The Brunelliaceae are characterized by a linear, sutural stigma that extends along the entire length of the curved or hooked style and that is composed of bands of papillae (Cuatrecasas 1970) (Fig. 4.3F).

The resolution of the especially puzzling question whether the Rosidae and Hamamelidae are primitively petaliferous or apetalous is of utmost importance to arriving at a satisfactory phylogeny. In the Cunoniaceae, petals are present in some genera, absent in other genera, or, present or absent in different species of the same genus (*Ceratopetalum*). Most phylogenists have assumed that the ancestral rosids bore flowers with a perianth of well-developed sepals and petals (Figs 4.1E, 4.2D). Cuatrecasas (1970), Ehrendorfer (1977), and Ehrendorfer *et al.* (1984), however, argued that petals are a secondarily derived feature within the Brunelliaceae and Cunoniaceae. In this regard it is intriguing that a corolla is absent in the apocarpous genera *Acsmithia* (Fig. 4.1A), *Spiraeanthemum*, and *Brunellia*, as well as the Davidsoniaceae. According to Ehrendorfer *et al.* (1984) a staminal origin of petals is supported by petal position, form, and the fact that both petals and stamens are one-trace structures. These authors interpret the lobed petals of the cunoniaceous genera *Schizomeria* (Fig. 4.2C) and *Gillbeea* as being staminodial in origin. *Gillbeea* is of interest because of glands at the lateral petal apices (Fig. 4.2G). Walker (1968) had earlier presented the viewpoint that apetally was probably primitive in the Hamamelidaceae, and that the more archaic genera of the family still retain this feature. Ehrendorfer (1977) subsequently incorporated this idea into a model of floral evolution that emphasized mutual affinities for apetalous Hamamelidae and Rosidae.

Nevertheless, in the absence of conclusive evidence to the contrary, I

gynoecium (G) (×167) (provided by R. Rutishauser). H, *Geissois pruinosa* (Endress 6005). Young flower with perianth removed showing numerous stamen primordia (A) arranged in a single ring and bicarpellate gynoecium (G) (×167) (provided by R. Rutishauser from material supplied by P. Endress). I, *Weinmannia pinnata* (Campbell 314). Pollen. Equatorial view (×1945).

believe the more compelling hypothesis assumes the primitive rosids possessed perfect flowers with well-developed corollas, and that the loss of petals in some taxa has been associated with shifts from entomophily toward anemophily. Like the Hamamelidales, much of the alteration in floral morphology and inflorescence type in the primitive rosids appears to be directly related to adaptive shifts in pollination biology. Accordingly, this interpretation takes exception to generalizations that describe the Cunoniaceae as entirely entomophilous (see Willemstein 1987). Although little is known of the floral biology of cunoniaceous plants, putative evolutionary trends in floral structure can be postulated. It is evident that many of these trends closely approximate transformations in the Hamamelidales.

In my opinion the probable pre-hamamelid–rosid was characterized by largely unspecialized, insect-pollinated flowers. Flowers of this organization were bisexual, of comparatively small size, tetramerous or pentamerous, with a diplostemonous androecium, an apocarpous gynoecium of few carpels and low or intermediate ovular number per carpel, nectariferous, and had a perianth composed of conventional sepals and petals. This generalized floral form can be viewed as ancestral to both rosids and hamamelids, with many divergent morphologies having radiated from this basic condition.

A more specialized, presumably insect- or bird-pollinated, flower evolved from petalled ancestors in the cunoniaceous genera *Geissois* and *Lamanonia*. *Geissois* has visually attractive, somewhat bottlebrush form of pendulous, sometimes ramiflorous, inflorescence (Fig. 4.1G) in which the thick calyx members and stamen filaments of the comparatively large flowers are various shades of red. In both genera loss of petals has been accompanied by increases in stamen number by advanced proliferation (Fig 4.1H, I). Stamen filaments are rather long. Carpels are multiovulate and a disk is present. Apparently different types of insects are attracted to the petaliferous but non-showy, greenish flowers of *Cunonia macrophylla*, the reddish flowers of *C. atrorubens*, and the white or cream-coloured flowers of species like *C. capensis* and *C. balansae*.

Putative major trends toward wind pollination have occurred in a number of primitive Rosidae, resulting in reduced, or specialized, floral morphologies. The following features are scattered among different genera and are consistent with the hypothesis of independent shifts toward anemophily, or partial anemophily, in different lineages: dioecy, unisexual flowers, loss of petals, flowers small and inconspicuous, epigyny or partial epigyny, low ovule number, small pollen size, variation in stigma type including decurrent stigmatic surfaces, and flowers clustered in dense inflorescences. These features are sometimes accompanied by the abortion of ovules during fruit maturation resulting in reduced seed number. The diversity in disk morphology among

Cunoniaceae may be significant in this regard. Most species have a rather thick annular disk, although in some taxa it is reduced or only the lower part of the ovary is slightly differentiated. A condition of pronounced protandry occurs in some genera (*Caldcluvia*, *Weinmannia*).

Accordingly, I interpret the absence of petals in the small flowers of *Acsmithia*, *Spiraeanthemum*, and *Brunellia* to be a derived character state. These genera are isolated elements within the assemblage and, although they retain a primitive gynoecial condition, they appear to be at least partially adapted for wind pollination. In addition to apetally, exclusively unisexual flowers have evolved in *Spiraeanthemum*. Sexuality in *Brunellia* is variable. A few species have totally bisexual flowers, others have bisexual and female flowers on separate individuals, and other taxa have functional male and female flowers on different trees (Cuatrecasas 1970). *Brunellia* can be seen as showing transitional stages to imperfect, anemophilous flowers. Although Cuatrecasas (1985) reaffirmed his belief that the extended sutural stigma on the elongated style of *Brunellia* is a relictual feature, I believe it is more likely related to an anemophilous floral biology. The capitate or subcapitate stigmas of *Acsmithia* and *Spiraeanthemum* can also be seen as specializations. The low ovular number of these plants (ovules reduced to one per carpel in some species of *Acsmithia*) correlates with the anemophilous syndrome. The fact that the rosids were, doubtless, primitively biovulate seemingly conferred a certain degree of preadaptation that may have led to this adaptive shift. The possibility that a decrease in flower size is associated with an increase in flower number per inflorescence by the development of more complex, branched inflorescences, as described for Hamamelidaceae (Endress 1977), needs exploration.

Wind pollination is also strongly inferred in the monotypic Australian genus *Vessellowskya*. This is evidenced by the modified decurrent stigma with an expanded surface, coupled with the dioecious condition. Other interesting features are that the flowers are nearly all trimerous and that the male flowers have sepals and petals but the female flowers only sepals (Hoogland, unpublished). The retention of a disk and high ovule numbers, however, indicate that the trend toward wind pollination is incomplete and that a continued partial insect pollination perhaps occurs as well.

The cunoniaceous genera *Callicoma*, *Codia*, and *Pancheria* have flowers clustered in condensed, ball-like heads (Fig. 4.1J). Engler (1928) gave considerable emphasis to this feature in characterizing his tribe Pancherieae, although this appears to be an artificial grouping showing convergence. Specializations are again evident in the absence of petals in *Callicoma* and the dioecious condition of *Pancheria*.

The absence of petals in different rosids leads to the supposition that corollas have been secondarily lost independently in different lineages.

This conclusion is supported by the presence of different floral groundplans among apetalous taxa. *Acsmithia* and *Spiraeanthemum* have an apocarpic construction coupled with eight to ten stamens arranged in two whorls (Fig. 4.1D). The different type of apetalous flower of *Geissois* is syncarpic and has a secondarily polyandrous androecium composed of up to 26 stamens arranged in a single ring (Figs 4.1I, 3H). I agree with R. D. Hoogland (personal communication) that the transformation of the alternisepalous stamens in *Acsmithia* and *Spiraeanthemum* into petals, and the subsequent duplication and relocation of the antisepalous stamens to form a diplostemonous androecium (Fig. 4.1M), is difficult to imagine. Similarly, if petals have arisen secondarily from apetalous ancestors by the sterilization of stamens, it is hard to explain the retention of apetally in five species of *Ceratopetalum*, with one species independently evolving petals from alternisepalous stamens and then subsequently changing back to the diplostemonous condition. The rather adavanced morphology of the branched petals and specialized fruit type of *Ceratopetalum* supports a reduction series from the petalled condition in *C. gummiferum* to the apetalous condition in the other species (Fig. 4.2F).

Cuatrecasas (1985) was an advocate of the multistaminate condition being plesiomorphic in Brunelliaceae. I believe it is more probable that a stable, intermediate number of stamens is basic within the rosids and hamamelids, with a subsequent increase and decrease in stamen number. In the Cunoniaceae, haplostemonous androecia have formed in a few species of *Cunonia* and *Weinmannia*, whereas a contrasting polyandrous condition has evolved in *Bauera*, along with a six-merous calyx and corolla and a bicarpellate gynoecium. Preliminary observations show that the stamen primordia in *Bauera* are initiated more or less simultaneously or with a slight centrifugal tendency (Rutishauser, personal communication) (Fig. 4.3G).

The following differences and resemblances between the two groups in stamen morphology can be noted: stamens with uniformly small anthers in rosids, staminal structure diverse in Hamamelidales with large anthers in wind-pollinated taxa; anthers basifixed in Hamamelidales, anthers dorsifixed (versatile) in Cunoniaceae and Brunelliaceae (basifixed in *Tetracarpaea*); anther dehiscence by means of unusual and uncommon valves in Trochodendrales, zoophilous Hamamelidaceae, and Platanaceae, anther dehiscence by simple longitudinal slits in all rosids and wind pollinated Hamamelidaceae, Myrothamnaceae, and Fagales. Anthers of all Hamamelidales, as well as some Cunoniaceae (*Ceratopetalum, Vesselowskya, Schizomeria, Platylophus, Caldcluvia paniculosa*) share a sterile apical extension of the connective (Figs 4.1E, 4.2E) and filaments are typically long.

In contrast to the primitive tricolpate pollen of the lower hamamelids

(Trochodendrales, Hamamelidales) (Endress 1977; Bogle and Philbrick 1980; Zavada and Dilcher 1986), pollen of the primitive rosids is tricolporate (Fig. 4.3I). The small, basically reticulate or tectate perforate grains of the Cunoniaceae (Fig. 4.3I) were regarded by Hideux and Ferguson (1976) to be of a low level of advancement as compared with other woody saxifragalean taxa. The fact that higher hamamelids have basically triporate pollen led Doyle and Donoghue (1987) to conclude that the triporate Amentiferae are actually rosids, only distantly related to the tricolpate groups. The granular type of infrastructure characteristic of Urticales, Juglandales, Myricales, and Casuarinales is unknown in the Rosidae.

Fruit and seed morphology

Stone (1973) described the many independent and parallel adaptations of the fruit and seed in taxonomically diverse groups of Hamamelidae, and cautioned against using mature fruit morphology in assessing phylogenetic relationships without careful study. However, a consideration of fruit diversity and trends in dispersal mechanisms can be useful in evaluating suggested familial affinities (Tiffney 1986). Stone (1973) described the diversity of fruit types in the Hamamelidales, including follicles (Cerecidiphyllaceae), capsules (Hamamelidaceae, Myrothamnaceae), winged-nuts (Eupteleaceae), achenes (Platanaceae), and drupes (Didymelaceae), and including forms with one to many carpels, one to many ovules, and the dehiscent and indehiscent condition. Stone further observed that this variation is theoretically capable of deriving essentially all the fruit types in the more advanced families comprising the Amentiferae, where few-seeded, indehiscent, wind- or animal-dispersed fruits predominate.

Dispersal in the Hamamelidaceae is by wind, by water, and more commonly by specialized short-range ejection (Endress 1977). It appears that the ancestral hamamelids had loculicidal fruits with wind-dispersed, thin-walled, winged, albuminous seeds. All of the multiovulate genera of Hamamelidaceae have retained this syndrome. In contrast, the large nuts of all modern Fagaceae are animal dispersed, although the family may be primitively abiotically dispersed (Tiffney 1986).

The primitive Rosidae nearly match the hamamelids in the range of fruit structure and dispersal biology. The considerable diversity among primitive rosids in fruit and seed type represents a wide range of different levels of evolutionary specialization and dispersal methods (Dickison 1984). Fruit types are also very diverse in the Rosaceae. The Cunoniaceae, in particular, are an excellent example of a family within which major transitions in floral and dispersal biology can be traced, and from which the various hamamelid syndromes could potentially have evolved. A summary of fruit types in the family is presented in Fig. 4.4.

The initial evolutionary step in fruit specialization involved the conversion of a cluster of follicles into the more common bilocular, ventrally dehiscent capsule or a spherical head of capsules. Ultimately, various forms of indehiscent fruit evolved. The dehiscent fruits of both Cunoniaceae and Hamamelidaceae have a mature pericarp that is differentiated into an outer parenchymatous region and an inner zone of rigid, thick-walled, and lignified cells. The absence of a woody endocarp is an uncommon and clearly apomorphic character state in the

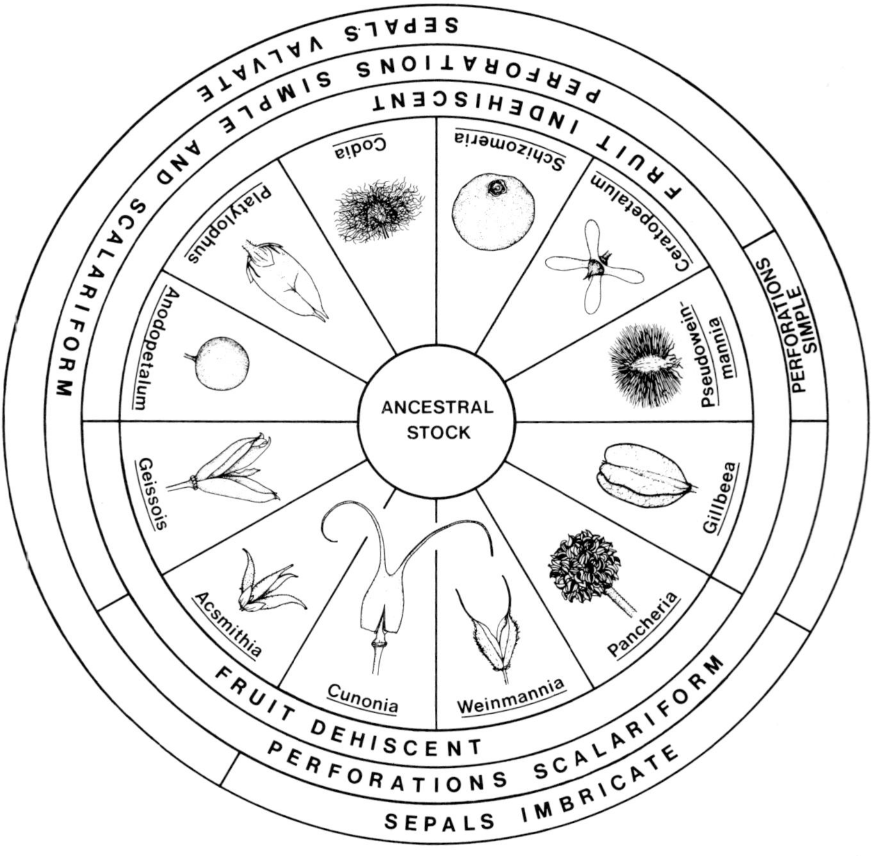

Fig. 4.4 Diagram showing diversity of fruit types in the Cunoniaceae. Fruits not drawn to scale. (Illustrations used with the permission of the President and Fellows of Harvard College.)

Cunoniaceae. The primitive cunoniaceous seed is winged, with abundant endosperm, thin-walled, has a single fibrous layer, and like the less specialized hamamelids is abiotically dispersed.

Subsequent evolutionary advancement has resulted in major shifts in dispersal methods (Dickison 1984). In more advanced Cunoniaceae the entire fruit has become modified for dispersal and protection, and there have been reductions in the number of seeds per fruit. District trends of structural specialization have produced densely pubescent and winged fruit types that are associated with anemochory. *Pseudoweinmannia* is interesting in that its densely hairy, indehiscent fruits are undoubtedly scattered by wind, but its elaiosomelike seed appendages suggest that a secondary agent (perhaps ants) may also be involved in seed dispersal. Trends have also occurred toward formation of drupaceous and berrylike fruits that probably promote seed dispersal by animals. For *Platylophus* the inflated fruit wall and the absence of dispersal structures on the small, thin-walled seeds may be reflections of dispersal by flowing water. The Cunoniaceae, therefore, show major adaptive shifts in dispersal agents—from wind to animals, and even water.

Despite the wide variability that exists among fruit types in the Cunoniaceae, and the potential importance ascribed to fruit morphology in intrafamilial classification (Dickison 1984), specific trends of specialization cannot be readily outlined. I find it impossible to satisfactorily hypothesize fruit homologies and, therefore, to score and utilize fruit types in a manner that would group genera in a phylogenetically logical way. Godley (1983) has also drawn attention to this problem by suggesting that not all dehiscent, capsular fruits in the Cunoniaceae may be homologous.

In the Brunelliaceae the carpels undergo asymmetric growth during fruit formation, displacing the style to the abaxial side in a manner illustrated by Cuatrecasas (1970). Mature follicles dehisce along the displaced ventral suture and seeds are forced out of the fruit but remain attached to the exocarp by a ribbon of tissue that represents a continuation of the funiculus. This unique, specialized fruit form is a distinguishing feature of the family.

Vegetative morphology and anatomy

Stipulate leaves exist in all primitive Rosales, Hamamelidales, and Fagales. As compared with the mostly simple, alternate foliage of the hamamelids, mostly opposite, pinnately compound leaves occur in the early rosids. Exceptions are seen in the alternate leaves of Davidsoniaceae and most Rosaceae. Stipules of the Cunoniaceae and allies are borne laterally on the branches on each side of the petioles, with the two stipules on each side of the branch usually united to form the characteristic interpetiolar stipular condition. Unfused stipules do, however, occur. Whorled leaves and palmately compound (three- to five-foliolate) and simple (unifoliolate) leaf types are derived in the Rosidae. In this regard the whorled leaves with interpetiolar stipules of *Trigonobalanus verticillata* (Fagaceae) may be significant (Nixon, personal communication). Some Cunoniaceae (*Ceratopetalum, Weinmannia, Pancheria*) and Brunelliaceae have both compound and simple leaves, whereas *Acsmithia* and *Spiraeanthemum* have only simple foliage. All of the Cunoniaceae and Brunelliaceae are evergreen in contrast with the deciduous habit of many hamamelids and other wind-pollinated plants.

The Hamamelidales and Fagales are usually described as having palmate or pinnate venation, although the difficulties in applying this terminology to these plants has been discussed by Spicer (1986). The non-entire margin is regarded as basic, the entire margin derived. Hickey and Wolfe (1975) considered the pinnate (craspedodromous) venation pattern of some Hamamelidaceae and Fagaceae to be derived from the palmate (actinodromous) condition by the suppression of the lateral primary veins.

Leaf venation in the rosids is pinnate, with either camptodromous or craspedodromous secondaries. The presumably primitive pattern is semicraspedodromous, in which the secondaries branch just within the margin, one of the branches terminating at the toothed margin, the other joining the superadjacent secondary after following a marginal course (Fig. 4.5A, B). Striking resemblances can be noted between the simple craspedodromous venation of certain Fagaceae and some coarsely toothed Cunoniaceae, such as *Callicoma serratifolia* (Fig. 4.5I), *Caldcluvia paniculata* (Fig. 4.5H), and *Acrophyllum australe* (Fig. 4.5G). The secondary veins in these leaves are straight and all terminate at the margin. Entire-margined cunoniaceous foliage possess secondary veins that curve upward just inside the margin forming a eucamptodromous (Fig. 4.5D) or irregular combination of eucamptodromous and brochidodromous configuration (Fig. 4.5C, E, F).

Hickey and Wolfe (1975) described a cunonioid tooth type, distinguished by a small, clear, glandular apex, with the principal vein to the tooth branching near the sinus, and sending one branch to the superadjacent secondary vein or to the area of the sinus and the other branch to the tooth apex along the apical side of the tooth (Fig. 4.6A). These authors suggested that the marginal teeth of *Trigonobalanus* are perhaps a 'modified cunonioid' type (see also Wolfe, Vol. 40A, Ch. 5). Teeth of *Callicoma* are vascularized by a major vein that extends along a medial course to the tooth apex (Fig. 4.6B). Cunonioid teeth are also present in Juglandales (Hickey and Wolfe 1975).

The nodal anatomy of the Cunoniaceae is unusually diverse,

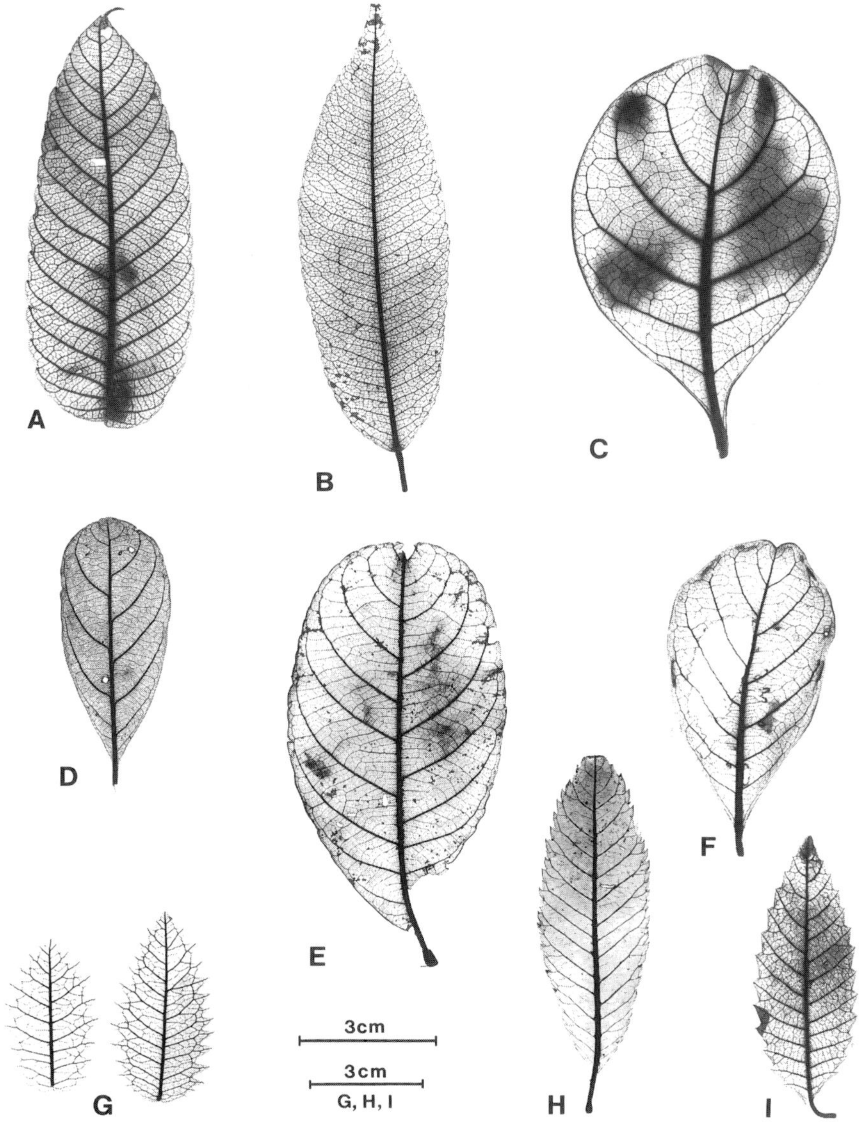

Fig. 4.5 Cleared leaves of Brunelliaceae and Cunoniaceae showing major venation patterns. A, *Brunellia comocladifolia* subsp. *domingensis* (Dickison 305). B, *Ceratopetalum apetalum* (Dickison 194). C, *Codia incrassata* (Dickison 289). D, *Acsmithia undulata* (Dickison 262). E, *Gillbeea adenopetala* (Dickison 210). F, *Codia discolor* (Dickison 269). G, *Acrophyllum australe* (Dickison 210). H, *Caldcluvia paniculata* (Bennett 1723). I, *Callicoma serratifolia* (Dickison 198).

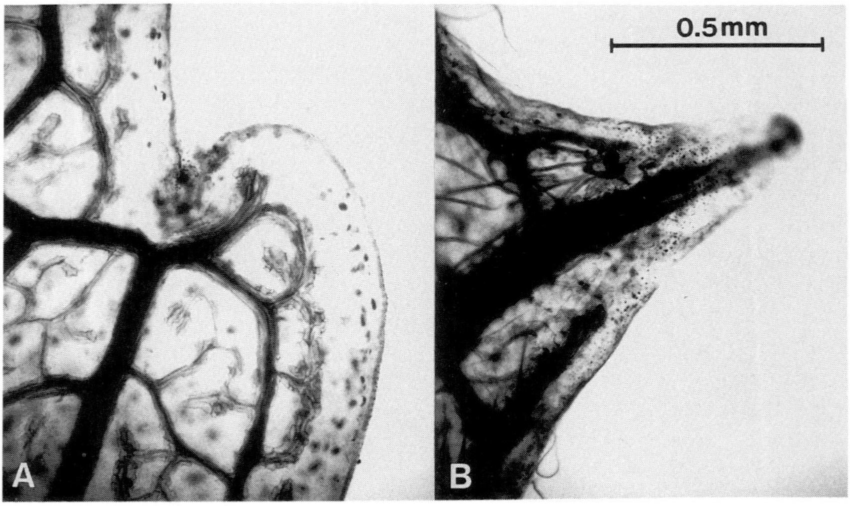

Fig. 4.6 Tooth morphology of Cunoniaceae. A, *Cunonia pulchella* (Dickison 287). B, *Callicoma serratifolia* (Dickison 198).

although the generalized trilacunar, three-trace pattern predominates and is considered basic (Dickison 1980*b*). The frequent evolutionary transition from opposite to whorled leaves has been accompanied by the formation of specialized and distinctive split-lateral traces (in the sense of Howard 1970).

Trilacunar, three-trace nodes predominate in other primitive rosalean plants and suggests that this is undoubtedly the ancestral pattern for the entire assemblage. It is significant that this nodal condition is found in species of *Brunellia* with whorled leaves. The nodes of such important hamamelidaceous families as Hamamelidaceae, Fagaceae, Betulaceae, Juglandaceae, and Moraceae are also of the trilacunar, three-trace type (Sinnott 1914), and thus could have been derived from the rosalean structural condition. Interestingly, the relatively uncommon condition of split-lateral leaf traces occurs in the Myrothamnaceae, a family placed very near Cunoniaceae in some systems. However, in the Cunoniaceae this condition is related only to whorled leaves, whereas in *Myrothamnus* leaves are opposite in arrangement.

The diversity of xylem structure in the Hamamelidae is enormous, ranging from the primitively vesselless Trochodendrales to moderately advanced families (Tippo 1938; Moseley 1973; Dickison 1975*a*). The absence of vessels in the Trochodendrales makes this an extremely primitive order, and implies that the group could only have been derived from a vesselless ancestor. Wood anatomy does not necessarily imply a vesselless ancestry for the porous families of hamamelids.

Unlike the often specialized floral morphology, wood structure of the Hamamelidaceae has remained at a uniformly very low level of advancement. The fact that they are an entirely scalariform perforation plate family makes them structurally more primitive than Fagaceae and other amentiferous groups. Exclusively scalariform perforations also occur in Cercidiphyllaceae, Eupteleaceae, and Myrothamnaceae.

Woods of the Cunoniaceae exhibit a wide range of anatomical variation, representing quite different levels of phylogenetic advancement (Dickison 1980*a*). Wood structure reinforces the basal position of the Cunoniaceae within the rosalean complex, since all rosoid structural types can potentially be derived from the primitive cunoniaceous condition. None of the rosids are vesselless. The Rosales, as a group, have been considered more structurally advanced than Hamamelidaceae (Tippo 1938), although wood- anatomically the Hamamelidaceae and the more unspecialized genera of Cunoniaceae are very comparable; viz., mostly solitary vessel elements with many-barred (fewer than 20) scalariform perforation plates, scalariform to opposite intervessel pitting, imperforate tracheary elements of the tracheid or fibre–tracheid type with distinctly bordered pits, heterogeneous rays (including Krib's Types I, IIA, and IIB), and axial parenchyma that is diffuse or diffuse-in-aggregates. Some genera of Hamamelidaceae have slightly longer perforate and imperforate elements, supporting a lower level of advancement. It must be emphasized, however, that these shared xylem features are primitive character states and are of little value in establishing cladistic relationships.

Discussion

When comparisons are made, a direct phenetic and probable close phylogenetic relationship between the primitive members of the woody Rosidae and some groups of the Hamamelidae gains considerable support. Confidence in this hypothesis increases as a result of the many general resemblances and parallel evolutionary trends in both vegetative and reproductive morphology. Key traits and conditions shared by the two groups of families include: a primitive wood anatomy of similar construction; trilacunar, three-trace nodes; stipulate leaves with often toothed margins and pinnate (craspedodromous) venation; a similar floral groundplan with transitions from apocarpous to syncarpous and superior to inferior gynoecia; trends toward both the increase and decrease in numbers of floral parts; strong tendencies for both unisexuality and floral reduction, including the frequent loss of petals; formation of capitate inflorescences; frequent shifts from entomophily presumably to anemophily; small, reticulate, triaperturate pollen with tectate–columellate structure; follicular and capsular fruit with lignified endocarps; and winged albuminous seeds with abiotic dispersal. Additional similarities include S-type plastids and laterocytic stomata. The two groups are held apart primarily by the possession of fundamentally pinnately compound leaves with pinnate venation in the rosids versus basically simple leaves with palmate venation in the hamamelids. The lower Hamamelidae have retained significantly more primitive tricolpate pollen and a less advanced wood anatomy (vesselless in Trochodendrales), although the distinctiveness of the above-mentioned groups is not sharp and I see insufficient grounds for regarding them as separate phyletic lines.

Few, if any, Rosidae have reached the level of specialization of the catkin-bearing Amentiferae, with extremely simple, often perianthless flowers clustered into spikes or aments. Nonetheless, many primitive Rosales are similar to the Hamamelidales in that they have undergone extensive radiations and adaptations that have presumably often been tied to apparent shifts toward anemophily and dispersal by various means. Unlike the situation in the Hamamelidales–Fagales complex, it has generally not been widely appreciated that various insect or other animal pollinators, along with transitions from insects to wind as the agent of pollination, have also accounted for profound changes in floral and inflorescence morphology among basal rosid taxa. Selection has resulted in an increase in number of floral parts and also frequent reduction and loss of parts. In some Hamamelidales and other amentiferous forms both sepals and petals have been secondarily lost, whereas floral simplification in the rosids has culminated in the loss of petals only. Of approximately 21 genera now recognized in the Cunoniaceae, nearly one-half are apetalous or have apetalous members. Apetally also distinguishes all Brunelliaceae and Davidsoniaceae. Clearly, floral reduction is not an isolated and uncommon trend among the primitive rosid taxa. The combination of primitive xylem coupled with apetally, valvate aestivation, and an inferior ovary also occurs in the Grubbiaceae, a family often regarded as having rosoid affinities (Carlquist 1977).

The Cunoniaceae are predominantly a tropical forest family of middle elevations, with subsequent radiations into more xeric or exposed sites, and into higher latitudes with increased seasonality (Dickison 1980a). This overall pattern of migration into generally drier and more exposed habitats supports postulated changes in pollination biology, since wind pollination is most likely to occur in open, arid situations (Whitehead 1969; Stebbins 1974). The large rainforest trees *Ceratopetalum* and *Pseudoweinmannia* are perhaps exceptions to this generalization. *Brunellia* also grows in greatest numbers in humid, wet, montane regions, with diversification leading to the establishment of species in open, less humid, and dry situations with seasonal periods (Cuatrecasas 1985).

Exceptions to the general rule of insect pollination in the Rosaceae are reported for the tribe Poterieae, where wind-pollinated genera with apetalous, inconspicuous, unisexual flowers in dense inflorescences are found (Stebbins 1974). The genus *Sanguisorba* is noteworthy since some species lack petals but their flowers produce nectar and are reportedly insect pollinated. This may represent a case of secondary reversion from wind to insect pollination (Stebbins 1974). It is possible that similar secondary reversions have occurred among certain Cunoniaceae.

The occurrence of floral specializations associated with wind pollination among primitive Rosidae implies that these were early adaptations within the rosalean line. The recognition that both the primitive Rosidae and Hamamelidae exploited this adaptive syndrome clarifies patterns of radiation and morphological change in both groups. The more extensive development of the anemophilous condition within the hamamelids, coupled with the deciduous habit, permitted their extensive diversification and dominance in temperate latitudes (Endress 1977. The occurrence of similar evolutionary trends in pollination biology among both groups are matched by concomitant changes in dispersal biology, from primitive abiotic to a more derived biotic dispersal mode. Wind pollination and correlated floral advancements have evolved independently many times within the dicotyledons and have been regarded as of little value, by themselves, in assessing relationships (Cronquist 1968; Thorne 1976). Nevertheless, I believe the totality of vegetative and reproductive similarities among the lower Hamamelidae and Rosidae are indicative of their close affinities and not just the result of convergence.

The interpretation presented here concludes that the main rosid–hamamelid radiations probably descended from common ancestral stock. The morphological series in both groups can be traced back to a basic floral type that was comparatively small, bisexual, apocarpic, with a stable, intermediate number of parts and a clearly differentiated perianth. Pollen was of the tricolpate type. Flowers of this hypothetical gross morphology were perhaps reminiscent in general appearance of the extant genus *Tetracarpaea*, although certainly no modern taxon combines all of the presumed primitive features. The importance of the perianth as a marker trait in angiosperm evolution has been emphasized by Endress (1986). In contrast to the Magnoliales, where sepals and petals are often not distinguishable, the perianth of the Rosales and Hamamelidales has reached a higher level of advancement by being fundamentally divisible into calyx and corolla.

I view the rosid–hamamelid complex as an ancient aggregation of many independent families that have unquestioned ancestral relationships and that have come under similar selective pressures. Carlquist (1982) called attention to the very unspecialized structure of the woods in this association and suggested that the magnolialean radiation was perhaps not the only early radiation of primitive angiosperms. If the Trochodendrales are regarded as an isolated order intermediate between the Magnoliidae and Hamamelidae (Endress 1986), then the hamamelid–rosoid complex stands as a major plexus from which many important phyletic lines diverged. Systems of classification that regard the Rosidae and Hamamelidae to constitute natural groups independently derived from the Magnoliidae appear to obscure phylogenetic relationships. Sorting out the many complex relationships that exists among families of these groups present difficult future challenges.

Acknowledgements

This contribution was supported by National Science Foundation grant BSR 84-01642. I especially wish to thank Dr R. D. Hoogland (Paris) for his continuous support and for generously sharing his extensive knowledge of the Cunoniaceae. Appreciation is extended to Mr Brad Bennett (Chapel Hill) for useful discussions and careful reading of the manuscript.

References

Baillon, H. (1871). Nouvelles notes sur les Hamamelidées. *Adansonia* **10,** 120–37.

Bange, G. G. J. (1952). A new family of dicotyledons: Davidsoniaceae. *Blumea* **7,** 293–6.

Bentham, G. and Hooker, J. D. (1865). Hamamelideae. *Genera Plantarum* **1,** 664–9.

Bessey, C. E. (1915). The phylogenetic taxonomy of flowering plants. *Annals of the Missouri Botanical Garden* **2,** 109–64.

Bogle, A. L. (1970). Floral morphology and vascular anatomy of the Hamamelidaceae: the apetalous genera of Hamamelidoideae. *Journal of the Arnold Arboretum* **51,** 310–66.

Bogle, A. L. (1986). The floral morphology and vascular anatomy of the Hamamelidaceae: subfamily Liquidambaroideae. *Annals of the Missouri Botanical Garden* **73,** 325–47.

Bogle, A. L. and Philbrick, C. T. (1980). A generic atlas of hamamelidaceous pollens. *Contributions from the Gray Herbarium* **210,** 29–103.

Carlquist, S. (1977). Wood anatomy of Grubbiaceae. *Journal of South African Botany* **43,** 129–44.

Carlquist, S. (1982). Wood anatomy of Daphniphyllaceae: ecological and phylogenetic considerations, review of pittosporalean families. *Brittonia* **34,** 252–66.

Crepet, W. L. and Nixon, K. C. (1987). Tricolporate angiosperms: success, phylogeny and early radiation. *XIV International Botanical Congress (Berlin) Abstracts,* 284.

Cronquist, A. (1968). *The evolution and classification of flowering plants.* Houghton Mifflin, Boston.

Cronquist, A. (1981). *An integrated system of classification of flowering plants.* Columbia University Press, New York.

Cuatrecasas, J. (1970). Brunelliaceae. *Flora neotropica* **2,** 1–189.

Cuatrecasas. J. (1985). Brunelliaceae. *Flora neotropica* **2,** (supplement), 1–103.

Dahlgren, R. M. T. (1977). A commentary on a diagrammatic presentation of the angiosperms in relation to the distribution of character states. *Plant Systematics and Evolution Supplement* **1,** 253–83.

Dahlgren, R. M. T. (1980). A revised system of classification of the angiosperms. *Botanical Journal of the Linnean Society* **80,** 91–124.

Dickison, W. C. (1975a). The bases of angiosperm phylogeny. *Annals of the Missouri Botanical Garden* **62,** 590–620.

Dickison, W. C. (1975b). Studies on the floral anatomy of Cunoniaceae. *American Journal of Botany* **62,** 433–47.

Dickison, W. C. (1978). Comparative anatomy of Eucryphiaceae. *American Journal of Botany* **65,** 722–35.

Dickison, W. C. (1980*a*). Comparative wood anatomy and evolution of the Cunoniaceae. *Allertonia* **2,** 281–321.

Dickison, W. C. (1980*b*). Diverse nodal anatomy of the Cunoniaceae. *American Journal of Botany* **67,** 975–81.

Dickison, W. C. (1984). Fruits and seeds of the Cunoniaceae. *Journal of the Arnold Arboretum* **65,** 149–90.

Doyle, J. A. and Donoghue, M. J. (1987). The importance of fossils in elucidating seed plant phylogeny and macroevolution. *Review of Palaeobotany and Palynology* **50,** 63–95.

Ehrendorfer, F. (1977). New ideas about the early differentiation of angiosperms. *Plant Systematics and Evolution, Supplement* **1,** 227–34.

Ehrendorfer, F., Morawetz, W., and Dawe, J. (1984). The neotropical angiosperm families Brunelliaceae and Caryocaraceae: first karyosystematical data and affinities. *Plant Systematics and Evolution* **145,** 183–91.

Eichler, A. W. (1878). *Blüthendiagramme,* Vol. 2. Wilhelm Engelmann, Leipzig.

Endress, P. K. (1967). Systematische Studien über die verwandtschaftlichen Beziehungen zwischen den Hamamelidaceen und Betulaceen. *Botanische Jahrbucher* **87,** 431–525.

Endress, P. K. (1977). Evolutionary trends in the Hamamelidales-Fagales group. *Plant Systematics and Evolution, Supplement* **1,** 321–47.

Endress, P. K. (1986). Floral structure, systematics, and phylogeny in Trochodendrales. *Annals of the Missouri Botanical Garden* **73,** 297–324.

Engler, A. (1897). Erläuterungen zu der Übersicht über die Embryophyta siphonogama. In *Die Natürlichen Pflanzenfamilien,* Nachträge zum 2 bis 4 Teil, Nachtrag 1, Lieferung 165, (ed. A. Engler and K. Prantl), pp. 358–66. Wilhelm Engelmann, Leipzig.

Engler, A. (1928). Cunoniaceae. In *Die Natürlichen Pflanzenfamilien*, Ed. 2, 18a, (ed. A. Engler and K. Prantl), pp. 229–62. Wilhelm Engelmann, Leipzig.

Eyde, R. H. (1970). Anatomy. In J. Cuatrecasas, *Brunelliaceae,* Flora Neotropica Monograph 2, pp. 32–43. Hafner, Darien, Connecticut.

Faegri, K. and Pijl, L. van der. (1979). *The principles of pollination ecology,* Ed. 3. Pergamon, New York.

Godley, E. J. (1983). The fruit in *Ackama, Caldcluvia,* and *Weinmannia. New Zealand Journal of Botany* **21,** 455–6.

Hallier, H. (1903). Ueber den Umfang, die Gliederung und die Verwandschaft der Familie der Hamamelidaceen. *Botanisches Zentralblatt Beihefte* **14,** 247–60.

Hallier, H. (1912). L'origine et le système phylétique des Angiospermes exposés à l'aide de leur arbre généalogique. *Archives Neerlandaises des Sciences Exactes et Naturelles* Ser. 3B. **1,** 146–234.

Harms, H. (1930). Hamamelidaceae. In *Die Natürlichen Pflanzenfamilien,* Ed. 2, 18a (ed. A. Engler and K. Prantl), pp. 303–45. Wilhelm Engelmann, Leipzig.

Hickey, L. J. and Wolfe, J. A. (1975). The bases of angiosperm phylogeny: vegetative morphology. *Annals of the Missouri Botanical Garden* **62,** 538–89.

Hideux, M. J. and Ferguson, I. K. (1976). The stereostructure of the exine and its evolutionary significance in Saxifragaceae *sensu lato.* In *The evolutionary significance of the exine* (ed. I. K. Ferguson and J. Muller), pp. 327–77. Academic Press, London.

Hils, M. H., Dickison, W. C., Lucansky, T. W., and Stern, W. L. (1988). Comparative anatomy and systematics of woody Saxifragaceae: *Tetracarpaea. American Journal of Botany.* **75,** 1687–700.

Hoogland, R. D. (1960). Studies in the Cunoniaceae. I. The genera *Ceratopetalum, Gillbeea, Aistopetalum,* and *Calycomis. Australian Journal of Botany* **8,** 318–41.

Hoogland, R. D. (1979). Studies in the Cunoniaceae. II. The genera *Caldcluvia, Pullea, Acsmithia,* and *Spiraeanthemum. Blumea* **25,** 481–505.

Howard, R. A. (1970). Some observations on the nodes of woody plants with special reference to the problem of the 'split-lateral' versus the 'common gap'. In *New research in plant anatomy* (ed. N. K. B. Robson, D. F. Cutler, and M. Gregory), pp. 195–214. *Botanical Journal of the Linnean Society, Supplement* 1, **63.**

Hutchinson, J. (1926). *The families of flowering plants,* Vol. 1. Macmillan, London.

Hutchinson, J. (1959). *The families of flowering plants,* (2nd edn.), Vol. 1. Clarendon Press, Oxford.

Kennedy, M. J. and Prakash, N. (1981). A morphological and embryological study of *Callicoma serratifolia* Andr. (Cunoniaceae). *Australian Journal of Botany* **29,** 721–31.

Moseley, M. F. Jr. (1973). Vegetative anatomy and morphology of Amentiferae. *Brittonia* **25,** 356–70.

Robertson, K. R. and Rohrer, J. R. (1987). Are apples pears? *Illinois Natural History Survey Reports,* No. 267. Champaign, Illinois.

Sinnott, E. W. (1914). Investigations on the phylogeny of the angiosperms. 1. The anatomy of the node as an aid in the classification of angiosperms. *American Journal of Botany* **1,** 303–22.

Smith, A. C. (1985). *Flora vitiensis nova*, Vol. 3. Pacific Tropical Botanical Garden, Honolulu.

Spicer, R. A. (1986). Pectinal veins: a new concept in terminology for the description of dicotyledonous leaf venation patterns. *Botanical Journal of the Linnean Society* **93,** 379–88.

Stebbins, G. L. (1974). *Flowering plants. Evolution above the species level.* Belknap Press, Cambridge.

Sterling, C. (1953). Developmental anatomy of the fruit of *Prunus domestica L. Bulletin of the Torrey Botanical Club* **80,** 457–77.

Sterling, C. (1969). Comparative morphology of the carpel in the Rosaceae. X. Evaluation and summary. *Österreichische Botanische Zeitschrift* **116,** 46–54.

Stone, D. E. (1973). Patterns in the evolution of amentiferous fruits. *Brittonia* **25,** 371–84.

Takhtajan, A. L. (1969). *Flowering plants: origin and dispersal* (trans. C. Jeffrey). Oliver & Boyd, Edinburgh.

Takhtajan, A. L. (1980). Outline of the classification of flowering plants (Magnoliophyta). *Botanical Review* **46,** 225–359.

Takhtajan, A. L. (1987). *Systema magnoliophytorum.* Nauka, Leningrad.

Tang, Y. (1943). Systematic anatomy of the woods of the Hamamelidaceae. *Bulletin Fan Memorial Institute Biology* **1,** 8–63.

Thorne, R. F. (1976). A phylogenetic classification of the Angiospermae. *Evolutionary Biology* **9,** 35–106.

Tiffney, B. H. (1986). Fruit and seed dispersal and the evolution of the Hamamelidae. *Annals of the Missouri Botanical Garden* **73,** 394–416.

Tippo, O. (1938). Comparative anatomy of the Moraceae and their presumed allies. *Botanical Gazette* **100,** 1–99.

Walker, J. W. (1968). Review, The evolution and classification of flowering plants by A. Cronquist. *Rhodora* **70,** 422–8.

Whitehead, D. R. (1969). Wind pollination in the angiosperms: evolutionary and environmental considerations. *Evolution* **23,** 28–35.

Willemstein, S. C. (1987). An evolutionary basis for pollination biology. *Leiden Botanical Series* **10,** 1–425.

Willis, J. C. (1973). *A dictionary of the flowering plants,* (8th edn) (revised by H. K. Airy Shaw). Cambridge University Press.

Zavada, M. S. and Dilcher, D. L. (1986). Comparative pollen morphology and its relationship to phylogeny of pollen in the Hamamelidae. *Annals of the Missouri Botanical Garden* **73,** 348–81.

5. Leaf-architectural analysis of the Hamamelididae

JACK A. WOLFE

Branch of Paleontology and Stratigraphy, United States Geological Survey, Denver, CO 80225, USA

Abstract

Analysis of leaf architecture of extant putative Hamamelididae indicates that the subclass is strongly polyphyletic. Among the basal hamamelidids, two distinct groups are present: the Trochodendrales–Hamamelidales and the Eupteleales (including Platanaceae). The trochodendralean–hamamelidalean group probably has a sister-relation to a line from which, in turn, the Dilleniidae, Eupteleales, and Rosidae have been derived. The higher hamamelidids appear to be a monophyletic group, comprising the pre-Normapolles Fagales and the Normapolles Betulales, Myricales, and Juglandales. The higher hamamelidids have leaf architecture that indicates derivation from primitive Rosidae. Probable dilleniid hamamelidids include i) Urticales, whose leaf architecture is best interpreted as indicating a relationship to Malvales, ii) *Eucommia*, which appears to have a close relationship with Stachyuraceae, and iii) an undescribed Costa Rican taxon of suggested hamamelidid affinity, which is closely related to Dilleniaceae.

Introduction

An earlier analysis (Wolfe 1974) of comparative foliar morphology (leaf architecture) of groups then included in the Hamamelididae suggested that foliage of most groups appeared to have levels of specialization that generally paralleled levels of specialization based on other organs and generally supported putative relationships inferred for these groups. On

Evolution, Systematics, and Fossil History of the Hamamelidae, Volume 1: Introduction and 'Lower' Hamamelidae (eds P. R. Crane and S. Blackmore), Systematics Association Special Volume No. 40A, pp. 75–104. Clarendon Press, Oxford, 1989.

the other hand, some groups appeared to have leaf architecture that allied them more closely to Rosidae or Dilleniidae than to Hamamelididae. A somewhat later survey of leaf architecture in the dicotyledons (Hickey and Wolfe 1975) broadly followed the earlier analysis with respect to groups within the Hamamelididae and added a new analytical dimension to leaf architecture, that of tooth morphology. Subsequently, knowledge of the distribution of various leaf architectural characters in the dicotyledons has grown. Much of this knowledge is still unpublished, with some notable exceptions (e.g. Rury and Dickison 1977, on the dilleniaceous *Hibbertia*; Levin 1986, on Subfamily Phyllanthoideae of Euphorbiaceae). In addition, subsequent studies, particularly of Early Cretaceous angiosperm leaves, have documented the relative times of appearances of various grades of foliar specialization in both megascopic (Doyle and Hickey 1976; Hickey and Doyle 1977) and microscopic (Upchurch 1985) characters. Studies of reproductive structures of Cretaceous and Paleogene members of Hamamelididae (e.g. Crane *et al.* 1988, in press; Crane and Stockey 1985; Manchester and Crane 1983, 1987; Manchester 1986, 1987; Wing and Hickey 1984) also allow evaluation of attached or associated foliage; these evaluations, along with studies of Paleogene taxa known only from foliage (Wolfe and Wehr 1987), add much new data to the analysis undertaken here. I will concentrate on foliage, but relevant palynological information will also be incorporated.

Basal Hamamelididae

Included here are those orders that Takhtajan (1980) and Cronquist (1981) considered to be the most primitive in Hamamelididae (Cronquist's 'Hamamelidae') and from which other orders of Hamamelididae were considered to be derived. These basal hamamelidid orders are Trochodendrales, Hamamelidales, and Eupteleales. Cercidiphyllaceae are, following Cronquist (1981), included in Hamamelidales, but Platanaceae are related to Eupteleaceae (see below) and are placed in Eupteleales. Myrothamnaceae are, as suggested by Endress (Vol 40A, Ch. 10), allied to Trochodendrales and Hamamelidales. Also included in the analysis is a fossil taxon from the Eocene of British Columbia (Fig. 5.1), which probably represents an extinct order.

Trochodendrales and Hamamelidales

As discussed in my earlier analysis (Wolfe 1974), leaf architecture is consistent with a close relationship between Trochodendrales (Trochodendraceae and Tetracentraceae) and Hamamelidales (in part, Cercidiphyllaceae and Hamamelidaceae). When extinct taxa are added to the analysis, the phenetic similarities are even more indicative of close relationships and indicate that an orbicular to ovate, palmately veined leaf is basic to both Trochodendrales and Hamamelidales.

The pinnately veined leaf of extant *Trochodendron* was suggested (Wolfe 1974) to be primitive for Trochodendrales and the palmately veined (actinodromous) leaf of *Tetracentron*, *Cercidiphyllum*, and primitive Hamamelidaceae to be specialized. Although the actinodromous grade is of later occurrence than the pinnately veined grade (Doyle and Hickey 1976; Hickey and Doyle 1977), reversals must have occurred. Thus, in Hamamelidaceae, all five subfamilies (excluding the debatable Rhodoleioideae; see below) have some or all taxa that have palmately veined leaves. Pinnately veined leaves occur in some species of *Altingia* in Altingioideae and in some Hamamelidoideae (e.g. *Parrotia*). In many pinnately veined Altingiodeae and Hamamelidoideae, secondary veins in the basal part of the lamina have a 'crowded' aspect or two relatively strong secondary veins originate at the base. These instances are inferred to represent reversals from a palmately veined leaf. If pinnately veined leaves are derived in Hamamelidaceae, the pinnately veined leaf of extant *Trochodendron* could also be derived.

An undescribed fossil taxon from the early middle Eocene (ca. 48–50 Ma) Republic flora (Wolfe and Wehr 1987) indicates that Trochodendraceae probably are primitively palmately veined (Fig. 5.1). The Republic taxon (RT) has characters possessed by both *Trochodendron* and *Tetracentron*: intercostal venation composed of an intersecondary vein that is joined by widely and irregularly spaced, orthogonal tertiary veins; festooned brochidodromy; well-developed admedially ramified bracing of the axils of the secondary veins; chevron-like braces between the major basal veins; and numerous chloranthoid teeth (Fig. 5.2). The RT, like the extant *Trochodendron*: has small, appressed teeth that are typically confined to the apical half of the lamina; has an almost rhomboidal shape; and was probably evergreen (the fossils are coriaceous and have a thick carbon residue). However, unlike *Trochodendron* but like *Tetracentron*, the RT is stipulate (a presumed primitive condition) and palmately veined. Xylem of the RT is unknown, but vesselless wood assigned to *Trochodendron* is known from the Eocene (Scott and Wheeler, 1982) and Oligocene of Oregon (Hergert and Phinney 1954). Leaves of *Trochodendron* are unknown in the Paleogene of the Pacific Northwest, but, in addition to the Eocene RT, one late Oligocene assemblage has leaves of a species congeneric with the RT. The woods may represent the same taxa as the RT and its Oligocene congener.

The extinct *Nordenskioldia* has vesselless wood and reproductive structure most closely allied to *Trochodendron* (Crane *et al.* in press). The

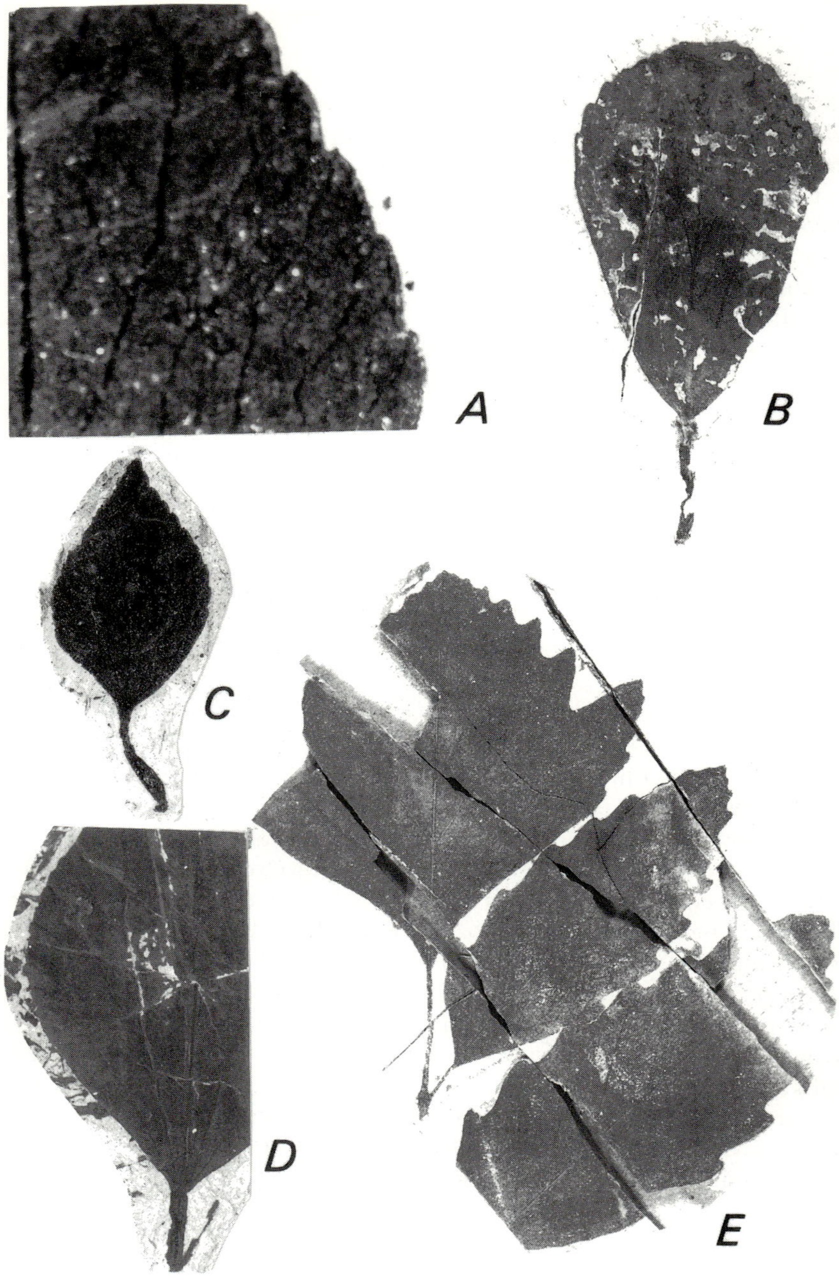

Fig. 5.1 Fossil leaves of some middle Eocene basal Hamamelididae. A–D, An undescribed taxon from Republic, Washington, related to *Trochodendron*; unlike *Trochodendron*, the fossil is stipulate and actinodromous. E, An undescribed taxon from Princeton, British Columbia; this taxon probably represents an extinct order that has chloranthoid teeth and pinnately compound foliage. A, ×4. B–D, ×1. E, ×0.6.

leaves associated at many localities are typically entire-margined, but some have a few, large chloranthoid teeth. As in the RT, the presumed leaves of Paleogene *Nordenskioldia* are palmately veined and probably evergreen, although they were estipulate or had deciduous stipules.

If the pinnately veined leaf was basic to Trochodendrales, then the palmately veined leaf would have had to originate four times in this order and Hamamelidales: in the lines leading to the RT, to *Nordenskioldia*, to *Tetracentron*, and to Cercidiphyllaceae–Hammameli-daceae. On the other hand, if the palmately veined leaf were basic to Trochodendrales, then the pinnately veined condition needs to have originated three times: in the line leading to *Trochodendron*, within *Altingia*, and within Hamamelidoideae. Parsimony (although not marked), as well as the association of palmate venation with the stipulate condition in the RT, suggest that actinodromy is basic to the line leading to Trochodendrales and Hamamelidales *sensu stricto*.

All members of Trochodendrales have clear-cut chloranthoid teeth and are vesselless. In both characters, Trochodendrales are divergent from the line leading to Cercidiphyllaceae and Hamamelidaceae. *Cercidiphyllum* and the extinct confamilial *Joffrea* (Crane and Stockey 1985) have modified chloranthoid teeth (although possible sucker-shoot leaves of *Joffrea* have clear-cut chloranthoid teeth: Fig. 5.2), in which the accessory veins in the apical part of the tooth are weakly developed. *Joffrea* has well-developed, admedially oriented bracing of the secondary veins and a more apparent chevron pattern of the bracing of the primary veins than in *Cercidiphyllum*; in these characters, *Joffrea* is less specialized than *Cercidiphyllum*.

In Hamamelidaceae, lateral bracing of the medial vein in the tooth is weakly developed in both apical and basal parts of the tooth, although as in the chloranthoid tooth, the medial vein thickens below a small, triangular gland (Fig. 5.2, the 'fothergilloid tooth' of Hickey and Wolfe 1975). Most Hamamelidaceae, however, have either entire margins or an aglandular spinose tooth. The palmately veined, cordate-based, ovate leaves of *Disanthus* (Disanthoideae) and *Chunia* (Exbucklandio-deae) have the greatest phenetic similarity to leaves of Trochodendrales and Cercidiphyllaceae, including festooned brochidodromy, chevron-like primary braces, and admedially oriented secondary braces. Like Trochodendrales and Cercidiphyllaceae, *Disanthus* and *Chunia* have lateral primary veins that extend into the apical third of the lamina and medial secondary veins that are restricted to the apical third of the lamina and are at a much higher angle to the midrib than are the major pair of lateral primary veins. Although *Disanthus* and *Chunia* are specialized in having an entire margin, in other characters these taxa have the least specialized leaf architecture in Hamamelidaceae.

In other palmately veined Hamamelidaceae, lateral primary veins

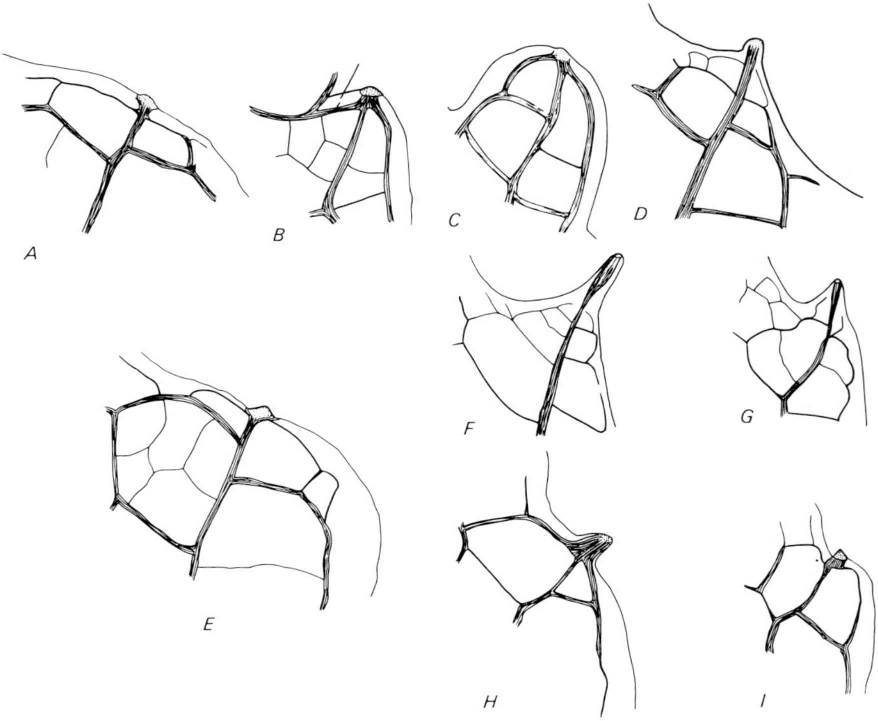

Fig. 5.2 Chloranthoid (A–C, E) and presumably derived teeth (D, fothergilloid; F, G, platanoid; H, dillenioid; I, protocunonioid). A, *Amborella trichopoda* (Amborellaceae, Laurales). B, *Tetracentron sinensis* (Tetracentraceae, Trochodendrales). C, *Joffrea* sp. (Cercidiphyllaceae, Hamamelidales). D, *Molinadendron guatemalense* (Hamamelidaceae, Hamamelidales). E, The Princeton taxon (PT). F, *Euptelea franchetii* (Eupteleaceae, Eupteleales). G, *Platanus racemosa* (Platanaceae, Eupteleales). H, *Hibbertia dentata* (Dilleniaceae, Dilleniales). I, *Ixerba brexioides* (Brexiaceae, Saxifragales). All figures × 15.

extend only about half the distance to the apex, and medial secondary veins occur in the basal half of the lamina and are approximately parallel to (or typically converge toward) the major pair of lateral primary veins. In Hamamelidoideae, at least one line developed craspedodromous secondary and pectinal veins (cf. Spicer 1986). Most pinnately veined Hamamelidoideae have secondary veins in the basal part of the lamina that are parallel to (or slightly converge with) one another; a notable exception is *Hamamelis*, in which these secondary veins markedly diverge.

Leaves of *Myrothamnus flabellifolia* (Myrothamnaceae) examined are consistent with, but not mandatory of, a basal hamamelidid relationship. Although the teeth on these specimens are aglandular, the actinodromy and chevron-shaped primary braces of this species suggest a relationship with the trochodendralean–hamamelidalean clade. However, Endress (Vol. 40A, Ch. 10) illustrates glandular teeth in *M. flabellifolia*, but the accessory veins join the major craspedodromous veins well below the glands, which is not a chloranthoid pattern. In contrast, the leaf of *M. moschata* illustrated by Endress (Vol. 40A, Ch. 10) has the typical vein–tooth–gland relations of the chloranthoid tooth and, concomitant with the other architectural features discussed above, places Myrothamnaceae in the trochodendralean–hamamelidalean clade. More precise relationships of Myrothamnaceae with other members of this clade are obscure.

Eupteleales

Although generally included in Hamamelidales, Platanaceae (with *Euptelea*) appear to represent a sister-group to the trochodendralean–hamamelidalean line. Three foliar specializations ally Platanaceae with Eupteleaceae (Fig. 5.3): 'stitched' intercostal venation (i.e. the tertiary veins tend to branch at almost right angles midway between secondary veins, with one branch extending to the adjacent tertiary vein and thus forming a 'composite intersecondary vein'; Hickey 1973); scalloped (i.e. concave) sinuses; and platanoid teeth (the craspedodromous veins thin and then rethicken before entering the teeth, enter the teeth medially, have a weak series of supporting loops, and have an apical foramen; Fig. 5.2). Certain other features, found also in Hamamelidaceae, are common to some Platanaceae (including the extinct *Macginitiea*; cf. Manchester 1986; Wolfe and Wehr 1987) and Eupteleaceae: pinnate venation; craspedodromous secondary veins; strong and craspedodromous pectinal veins; and secondary veins that originate at a lower angle in the medial than in the basal part of the lamina. Because these last four characters represent derived conditions in Hamamelidaceae and because neither Platanaceae nor Eupteleaceae can on other grounds be derived from Hamamelidaceae, these characters cannot be used to suggest a close relationship between Hamamelidaceae and Platanaceae–Eupteleaceae (a probably analogous situation pertains to the fruiting 'balls' of *Liquidambar* and Platanaceae).

The palinactinodromy of Platanaceae is probably not homologous to actinodromy in Trochodendrales. In *Macginitiea*, in which the first pair of major lateral veins originates at the base of the lamina (and is thus technically actinodromous), the most basal axillary bracing veins are straight to arcuate, as opposed to the chevron-like bracing in Trochodendrales. Further, *Platanus* has an uncanalized (cf. Upchurch 1985) condition of the major lateral veins originating at either the base or more typically above the base, and thus the uniform condition in

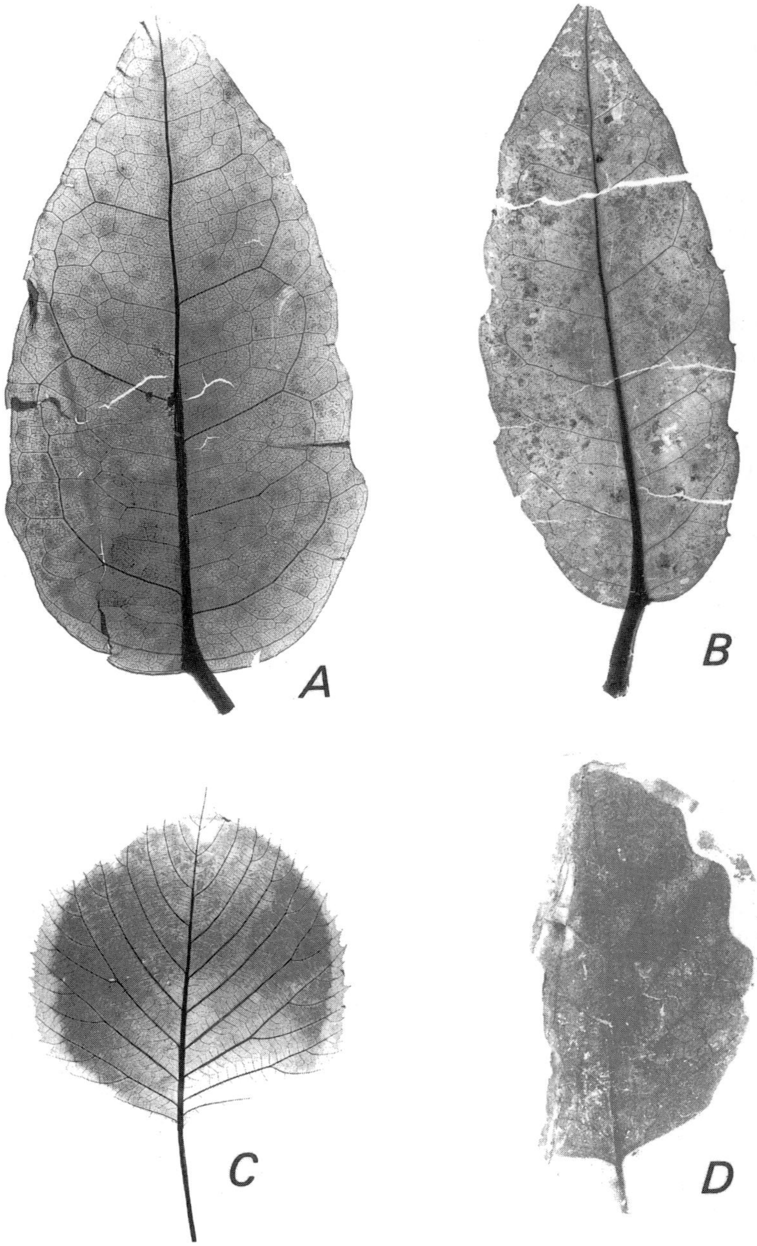

Fig. 5.3 Comparison of leaf architecture in Magnoliidae, Dilleniidae, and Hamamelididae. A, *Amborella trichopoda* (Magnoliidae). B, *Hibbertia dentata* (Dilleniidae). C, *Euptelea franchetii* (Hamamelididae). D, The Princeton taxon (Hamamelididae). All figures ×1.

Macginitiea is viewed as specialized. Because *Euptelea* is exclusively pinnately veined and because *Platanus* has both pinnate venation and palinactinodromy (basically a modified pinnate condition), pinnate venation is inferred to be basic for Eupteleales.

The 'stitched' intercostal venation of Eupteleales (Fig. 5.3) is interpreted as an intermediate morphologic stage between the basic magnoliid and higher hamamelidid stages. Weakening and shortening of the major intersecondary of the magnoliid stage would initially result in a compound intersecondary vein (see Wolfe and Wehr 1987, pl. 6, figs. 1, 2), and further weakening would leave isolated remnants of the vein midway between the secondary veins. In some *Platanus*, the 'stitched' venation has disappeared.

The probable primitive characters of Eupteleales occur in some magnoliids. Admedially ramified bracing is conspicuous in the axils of the secondary veins in *Euptelea* and most Platanaceae. The veins in the medial part of the lamina originate at lower angles than the veins in the basal part; this difference is apparent in the secondary veins of *Euptelea* (Wolfe 1974, fig. 24) and in the medial secondary and tertiary veins of *Macginitiea* (Wolfe and Wehr 1987, pl. 6, figs. 2, 4). Early platanoids either had a few, dentate teeth or are entire-margined (Upchurch 1985).

Euptelea has numerous teeth, but this may be a derived condition. In early platanoids teeth are few, and even in later Platanaceae the margin, except for lobing, is typically entire or with a few teeth. Entire-margined forms in Platanaceae are simple (rather than festooned) brochidrodromous or eucamptodromous, and one of the two extant species of *Euptelea* has some simple brochidodromous secondary veins. Uncanalized simple brochidodromy mixed with craspedodromy is probably primitive.

Presumably the platanoid tooth must be derived from the chloranthoid (Fig. 5.2). However, morphologic intermediates between the two tooth types are unknown. G. R. Upchurch (personal communication, February 1988) suggests that analysis of tooth morphology of some of the Early Cretaceous platanoids might be pertinent; although these leaves have the specializations in general architecture and cuticular anatomy to be assigned to the platanoid line (Upchurch 1985), the teeth of these leaves have not been carefully examined.

Princeton taxon

An unnamed and previously undescribed taxon from the early middle Eocene Princeton, B.C., flora (Fig. 5.1) occupies an important position in the current analysis. The Princeton taxon (PT), like Eupteleales, has: a leaf margin with scalloped sinuses; secondary veins in the medial part of the lamina that originate at lower angles than veins in the basal part; some (but not numerous) short intersecondary veins; craspedodromous secondary veins; and pinnate arrangement and venation. However, the teeth in the PT are chloranthoid and the leaf is clearly compound. Further, the intercostal venation does not have the 'stitched' aspect of Eupteleales. The PT has intercostal venation that is composed largely of widely, irregularly spaced tertiary veins that are orthogonal, compound, and percurrent; this venation pattern is interpreted to be derived from the euptelealean grade. The PT is best interpreted as representing a divergence separate from the platanoids, and, based on the compound leaf, a divergence leading to the 'higher hamamelidids' that have pinnately compound foliage.

However, the recently described Paleocene *Platanites hebridicus* (Crane *et al.* 1988) has clear platanoid teeth combined with a ternate lamina. This could be interpreted as indicating that a pinnately compound leaf is basic to the euptelealean line. If this interpretation is valid, then the pinnately compound lamina originated prior to the divergence of the euptelealean from the PT clade.

I emphasize, however, that the PT has some specializations that indicate a divergence from the main line leading to 'higher Hamamelididae'. In particular, the leaflets have long petiolules, and the abscission area is well developed, characters that would not be expected in a newly evolved compound leaf (e.g. as in *Sapindopsis*; cf. Doyle and Hickey 1976; Hickey and Doyle 1977). Further, the secondary veins are exclusively craspedodromous, as in higher hamamelidids that occur in the same assemblage as the PT.

Relationships of basal Hamamelididae

The foliar characters shared by the trochodendralean and the euptelealean clades all occur in some primitive magnoliids: stipulate leaves, alternate phyllotaxy, simple organization, numerous intersecondary veins, admedially ramified axillary bracing of the secondary veins, secondary veins not decurrent, low angles of origin of the secondary veins, and chloranthoid teeth. From the standpoint of leaf architecture, the two clades could have diverged at the magnoliid level; only by including characters such as the simple tricolpate pollen (and its exine structure) can the two clades be considered to have diverged above the magnoliid level. The trochodendralean clade could be derived from a magnoliid that, prior to becoming actinodromous, had secondary veins approximately parallel to one another in the central and basal parts of the lamina and numerous teeth (especially apically); the festooned brochidodromy and magnolialean intercostal venation of the trochodendralean line is found in many primitive magnoliids. In contrast, the euptelealean–PT clade could have risen from a magnoliid that had mixed simple brochidodromy and craspedodromy, secondary veins in the medial part of the lamina at higher angles than veins in the basal part, and a few coarse, dentate teeth (primarily in the medial and basal parts of the lamina).

In this context, the euptelealean–PT clade could be a sister-group to Dilleniidae, which primitively share these same specializations (Fig. 5.3). The divergence between Dilleniidae and the euptelealean–PT clade would be in the euptelealean–PT clade's tendencies to become dominantly craspedodromous and to regularize the intercostal venation by shortening the intersecondary veins and developing a system of orthogonal, percurrent, branched tertiary veins. The PT line maintained the chloranthoid tooth until after the development of a pinnately compound lamina. The dilleniid line generally maintained many primitive foliar characters (e.g. many members of Theales have low-rank venation) but early reduced the tooth to a narrow, elongated, glandular serration that has weak (or no) accessory veins. Early members of the euptelealean–PT clade are inferred to have had foliage like some species of *Hibbertia* (e.g. *H. dentata*; Fig. 5.3) but probably would have rapidly evolved the specializations in intercostal venation characteristic of later members of this clade. Whereas Dilleniidae are considered to be a sister-group of the line ancestral to the euptelealean–PT clade (this ancestral line, as discussed below, gave rise to other Rosidae and higher Hamamelididae), the trochodendralean–hamamelidalean clade is considered to be a sister-group of both Dilleniidae and Rosidae.

The origin of the pinnately compound condition near the divergence of Eupteleales is in accord with the fossil record. Early platanoid leaves have cuticular anatomy that has strong phenetic similarities to that of coeval *Sapindopsis* (Upchurch 1985), which a leaf type that has generally low-rank venation and is pinnatifid to pinnatisect (Doyle and Hickey 1976). The pinnatifid to pinnatisect condition is interpreted as a specialized stage that would give rise to truly pinnately compound foliage. Most early *Sapindopsis* have an entire margin, but some are sparsely toothed (Hickey and Doyle 1977; Upchurch 1985); tooth morphology of these leaves has not been described. Some younger species included in *Sapindopsis* have cunonioid teeth (Doyle and Hickey 1976; Hickey and Doyle 1977).

Higher Hamamelididae

This part of the analysis includes only the Fagales, Betulales, Myricales, and Juglandales (for other orders considered to be related, see 'Other Hamamelididae'). These orders represent the core of the old Amentiferae and are major floristic components of northern hemisphere

Cenozoic mesothermal and especially microthermal vegetation. These orders are considered by Takhtajan (1980) and Cronquist (1981) to be derived directly from Hamamelidales *sensu lato.* I suggested (Wolfe 1974) that foliage of Betulaceae and Fagaceae could be directly derived from hamamelidalean foliage from a type such as represented by *Corylopsis* and allied genera. On the other hand, I considered Juglandales to be of rosid affinities. As Cronquist (1981) validly pointed out, Juglandales are, on other than foliar characters, allied to Myricales, Betulales, and probably to Fagales; where one of these orders is placed, the others must follow.

Non-Normapolles higher Hamamelididae

Fagales

When most of us think of fagaceous foliage, we envision a chestnut-like leaf: numerous, parallel (and nearly straight), craspedodromous secondary veins that medially enter markedly serrate (or even spinose) teeth. Indeed, this morphotype occurs in *Nothofagus*, *Fagus*, *Castanea*, *Quercus*, *Lithocarpus*, and *Castanopsis*; an apparently reasonable inference would be that this morphotype is basic to Fagales. However, I suggest that this typical fagaceous leaf has evolved numerous times in response to similar environmental factors and that the leaf basic to Fagales is the type possessed by subgenus *Trisyngyne* of *Nothofagus* and by *Trigonobalanus sensu lato*.

The very high rank (i.e. high degree of rigid organization; cf. Doyle and Hickey 1976) of the typical fagaceous leaf should, in fact, argue against this being primitive, even for Fagales. High-rank foliage associated with a uniformly serrate to spinose margin has certainly arisen many times in the woody dicotyledons, being found in putatively advanced members of many families of Hamamelididae (e.g. Hamamelidaceae, Betulaceae, Juglandaceae, Ulmaceae) as well as in families of other subclasses (e.g. Dilleniaceae, Tiliaceae, Rosaceae, Hippocastanaceae, and Meliosmaceae). Doyle and Hickey (1976) and Hickey (1978) demonstrated a general increase in leaf rank in dicotyledonous foliage during the Early Cretaceous, indicating that high-rank foliage generally is a derived character. Considering that this type of foliage is typically found in woody plants that live in microthermal climates (Dilleniaceae being notable exceptions), the suggestions appear reasonable that this foliage has been selected for in many different lineages and that it could have evolved in different lineages of Fagaceae.

On the other hand, the leaves of two of the three species of *Trigonobalanus*, which Nixon (Vol. 40B, Ch. 3) segregates into three distinct genera, and of most species of *Nothofagus* (*Trisyngyne*) share certain characters. These leaves (Fig. 5.4) are typically mixed craspedodromous–brochidodromous, have widely and somewhat irregularly spaced tertiary veins that branch, have some (but not numerous) short intersecondary veins, have secondary veins in the medial part of the lamina that originate at lower angles than veins in the basal part, and have admedially ramified secondary braces. The secondary veins, if craspedodromous, tend to weaken as they approach the margin and are deflected at junctions with other veins. Most significantly, the secondary veins extend to the sinuses, where the veins typically end rather than entering the crenate teeth. The unusual association of crenate teeth with secondary veins ending at the sinuses is also found in more advanced fagaceous taxa (in some species of *Fagus* and *Nothofagus*) in association with higher-rank venation. Proximity of craspedodromous veins to sinuses is, by itself, a syndrome that characterizes the cunonioid tooth (Figs 5.2, 5.4) that is considered to be basic to Rosidae (Hickey and Wolfe 1975), although the typical cunonioid tooth has an apical branch that terminates at the sinus with the major branch extending along the apical side of the tooth.

In Cunoniaceae most taxa have either an aglandular cunonioid or a glandular protocunonioid tooth. In *Pullea* most species have a typical protocunoniod tooth, but some species have a crenate margin with the secondary veins (or their branches) ending at the sinuses, which are furnished with a gland (Fig. 5.4). I am not suggesting that these taxa of *Pullea* are in any way closely associated with the ancestry of Fagales, but the dental modification in *Pullea* demonstrates that vein termination at a sinus accompanied by crenate teeth can be derived from a more typical cunonioid (or in this instance protocunonioid) tooth, a demonstration of considerable relevance to inferred ancestry of Fagales.

In the present discussion, *Nothofagus* has been included in Fagaceae, although Nixon (Vol. 40B, Ch. 3) considers that *Nothofagus* contains more than one genus, represents a family distinct from Fagaceae, and is perhaps more closely allied to Betulaceae than to Fagaceae. While the first two concepts could be supported by leaf architecture, the third is uncertain. The doubly serrate margin, which occurs in some *Nothofagus* and in almost all Betulaceae, is probably derived in *Nothofagus* and would thus represent a convergence.

Another point emphasized is that, if the high-rank fagaceous leaf is not basic to Fagaceae, then similarities that I (Wolfe 1974) previously pointed out between the typical fagaceous leaf and similarly high-rank leaves in Hamamelidaceae are probably phylogenetically irrelevant. The trigonobalanid type of leaf that is basic to Fagaceae has almost no similarities to the leaf basic to Hamamelidales *sensu stricto* or to the trochodendralean–hamamelidalean line.

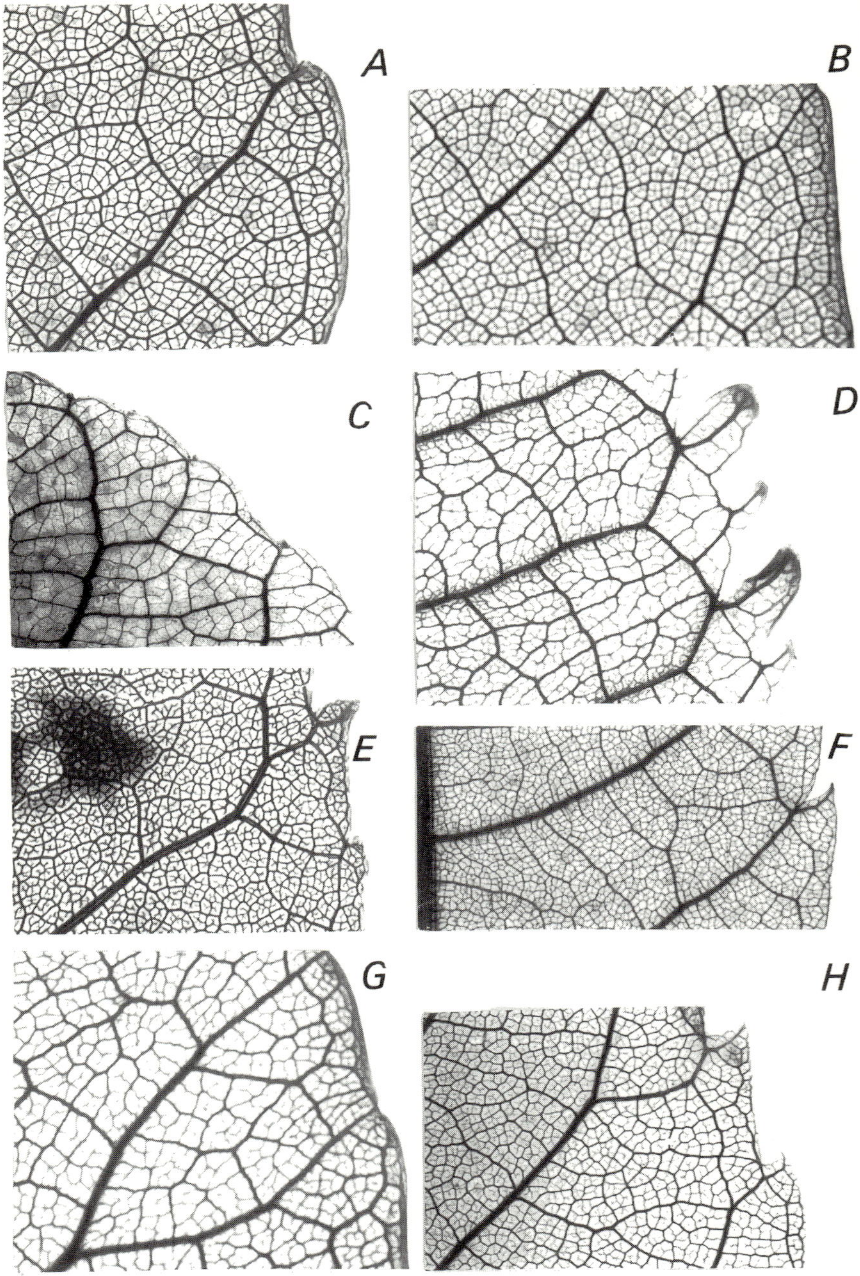

Fig. 5.4 Comparison of dental morphology in the higher Hamamelididae (A–F) with extant primitive Rosidae (G, H). A, *Trigonobalanus verticillata* (Fagaceae). B, *Nothofagus (Trisyngyne) baumanniae* (Fagaceae). C, *Alnus (Clethropsis) traberculosa* (Betulaceae). D. *Canacomyrica monticola* (Myricaceae). E, *Myrica mexicana* (Myricaceae). F, *Alfaroa costaricensis* (Juglandaceae). G, *Pullea glabra* (Cunoniaceae). H, *Lamanonia speciosa* (Cunoniaceae). All figures ×4.

The Normapolles Hamamelididae

All the groups discussed under 'Basal Hamamelididae' have fundamentally simple tricolpate pollen (Walker and Doyle 1975). In contrast, groups discussed below (Betulales, Myricales, Juglandales) have fundamentally triporate pollen that has '... clearly compound apertures (i.e. they are actually poroporate)' (Walker and Doyle 1975, p. 710). Some or all the compound-aperturate higher Hamamelididae have been suggested to be derivatives of the Normapolles complex (Góczán *et al.* 1967; Doyle 1969; Muller 1970; Wolfe 1974, 1976; Batten Vol. 40B, Ch. 2; Kedves, vol. 40B, Ch. 1), a group particularly characteristic of the Late Cretaceous of eastern North America and Europe.

Doyle's (1969; Doyle and Hickey 1976; Wolfe *et al.* 1975) work strongly indicates that the poroporate Normapolles complex was derived during the medial Cretaceous from simple tricolpates via the significant intermediate grade of a triangular tricolporate condition. In this context, Fagales, which have both tricolpate and tricolporate pollen, could have diverged during this intermediate stage; *Trigonobalanus* has pollen comparable to this intermediate stage (Walker and Doyle 1975, p. 710), and thus the colpate members of Fagales may represent one or more reversals from a tricolporate condition (Walker and Doyle 1975, inferred such reversals in advanced groups of Dilleniidae and Rosidae).

A stratigraphic–morphologic sequence of pollen from the Late Cretaceous of eastern North America (Wolfe 1974, 1976) strongly indicates the presence of both Rhoipteleaceae and Juglandaceae by the end of the Cretaceous, both families being derived from Campanian members of the Normapolles complex. Friis (1983) has fully confirmed this suggested transition and has provided a major datum for interpreting the phyletic relationships of this group. Flowers from the Santonian of Sweden represent juglandalean taxa, one of which has some features of Rhoipteleaceae and the other of Juglandaceae; each flower type contains a different palynomorph genus assignable to Normapolles. Other Normapolles-containing flowers may have some myricalean affinity. Clearly at least some of the Normapolles complex represent higher Hamamelididae.

Betulales

Of all the higher Hamamelididae, Betulaceae *sensu lato* have overall the most derived, high-rank foliage. The numerous, subparallel, craspedodromous secondary veins and craspedodromous pectinal veins with numerous, serrate teeth are found in all extant genera of the family. Intersecondary veins are absent, and the intercostal area is typically occupied by numerous, closely and uniformly spaced tertiary veins; secondary bracing is randomly reticulate.

In *Betula*, *Ostrya*, and *Carpinus* (and the extinct *Asterocarpinus*; cf. Manchester and Crane 1987), the secondary veins extend toward a sinus basal to the primary tooth but abruptly change course to swing apically into aglandular teeth. In *Corylus* and *Ostryopsis*, the secondary veins extend directly into aglandular teeth. Most species of *Alnus* have a vein–tooth relation like that of *Corylus*, but notable exceptions occur in both subgen. *Clethropsis* and subgen. *Alnus*. In inferred primitive taxa of subgen. *Alnus* (Furlow 1979), major veins typically branch before entering the subsidiary tooth, the apical branch extending to the sinus and the basal branch entering along the apical side of the serrate tooth; this is a typical cunonioid pattern, although the secondary veins typically enter the teeth medially. In some species of subgen. *Clethropsis* the venation–tooth relation is similar to primitive members of subgen. *Alnus*, except that even the primary teeth have the cunonioid pattern. In other *Clethropsis* (Fig. 5.4) a mixed craspedodromous–brochidodromous condition occurs and the margin can be crenate; from the brochidodromous loop, a thick vein extends to the sinus, in which is located a clear gland. The tooth morphology in subgenus *Clethropsis* recalls that inferred for primitive Fagales, and, because of the presence of a gland, this morphology resembles even more the protocunonioid pattern of the cunoniaceous *Pullea*. Furlow (1979) considered *Clethropsis* to be a sister-group to all other taxa of *Alnus*, and I suggest that *Clethropsis* has retained the foliar morphology primitive for Betulaceae *sensu lato*.

If the highly organized architecture of most Betulaceae is specialized for this family, then resemblances to leaves of specialized members of Hamamelidaceae previously noted (Wolfe 1974) are irrelevant. The leaf type here inferred to be basic to Betulaceae is similar to that inferred to be basic to Fagaceae and points to an ancestry among primitive Rosidae.

Myricales

Myricaceae typically do not display the high-rank venation that is developed in Fagales and Betulales. In general, Myricaceae have an irregular tertiary venation, although fourth- and fifth-order venation is well organized. Intersecondary veins are typically present. In toothed Myricaceae, secondary venation can be craspedodromous with the aglandular teeth typically entered medially. However, in some species of *Myrica*, the secondary veins weaken as they approach the margin and are deflected at junctions with other veins (cf. Fagales). As the branching secondary vein approaches the margin, the vein extends to the sinus and sends a branch along the apical side of the aglandular tooth; this pattern is typically cunonioid (Fig. 5.4). In the monotypic *Canacomyrica*, the brochidodromous arch of a secondary vein is placed against the sinus, where an abmedial secondary branch extends along the apical side of the tooth (Fig. 5.4), again a basically cunonioid pattern. Because *Canacomyrica* must on the basis of reproductive structure represent a sister-group to *Myrica*, the cunonioid tooth type is inferred to be primitive for Myricaceae.

Juglandales

Previous analyses of Juglandales (Wolfe 1974; Hickey and Wolfe 1975) concluded that the cunonioid teeth present in Rhoipteleaceae and Juglandaceae (e.g. *Alfaroa*), concomitant with the pinnately compound leaf organization, were anomalous in Hamamelididae but indicated a rosid affinity. The rosid affinity of Juglandaceae is further strengthened by the occurrence of glandular cunonioid (i.e. protocunonioid) teeth in early Eocene *Platycarya* (Wing and Hickey 1984).

Except for their pinnately compound leaf organization, Juglandales compare well in other leaf-architectural characters to characters inferred to be basic to Fagales, Betulales, and Myricales. That is, the leaflets of Juglandales primitively have protocunonioid and/or cunonioid teeth (Fig. 5.4), have widely and irregularly spaced tertiary veins, have intersecondary veins, have secondary veins in the medial part of the lamina that originate at lower angles than veins in the basal part, and have mixed craspedodromy–brochidodromy. Concomitant with the occurrence of glandular–peltate scales on the leaves, a strong case can be made for including Fagales, Betulales, Myricales, and Juglandales in the same supraordinal group.

The rosid connection

In each of the higher hamamelidid orders just discussed, the leaf architecture inferred to be primitive for the order has strong similarity to primitive Rosidae. The inclusion of Juglandales in the higher hamamelidid clade, moreover, indicates that the simple leaves of the other three orders probably represent reductions from a pinnately compound leaf (i.e. the simple leaves probably are homologous to a juglandalean leaflet), thus further strengthening the rosid connection. Glandular–peltate scales found in these higher hamamelidid orders, further, do not occur in the basal hamamelidids but do occur in Saxifragales (Escalloniaceae, Grossulariaceae), as well as in more derived rosids (e.g.

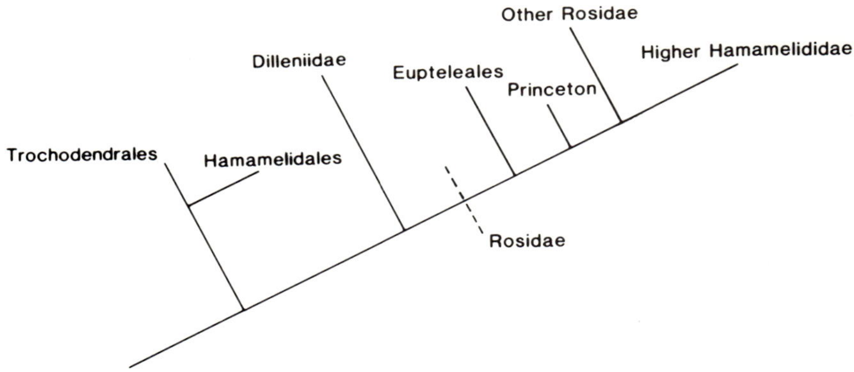

Fig. 5.5 Suggested phyletic relationships of Hamamelididae, Dilleniidae, and Rosidae.

Anacardiaceae, Sapindaceae, Meliaceae). As Walker and Doyle (1975) emphasized, the Normapolles-derived higher hamamelidids evolved from a pollen grade most characteristic of extant primitive Rosidae. Fagales, Betulales, Myricales, and Juglandales probably represent a clade derived from early members of Rosidae (Fig. 5.5). However, the suggestion by Nixon (Vol. 40B, Ch. 3) that Fagaceae might represent a clade derived separately from early rosids as a line distinct from the other three orders merits further research.

Does a rosid ancestry for these higher hamamelidid orders totally negate some relationships with the basal hamamelidid orders? When the discussion of the basal hamamelidids were adjourned earlier in this analysis, the last divergence (Fig. 5.5) represented the undescribed Princeton taxon that has a pinnately compound leaf combined with a chloranthoid tooth; individual leaflets have intersecondary veins, tertiary veins that are widely and irregularly spaced, and secondary veins that originate at a lower angle in the medial than in the basal part of the lamina. The canalized craspedodromy is specialized, but this might be expected in the microthermal environment occupied by the Princeton taxon.

The major leaf-architectural feature that separates the Princeton taxon from extant primitive rosids and higher hamamelidids is a chloranthoid versus a protocunonioid tooth. However, the protocunonioid tooth must almost certainly be derived from the chloranthoid tooth (Fig. 5.2). First, the chloranthoid tooth is the dominant specialized tooth type among early dicotyledons (Doyle and Hickey 1976; Hickey and Doyle 1977: Upchurch 1985), with the cunonioid (or protocunonioid?) appearing later. Second, the two tooth types differ primarily in the absence of accessory bracing veins along the apical side of the tooth and the positioning of the major vein along the apical side in the protocunonioid tooth. If the apical part of a chloranthoid tooth were condensed, the apical accessory bracing veins would probably disappear (or be greatly reduced in thickness) and the major vein would be displaced along the apical side; the strong apical branch that extends to the sinus in the typical protocunonioid/cunonioid tooth could, in fact, represent remnants of the apical accessory bracing veins of the chloranthoid tooth. In any case, derivation of a protocunonioid from a chloranthoid tooth would not involve major architectural changes. I suggest that extant basal hamamelidids such as Eupteleales and the PT clade represent divergences from a line that led to the early rosids; this line subsequently diverged, one line leading to generally acknowledged members of Rosidae and the other line to the higher hamamelidid clade (Fig. 5.5).

Other Hamamelididae

Several groups assigned by Takhtajan (1980) and Cronquist (1981) to Hamamelididae have been excluded from the above analysis. In the instances of Barbeyaceae and especially Casuarinaceae, reduction of foliage is severe and leaf architecture has little or nothing to say about affinities. Palynologically the Casuarinaceae probably are Normapolles derivatives (Walker and Doyle 1975). The entire-margined foliage of Barbeyaceae has a highly generalized venation pattern and offers no characters of taxonomic significance.

I have nothing more to offer than in my (Wolfe 1974) analysis regarding Leitneriaceae and Didymelaceae, except to point out that pollen morphology, as well as leaf architecture, does not support inclusion in either the basal hamamelidids or the higher hamamelidids (Walker and Doyle 1975) and may indicate a placement in Dilleniidae.

Balanopaceae are even more problematic. The general architecture of *Balanops* leaves is not suggestive of any hamamelidid group but is similar to that in some Dilleniidae, especially the thealean–violalean Ancistrocladaceae. The 'tooth' of *Balanops* is actually the basal end of an elongated foramen; I have been unable to determine whether the foramen was the base for a deciduous seta, as is characteristic of Theales. However, for some other hamamelidid groups, foliar evidence is less equivocal and indicates relationships to other groups.

Urticales

Previously I accepted the higher hamamelidid relationship of Urticales, based both on the possible derivation or urticalean pollen from the

Normapolles complex and on urticalean high-rank foliage (e.g. Ulmoideae) that has similarities to high-rank foliage in orders here accepted as higher hamamelidids. The predominantly porate pollen of Urticales might suggest Normapolles derivation, particularly in combination with the presence of arcoid (albeit faint) streaks. Urticalean pollen, however, does not show the clear-cut compound, pronounced aspidote apertures that are characteristic of Normapolles and derivative groups (Batten Vol. 40B, Ch. 2). In regard to leaf architecture, I have emphasized above that high-rank venation such as that of Ulmoideae has evolved independently in various hamamelidid groups (as well as in Dilleniidae and Rosidae) and should not be used uncritically as indicating close relationships between different orders.

My survey of urticalean foliage revealed no evidence of a primitive rosidean ancestry (i.e. no evidence of the glandular protocunonioid or of the aglandular cunonioid tooth). If the major tooth vein is asymetrically placed in an urticalean tooth, the displacement is toward the basal side of the tooth, rather than toward the aspical side as in primitive rosids. If a strong vein extends to or towards the urticalean sinus, this vein originates from the superjacent secondary vein (Fig. 5.6), rather than being an apical branch of the subjacent secondary vein as in primitive rosids (Fig. 5.4). If urticalean foliage is compound, it is palmately compound, rather than pinnately compound as in primitive rosids. From what is known of extant urticalean leaf architecture, no obvious relationships are apparent with primitive rosids or with basal hamamelidids.

On the other hand, Thorne (1983) has suggested and recently Takhtajan (1987) has accepted that Urticales are related to Malvales, basing this suggestion in part on some aspects of xylem anatomy (e.g. the presence of storied elements in both orders). Does leaf architecture support this alternative placement of Urticales?

Hickey and Wolfe (1975) defined both an 'urticoid' tooth and a 'malvoid' tooth. As defined, both tooth types are aglandular and have a medial vein (Fig. 5.6), but the urticoid is supposedly differentiated from the malvoid by the presence of lower-order accessory bracing veins in the urticoid. In fact, Malvales can also have teeth that have lower-order accessory bracing veins, and I can see no general difference between malvoid and urticoid teeth. However, because the malvoid–urticoid tooth is structurally simple, little weight should be placed on the occurrence of this tooth type in both Urticales and Malvales.

As shown in Table 5.1, a number of other foliar characters are shared by Urticales and Malvales, but most of these characters are not found in Hamamelididae. Of particular significance is that the palmately veined condition appears to be basic to Urticales, as it surely is to Malvales (Fig. 5.6). Unlike the chevron-like braces associated with trochodendralean

Table 5.1 Comparison of some leaf architectural characters of Urticales with Malvales and other Hamamelididae. Bo, Bombacaceae; Ca, Cannabinaceae; Ha, Hamamelidaceae; Had, higher Hamamelididae; Has, Hamamelidales; Ma, Malvaceae; Mo, Moraceae; St, Sterculiaceae; Ti, Tiliaceae; Trs, Trochodendrales; Ul, Ulmaceae; Ur, Urticaceae.

Character	Urticales	Malvales	Hamamelididae
Actinodromy	Ca, Mo, Ul, Ur	Ma, St, Ti	Has, Trs
Palmately compound	Ca	Bo	—
Primary vein braces straight or arcuate	Ca, Mo, Ul, Ur	Ma, St, Ti	—
Serrate margin with brochidodromy	Mo, Ul	Ti	—
Flattened brochidodromous arches	Mo	Bo	—
Orthogonal fourth-order veins	Mo, Ul, Ur	Ma, St, Ti	—
Admedially ramified intercostal venation	Mo	Bo	Ha
Superjacent branches to sinuses	Mo, Ul, Ur	St, Ti	—
Peltate glandular hairs	Ur	Bo, Ma, St, Ti	Had

actinodromy, the primary braces in Urticales and Malvales are straight to bowed. Orthogonal origination of fourth-order venation is characteristic of Malvales, as in some other palmate dilleniids; although Urticales do not have the strong percurrency of fourth-order venation developed in many palmate dilleniids, the urticalean fourth-order veins also originate orthogonally. The high-rank urticalean foliage previously mentioned is paralleled by equally high-rank foliage in a genus such as *Tilia*. Some Ulmaceae and Moraceae develop an almost unique type of leaf that is actinodromous, strongly brochidodromous close to the margin, and finely serrate: outside Urticales, only Tiliaceae have this type of leaf (Fig. 5.6). Overall, the syndrome of foliar characters present in Urticales is far more similar to that in Malvales than to that in Hamamelididae. This leaf-architectural analysis is not mandatory for a close relationship between Urticales and Malvales, but this appears much more satisfactory than a close relationship between Urticales and Hamamelididae.

Costa Rican Taxon

This still undescribed taxon has been suggested to be of higher hamamelidid affinity, possibly related to Fagales, Juglandales, and/or

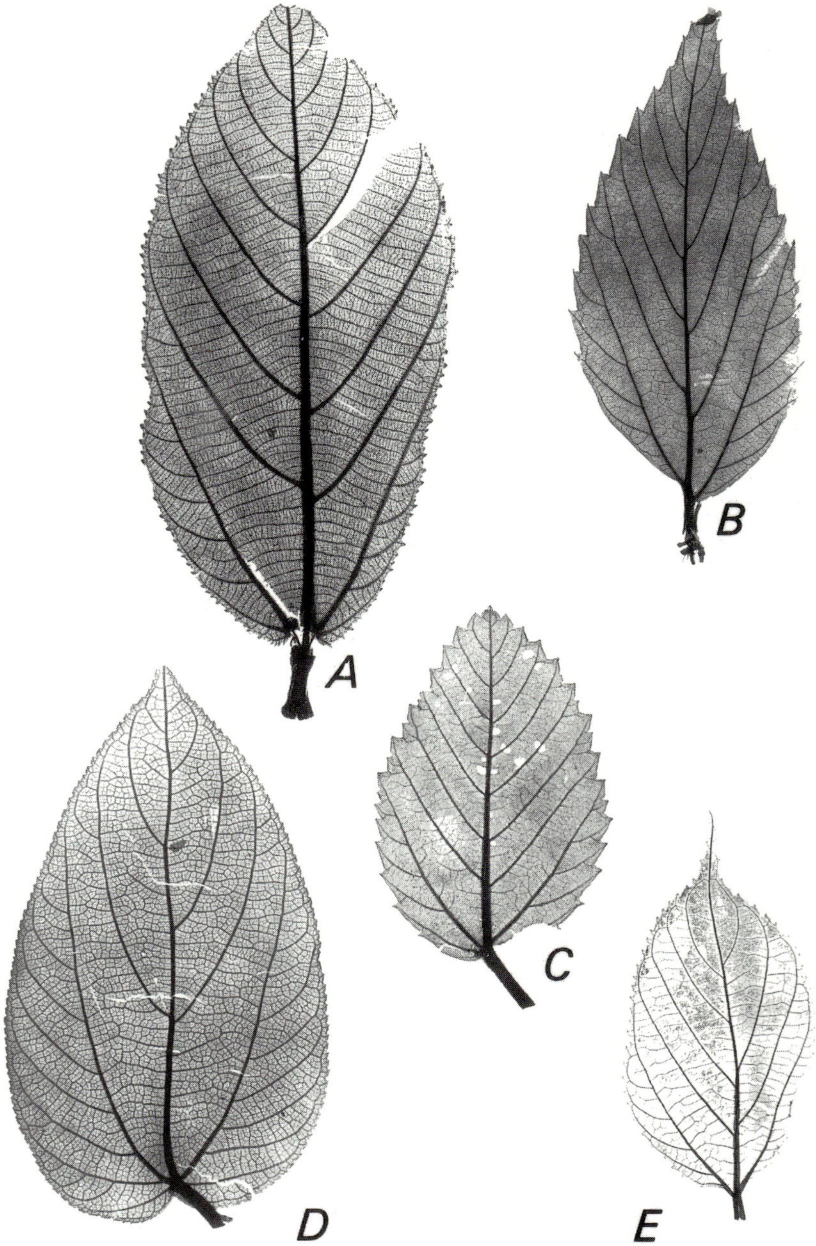

Fig. 5.6 Comparison of Urticales (B, D, E) and Malvales (A, C). All figures ×1.

Hamamelidales (Gómez-Laurito and Gómez P. personal communication, 1989). Another suggested relationship is with *Picrodendron* of Euphorbiaceae. In the context of leaf architecture, the Costa Rican taxon (CRT) has almost nothing in common with *Picrodendron* other than characters that can be found in a wide array of only distantly related dicotyledonous families. On the other hand, some gross foliar similarities are present between the CRT and high-rank foliage of particularly Fagaceae. These similarities are, however, of little phyletic significance.

The CRT, on the other hand, shows strong similarities in foliage to Dillenioideae, all of which are high rank. Most high-rank foliage is characterized by straight tertiary veins that, if branching, tend to branch orthogonally. In contrast, both Dillenioideae and the CRT have tertiary veins that have a markedly sinuous course and that tend to branch acutely (Fig. 5.7). The medial vein entering the teeth in both the CRT and the Dillenioideae expands and then ends in a clear glandular apex; the accessory veins in the tooth are weakly developed in the CRT, thus conforming to the dillenioid tooth (Hickey and Wolfe 1975). Every significant foliar character possessed by the CRT is also possessed by almost all genera of Dillenioideae.

Whether the reproductive specializations of the CRT can be derived from the generally primitive reproductive structure of Dillenioideae is not in the provenance of this analysis. The CRT could be a derivative of Dillenioideae or could represent a sister-group to Dillenioideae.

Eucommiaceae

Although traditionally included in Hamamelididae and thought to be allied to basal Hamamelididae, *Eucommia* has foliage highly anomalous for such a relationship. The secondary veins, although technically brochidodromous, extend apically in a basically eucamptodromous manner, and two sets of intramarginal loops are present (Fig. 5.8). Teeth are entered by veins that originate from the abmedial side of the loops; this major vein is placed somewhat apically in the tooth and expands before terminating in a clear, triangular gland (not a platanoid tooth as misinterpreted by Hickey and Wolfe, 1975). Neither the secondary venation pattern (including the extensive brochidodromous–eucamptodromous loops) nor the dental morphology is similar to any other taxon assigned to Hamamelididae. I (Wolfe 1974) previously suggested a possible relationship somewhere among primitive rosids, but now a relationship to the violalean–thealean Stachyuraceae appears more probable.

The basic pattern of secondary venation and extensive looping is found in both *Eucommia* and *Strachyurus* (Fig. 5.8). In both taxa the

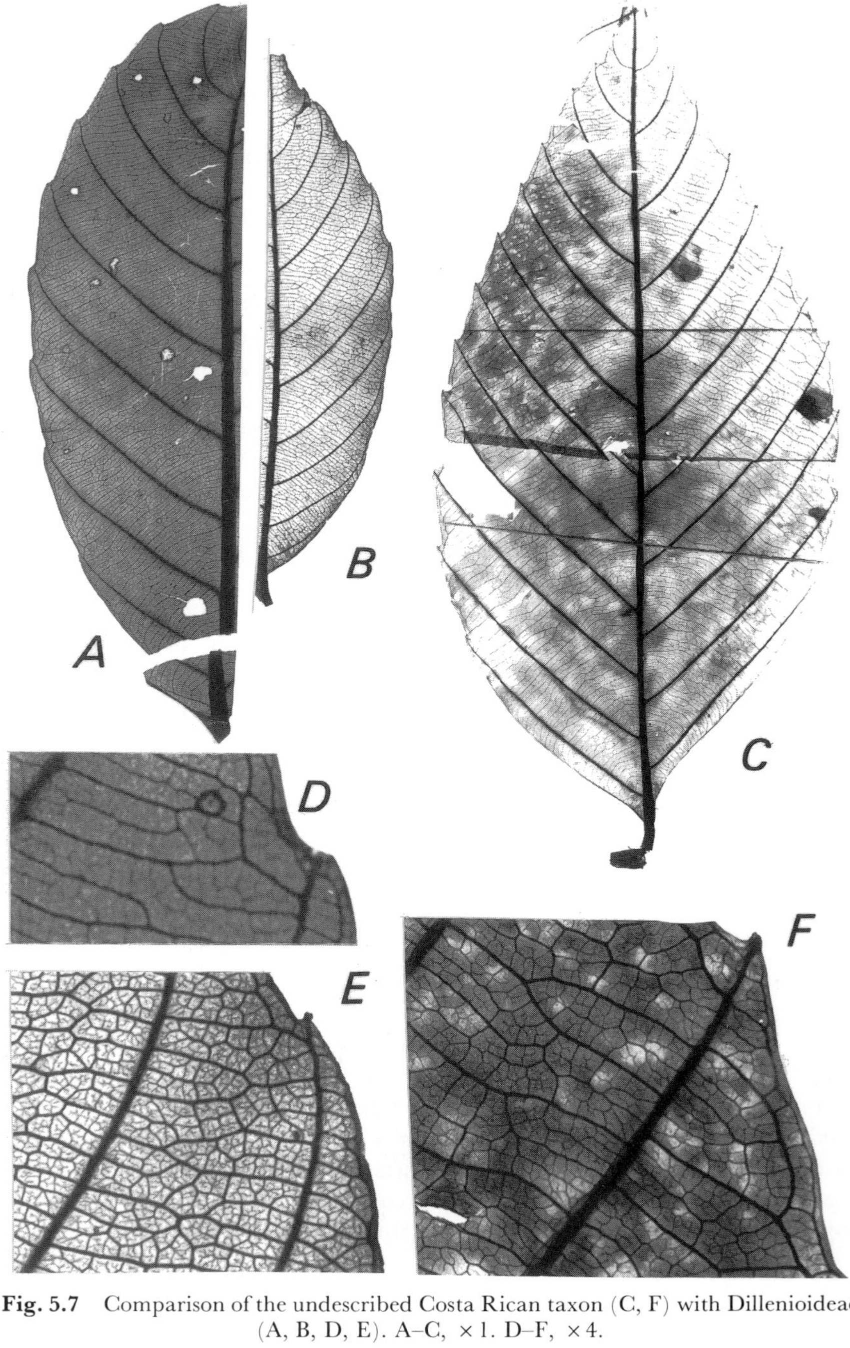

Fig. 5.7 Comparison of the undescribed Costa Rican taxon (C, F) with Dillenioideae (A, B, D, E). A–C, ×1. D–F, ×4.

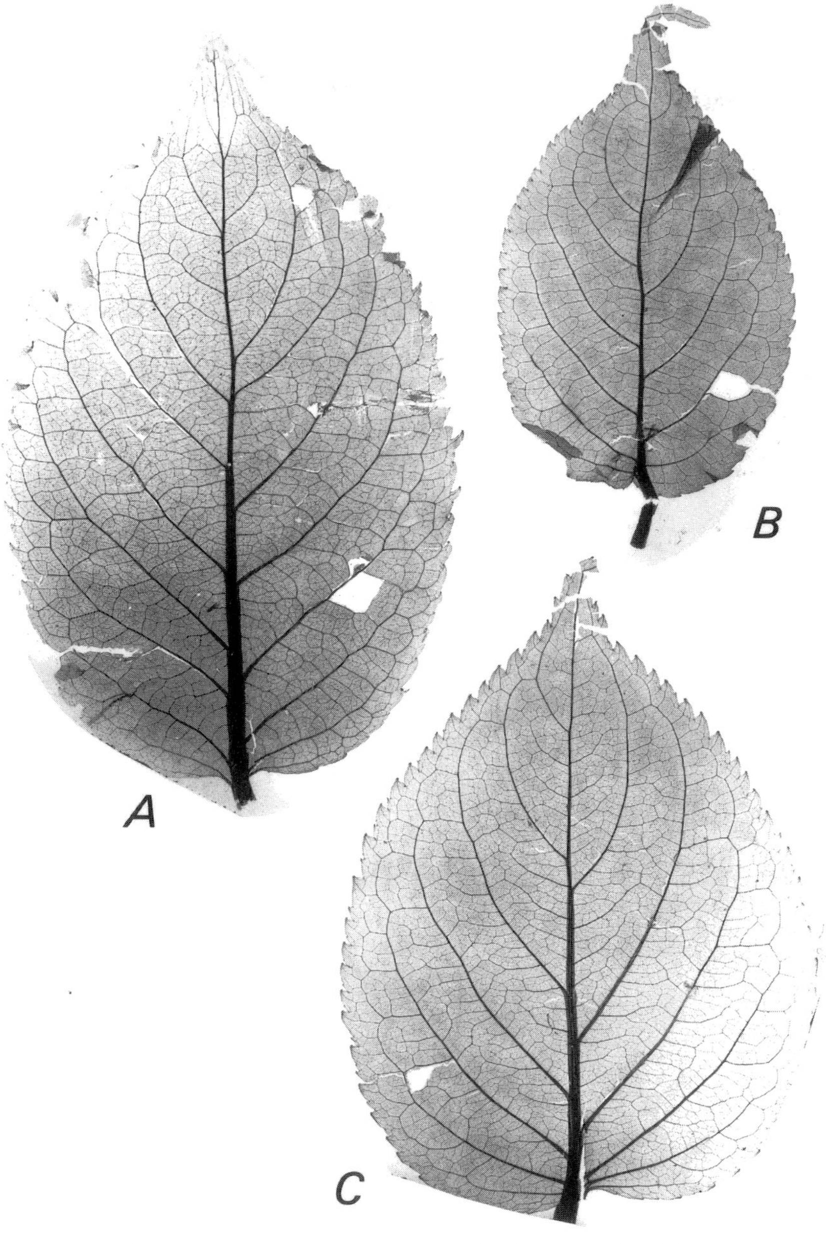

Fig. 5.8 Comparison of *Eucommia* (A) with *Stachyurus* (B, C). All figures ×1.

intercostal venation is formed by widely spaced, compound tertiary veins that originate acutely from the superjacent secondary vein and orthogonally from the subjacent secondary vein. Numerous inter-secondary veins are present, particularly in the apical third of the lamina. Both genera have a finely and sharply serrate margin. Indeed, were it not for some difference in dental morphology, differentiating foliage of *Eucommia* from foliage of *Stachyurus* would be impossible on architectural grounds.

Although the teeth of both *Eucommia* and *Stachyurus* have a thickened vein that terminates in a triangular, clear gland (Fig. 5.9), in *Stachyurus* the thickening is more conspicuous, the major vein is medial, and accessory veins are typically present. However, little morphologic change would be involved to derive the tooth of *Eucommia* from that of *Stachyurus*; if the apical part of a *Stachyurus* tooth were reduced, a tooth very much like that of *Eucommia* would result.

Considering that *Eucommia* and *Stachyurus* both inhabit mesothermal to warm microthermal forests of eastern Asia, one interpretation of the foliar similarities might be that both have adapted to the same environment. However, hamamelidids and Urticales that have adapted to this same environment typically have the high-rank foliage characterized by straight, craspedodromous secondary veins and well-organized intercostal venation. That is, the syndrome of foliar characters displayed by *Eucommia* is like the syndrome developed in microthermal member of Theales (e.g. *Stewartia*) and Violales and not like the syndrome developed in Hamamelididae or in Urticales.

Similarities in xylem anatomy are also apparent between *Eucommia* and *Stachyurus*. Both have small, solitary vessels (Metcalfe and Chalk

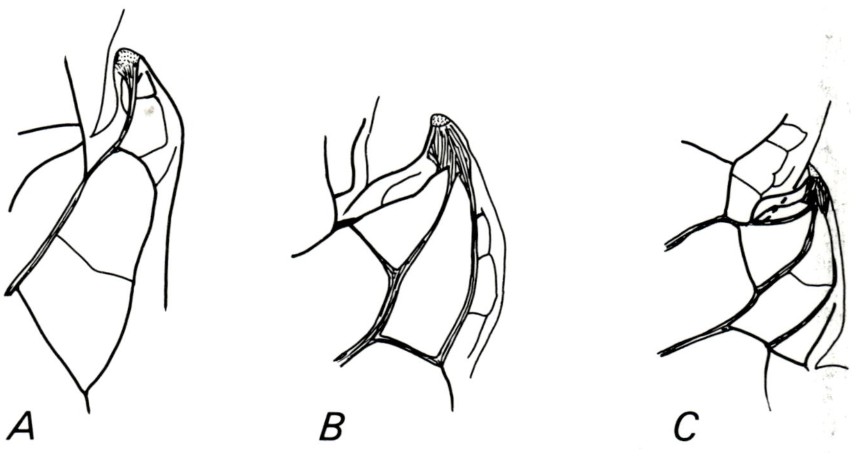

A B C

Fig. 5.9 Comparison of teeth of *Eucommia* (A) with teeth of *Stachyurus* (B, C). All figures ×12.

1972) and are generally very similar in most other aspects of xylem anatomy. Although *Stachyurus* has few-barred perforation plates and *Eucommia* has typically simple plates, scalariform plates occur next to the primary xylem in *Eucommia*. The longer vessels and fibres of *Stachyurus* could be related to the viney habit of some species. I suggest that *Eucommia* and *Stachyurus* have diverged markedly in reproductive structure but have retained a high degree of similarity in both foliar architecture and xylem anatomy.

Conclusions

As interpreted here, Hamamelididiae cannot be maintained as an even approximately monophyletic group (Fig. 5.5). Even rejecting from the hamamelidids taxa such as *Eucommia* and Urticales still leaves a large residue that is clearly polyphyletic. Certainly all the 'core' Hamamelididae ultimately have a common ancestry, but this ancestry is also involved in the derivation of Dilleniidae and Rosidae. If the remnant higher hamamelidids (Fagales, Betulales, Myricales, Juglandales) are included (perhaps as a superorder Juglandanae) within Rosidae, and if Eupteleales are also included within Rosidae, then we have a monophyletic Rosidae as a sister-group to the presumably monophyletic Dilleniidae. As pointed out above, a restricted Hamamelididae (i.e. Trochodendrales, Hamamelidales *sensu stricto* Myrothamnaceae) are probably a sister-group to the dilleniids and the rosids together.

Numerous problems remain with the remnants of Hamamelididae. Since Hamamelididae was proposed, some groups, such as Buxaceae and Daphniphyllaceae, have been squeezed into the subclass; from the standpoint of both pollen morphology and leaf morphology, neither family appears related to either basal or higher hamamelidids. The problem of *Rhodoleia* still remains. Leaf morphology of *Rhodoleia* is very primitive (Wolfe 1974), whereas the subspheroidal, tricolporate pollen is relatively specialized; neither leaf nor pollen morphology support a placement in or near Hamamelidaceae.

In this survey of putative hamamelidids, the necessity for examining as many species of a family as possible has become apparent. What might appear at first glance to be basic architectural patterns for a given family very typically turn out to be specializations, and primitive characters are restricted to a few taxa. The probable parallelism in many leaf-architectural features is also apparent; such parallelism can be used as secondary evidence of relationship if other characters indicate a primary relationship. But, if characters substantiating a primary relationship are absent, then the supposed parallelisms may actually be convergences.

I have relied heavily on morphology of teeth, including characters of glandularity and relations of veins to the teeth and sinuses. This needs to be researched in much more detail (including sectioning) to determine if the homologies assumed are valid. The classification of teeth that Leo Hickey and I proposed (Hickey and Wolfe 1975) was perhaps overly typological and needs to be refined and/or expanded to take into account the considerable variation and intergradations that can be demonstrated. Some species, moreover, have a considerably different tooth morphology in immature and/or sapling leaves than in mature leaves; glands that can be absent in mature leaves may well occur in immature and/or sapling leaves. Problems exist, particularly with families that have reduced leaves, but analogous problems occur if reproductive structures are reduced. However, leaf architecture can contribute significantly to understanding relationships between many dicotyledonous families.

Acknowledgments

For continuing discussions of the content of this paper, I particularly thank G. R. Upchurch, Jr. I have also benefited by discussions with P. R. Crane, J. A. Doyle, L. J. Hickey, R. A. Spicer, and S. L. Wing.

References

Crane, P. R. and Stockey, R. A. (1985). Growth and reproductive biology of *Joffrea speirsii* gen. et sp. nov., a *Cercidiphyllum*-like plant from the Late Paleocene of Alberta, Canada. *Canadian Journal of Botany*, **63**, 340–64.

Crane, P. R., Manchester, S. R. and Dilcher, D. L. (1988). Morphology and phylogenetic significance of the angiosperm *Platanites hebridicus* from the Palaeocene of Scotland. *Palaeontology* **31**, 503–17.

Crane, P. R., Manchester, S. R. and Dilcher, D. L. Fossil leaves and well preserved reproductive structures from the Fort Union Formation (Paleocene) near Almont, North Dakota, U.S.A. *Fieldiana, Geology (N.S.)* in press.

Cronquist, A. (1981). *An integrated system of classification of flowering plants.* Columbia University Press, New York.

Doyle, J. A. (1969). Cretaceous angiosperm pollen of the Atlantic Coastal Plain and its evolutionary significance. *Journal of the Arnold Arboretum* **50**, 1–35.

Doyle, J. A. and Hickey, L. J. (1976). Pollen and leaves from the mid-Cretaceous Potomac Group and their bearing on early angiosperm evolution. In *Origin and early evolution of angiosperms.* (ed. C. B. Beck), pp. 139–206. Columbia University Press, New York.

Friis, E. M. (1983). Upper Cretaceous (Senonian) floral structures of Juglandalean affinity containing Normapolles pollen. *Review of Palaeobotany and Palynology* **39**, 161–88.

Furlow, J. J. (1979). The systematics of the American species of *Alnus* (Betulaceae). *Rhodora* **81,** 1–121, 151–248.

Góczán, F., Groot, J. J., Krutzsch, W. and Pacltová, B. (1967). Die Gattungen des "Stemma Normapolles Pflug, 1953b" (Angiospermae). *Palaontologische Abhandlungen B* **2,** 427–633.

Hergert, H. L. and Phinney, H. K. (1954). *Trochodendroxylon beckii* gen. et sp. nov. from the Tertiary of Oregon. *Bulletin of the Torrey Botanical Club* **81,** 118–22.

Hickey, L. J. (1973). Classification of the architecture of dicotyledonous leaves. *American Journal of Botany* **60,** 17–33.

Hickey, L. J. (1978). Origin of the major features of angiospermous leaf architecture in the fossil record. *Courier Forschungsinstitut Senckenberg* **30,** 27–34.

Hickey, L. J. and Doyle, J. A. (1977). Early Cretaceous fossil evidence for angiosperm evolution. *Botanical Review* **43,** 3–104.

Hickey, L. J. and Wolfe, J. A. (1975). The bases of angiosperm phylogeny: Vegetative morphology. *Annals of the Missouri Botanical Garden* **62,** 538–89.

Levin, G. A. (1986). Systematic foliar morphology of Phyllanthoideae (Euphorbiaceae). I. Conspectus. *Annals of the Missouri Botanical Garden* **73,** 29–85.

Manchester, S. R. (1986). Vegetative and reproductive morphology of an extinct plane tree (Platanaceae) from the Eocene of western North America. *Botanical Gazette* **147,** 200–26.

Manchester, S. R. (1987). The fossil history of the Juglandaceae. *Monographs in Systematic Botany from the Missouri Botanical Garden* **21,** 1–137.

Manchester, S. R. and Crane, P. R. (1983). Attached leaves, inflorescences, and fruits of *Fagopsis*, an extinct genus of fagaceous affinity from the Oligocene Florissant flora of Colorado, U.S.A. *American Journal of Botany* **70,** 1147–64.

Manchester, S. R. and Crane, P. R. (1987). A new genus of Betulaceae from the Oligocene of western North America. *Botanical Gazette* **148,** 263–73.

Metcalfe, C. R. and Chalk, L. (1972). *Anatomy of the dicotyledons.* Oxford University Press, Oxford.

Muller, J. (1970). Palynological evidence on early differentiation of angiosperms. *Biological Reviews* **45,** 417–50.

Rury, P. M. and Dickison, W. C. (1977). Leaf venation patterns of the genus *Hibbertia* (Dilleniaceae). *Journal of the Arnold Arboretum* **58,** 209–241.

Scott, R. A. and Wheeler, E. A. (1982). Fossil woods from the Eocene Clarno Formation of Oregon. *International Association of Wood Anatomists Bulletin n.s.* **3,** 135–54.

Spicer, R. A. (1986). Comparative leaf architectural analysis of Cretaceous radiating angiosperms. In *Systematic and taxonomic approaches in palaeobotany* (ed. R. A. Spicer and B. A. Thomas), pp. 221–32. Oxford University Press, Oxford.

Takhtajan, A. (1980). Outline of the classification of flowering plants (Magnoliophyta). *Botanical Review* **46,** 225–359.

Takhtajan, A. (1987). *Systema Magnoliophytorum* [in Russian]. Nauka, Leningrad.

Thorne, R. F. (1983). Proposed new realignments in the angiosperms. *Nordic Journal of Botany* **3,** 85–117.

Upchurch, G. R., Jr. (1985). Cuticle evolution in Early Cretaceous angiosperms from the Potomac Group of Virginia and Maryland. *Annals of the Missouri Botanical Garden* **71,** 522–50.

Walker, J. W. and Doyle, J. A. (1975). The bases of angiosperm phylogeny: Palynology. *Annals of the Missouri Botanical Garden* **62,** 664–723.

Wing, S. L. and Hickey, L. J. (1984). The *Platycarya* perplex and the evolution of the Juglandaceae. *American Journal of Botany* **71,** 388–411.

Wolfe, J. A. (1974). Fossil forms of Amentiferae. *Brittonia* **25,** 334–55.

Wolfe, J. A. (1976). Stratigraphic distribution of some pollen types from the Campanian and lower Maestrichtian rocks (Upper Cretaceous) of the Middle Atlantic States. *United States Geological Survey Professional Paper* **977,** 1–18.

Wolfe, J. A. and Wehr, W. (1987). Middle Eocene dicotyledonous plants from Republic, northeastern Washington. *United States Geological Survey Bulletin* **1597,** 1–25.

Wolfe, J. A., Doyle, J. A. and Page, V. M. (1975). The bases of angiosperm phylogeny: Paleobotany. *Annals of the Missouri Botanical Garden* **62,** 801–24.

6. Sieve-element plastids, phloem proteins, and the evolution of flowering plants. IV. Hamamelidae

H.-DIETMAR BEHNKE

Zellenlehre, Universität Heidelberg, D-6900 Heidelberg 1, FRG

Abstract

Twenty-five species from 16 families of Hamamelidales (and allies), 18 species of five families of Fagales and Casuarinales, and 46 species from the Urticales have been studied by TEM methods for their sieve-element characters. Sieve-element plastids in the first group are of S-type, have an average diameter of 1.5 μm, and contain between five and ten starch grains. In the Fagales–Casuarinales sieve-element plastids are also of S-type but with an average diameter of only 1.2 μm, while in the Urticales sieve-element plastids are of P- or S-type, with four forms (Pc, Pcs, S, So) in Ulmaceae and the two S-forms (S, So) in the other families. The diameter of sieve-element plastids in the Urticales is considerably smaller (average of 0.9 μm) than in other groups examined. Non-dispersive phloem(P)-protein bodies occur in some taxa but are absent from Hamamelidales *sensu lato*: compound-spherical or spindle-shaped bodies are found in the Fagales, Casuarinales, and *Tetracentron*; globular bodies occur in the Urticales. Among the taxa covered here and judged on the basis of sieve-element plastids the Urticales are the more advanced group, not directly related to either Hamamelidae or Malviflorae. However, non-dispersive P-protein bodies common to Urticales are also found in Fagales and Malvales and might link these three taxa. In addition, similarities in P-type plastids would favour a closer relationship between Ulmaceae and Buxaceae.

Evolution, Systematics, and Fossil History of the Hamamelidae, Volume 1: Introduction and 'Lower' Hamamelidae (eds P. R. Crane and S. Blackmore), Systematics Association Special Volume No. 40A, pp. 105–128. Clarendon Press, Oxford, 1989.

Introduction

Ultrastructural characters from sieve elements have been used as an additional tool for the circumscription and putative phylogenetic arrangement of families and other taxa of angiosperms (e.g. Behnke 1976, 1981*a*; Behnke and Pop 1981; Behnke 1988). A first report on sieve-element plastids in Takhtajan's (1969) Hamamelididae considered 48 species in 16 families (Behnke 1973). Since then the number of hamamelid species studied has been increased and new information on plastids and non-dispersive P-protein bodies has become available. This, together with the comparative data available from over 2000 dicotyledon species, makes an evaluation of evolutionary trends within sieve-element plastids more feasible. While Cronquist's (1981) Hamamelidae define the extent of the present study, the dismemberment of this subclass and placement of some of its orders into the dilleniid (Takhtajan 1987), rosanaean (Dahlgren 1989), or malvanaean alliance (Thorne 1983, Dahlgren 1989) will be discussed with regard to the sieve-element data.

Material and methods

Eighty nine species of 27 families, either all placed within the Hamamelidae (Cronquist 1981) or partly assigned to Dilleniidae (Takhtajan 1987) or Malviflorae (Thorne 1983) and/or Rosanae (Dahlgren 1989), and 12 species of the Buxaceae affiliated with the hamamelidaean families by both Dahlgren (1989) and Takhtajan (1987) were investigated. Table 6.1 lists all species studied, their origin, herbarium voucher specimens, and the sieve-element data used.

Living material (recently removed from the plant or shipped under special care) is a prerequisite for fixation of sieve elements and investigation of their plastids with the transmission electron microscope (TEM). Thin hand sections were made with a razor blade from herbaceous shoots or end parts of tree branches. The sections were immersed into a fixing solution containing formaldehyde and glutaraldehyde and processed according to standard TEM methods (see Behnke 1982).

Results and discussion

The sieve-element characters used

Sieve-element plastids Using their ergastic contents (i.e. protein and starch accumulations), sieve-element plastids are categorized into types,

Table 6.1 The sieve-element characters of the investigated species of Hamamelidae arranged alphabetically according to the families covered (Fam is the family acronym after Weber 1982).

Fam	Species name	Origin	Herbarium	Sieve-element plastids			Non dispersive P-protein
				Type	Average diameter (μm)	Average Number of starch grains	
ALT	*Liquidambar styraciflua* L.	BG-HEID	HEID	S	1.6	7	
BET	*Alnus glutinosa* (L.) Gaertn.	BG-HEID		S	1.2		
BET	*Betula papyrifera* Sukatsch.	BG-HEID	HEID	S	1.2		gl
BET	*Carpinus betulus* L.	BG-HEID	HEID	S	1.2	5	
BET	*Ostrya japonica* Sarg.	BG-HEID	HEID	S	1.2	5	
BLN	*Balanops australiana* F. Muell.	BG-MO	MO	S	1.4	10	
BLN	*Balanops vieillandii* Baill.	BG-WU	HEID/WU	S	1.2	5	
BRB	*Barbeya oleoides* Schweinf.	T. Ebba (s.n.)		S	1.2	10	
BUX	*Buxus balearica* Lam.	BG-HEID	HEID	Pcs	0.8	2	
BUX	*Buxus glomerata* Muell.	E. Koehler		Pcs	0.8	1	
BUX	*Buxus harlandii* Hance	BG-BONN	HEID	Pcs	0.8	5	
BUX	*Buxus macowani* Oliver in Hook. (=*Notobuxus*)	BG-PRE	PRE/HEID	Pc	0.7		
BUX	*Buxus microphylla* Sieb. & Zucc. var. *japonica* Muel	BG-HEID	HEID	Pcs	0.8	5	
BUX	*Buxus natalensis* (Oliv.) Hutchinson (=*Notobuxus*)	BG-PRE	PRE/HEID	Pc	0.8		
BUX	*Buxus sempervirens* L.	BG-HEID	HEID	Pcs	0.7	5	
BUX	*Pachysandra terminalis* Sieb. et Zucc.	BG-HEID	HEID	Pc	0.8		
BUX	*Sarcococca humilis* (Rehd. et Wils.) Stapf ex Seal	BG-FI		Pc	0.7		

Table 6.1—*continued*

Fam	Species name	Origin	Herbarium	Sieve-element plastids			Non dispersive P-protein
				Type	Average diameter (µm)	Average Number of starch grains	
BUX	*Sarcococca pruniformis* Lindl. (*hookeriana* Baill.)	BG-WU	HEID	Pc	0.9		
BUX	*Sarcococca ruscifolia* Stapf	BG-HEID	HEID	Pc	0.9		
BUX	*Sarcococca saligna* (D. Don.) Muell.	BG-HEID	HEID	Pc	0.9		
CAN	*Cannabis sativa* L.	BG-HEID		S	1.0	1	
CAN	*Humulus lupulus* L.	BG-HEID	HEID	S	1.0	1	
CAS	*Casuarina cunninghamii* Miq.	BG-HEID		S	1.1	5	cs
CAS	*Casuarina equisetifolia* L.	BG-M		S	1.1	5	cs
CAS	*Casuarina stricta* Dryand	BG-HEID		S	1.0	2	cs
CDP	*Cercidiphyllum japonicum* Sieb. et Zucc.	BG-BONN	HEID	S	1.9	10	
CEC	*Cecropia pachystachya* Trec.	BG-BR	HEID	So	0.8		
CEC	*Cecropia peltata* L.	BG-HEID		S	1.1	1	
CEC	*Coussapoa schottii* Miq.	BG-P	HEID	S	0.9	3	gl
CEC	*Musanga cecropioides* R. Br.	BG-BR	HEID	S	1.2		
CRL	*Corylus colurna* L.	BG-HEID	HEID	S	1.1	10	
CRL	*Corylus maxima* Mill.	BG-HEID	HEID	S	1.1	8	
DPH	*Daphniphyllum macropodium* Miq.	BG-CP		S	1.3	10	
ECM	*Eucommia ulmoides* Oliv.	BG-HEID	HEID	S	1.8	10	
EPT	*Euptelea pleiosperma* Hook. f. et Thoms.	BG-BONN	HEID	S	1.3	10	
FAG	*Castanea sativa* Mill.	HDB7205/871018	HEID	S	1.3	5	cs
FAG	*Fagus sylvatica* L.	BG-HEID	HEID	S	1.4	6	sp

FAG	*Lithocarpus densiflora* Rehder	BG-E		S	0.8	5	cs
FAG	*Quercus edulis* Makino (=*Pasania*)	BG-E		S	0.8	3	cs
FAG	*Quercus glauca* Thunb. (=*Cyclobalanopsis*)	BG-E		S	0.9	2	cs
FAG	*Quercus robur* L.	BG-HEID	HEID	S	1.3	5	cs
FAG	*Trigonobalanus verticillata* Forman	BG-E		S	0.9	4	cs
HAM	*Corylopsis spicata* Sieb. et Zucc.	BG-HEID		S	1.0	5	
HAM	*Distylium racemosum* Sieb. et Zucc.	BG-HEID	HEID	S	1.3	5	
HAM	*Fothergilla monticola* Ashe	BG-HEID	HEID	S	1.3	5	
HAM	*Hamamelis mollis* Oliver ex Forb. & Hemsl.	BG-HEID		S	1.0	5	
HAM	*Parrotia persica* (DC.) C.A. Mey	BG-HEID	HEID	S	1.0	5	
HAM	*Parrotiopsis jaquemontiana* Rehder	BG-HEID	HEID	S	1.9	10	
JUG	*Carya ovata* (Mill.) K. Koch	BG-HEID		S	1.9	10	
JUG	*Juglans americana* hort.	BG-HEID		S	1.9	15	
JUG	*Juglans cordiformis* Maxim.	BG-HEID	HEID	S	1.7	5	
JUG	*Pterocarya fraxinifolia* (Poir.) Spach.	BG-HEID	HEID	S	1.4	10	
LTN	*Leitneria floridana* Chapm.	B.L. Turner		S	1.1	5	
MOR	*Artocarpus integrifolia* L.	BG-E		S	1.1	5	cs
MOR	*Brosimum alicastrum* Sw.	BG-HEID	HEID	S	0.8	3	cs
MOR	*Broussonetia papyrifera* Vent.	BG-CP		So	0.6		
MOR	*Cardiogyne africana* Bur.	BG-BR	HEID	So	1.3		
MOR	*Castilla ulei* Warb.	BG-RB-Kubitzki	HEID	So	0.8		cs
MOR	*Ficus altissima* Blume	BG-HEID	HEID	So	0.7		
MOR	*Ficus benjamina* L.	BG-HEID	HEID	So	0.6		tu
MOR	*Ficus carica* L.	BG-HEID	HEID	So	1.2		
MOR	*Maclura pomifera* (Raf.) Schneid.	BG-HEID	HEID	So	0.8		
MOR	*Morus alba* L.	BG-HEID	HEID	S	1.1	2	
MOR	*Pseudomorus brunoniana* Bur.	BG-K	HEID	S	1.3	8	
MOR	*Sorocea ilicifolia* Miq.	BG-RB-Kubitzki		S	0.9	3	
MOR	*Treculia africana* Decne. ex Trec.	BG-K	HEID	So	0.8		
MTM	*Myrothamnus flabellifolius* Welw.	BG-HEID	HEID	S		(5)	

Table 6.1—*continued*

Fam	Species name	Origin	Herbarium	Sieve-element plastids			Non dispersive P-protein
				Type	Average diameter (μm)	Average Number of starch grains	
MYR	*Myrica californica* Cham. & Schlecht.	BG-E		S	1.2	15	
NTH	*Nothofagus obliqua* (Mirb.) Oerst.	BG-HEID	HEID	S	1.8	10	sp
NTH	*Nothofagus procera* (Poepp. et Endl.) Oerst.	BG-CP		S	1.3	10	
PLT	*Platanus occidentalis* L.	BG-HEID	HEID	S	1.4	10	
RHD	*Rhodoleia championii* Hook.	BG-BO-Cibodas	HEID	S	1.2	10	
SMM	*Simmondsia chinensis* (Link) C.K. Schneid.	W.C. Oechel	HEID	S	1.0	8	
TRC	*Trochodendron aralioides* Sieb. et Zucc.	BG-CP/BONN	HEID	S	1.5	6	
TTR	*Tetracentron sinense* Oliv.	BG-BONN/DORT		S	1.5	5	sp
ULM	*Aphananthe aspera* Planch. in DC.	BG-BONN	HEID	S	1.0	3	gl
ULM	*Celtis australis* L.	BG-HEID	HEID	So	0.7		
ULM	*Celtis bungeana* Blume	BG-HEID	HEID	So	0.7		
ULM	*Chaetacme aristata* Planch.	BG-BR	HEID	Pcs	0.7	1	gl
ULM	*Gironniera subaequalis* Planch.	BG-BO	HEID	S	0.9	8	

ULM	*Hemiptelea davidii* (Hance) Planch.	BG-A	AA-5283	So	0.7		
ULM	*Holoptelea integrifolia* Planch.	BT-FTG/HBG	HEID	So	0.8		
ULM	*Planera aquatica* J. F. Gmel.	BG-BONN	HEID	Pc	0.8		
ULM	*Pteroceltis tatarinowii* Maxim.	BG-MOAR		S	1.2	5	gl
ULM	*Trema guinensis* Priemer	BG-FTG		S	1.2	10	gl
ULM	*Trema micrantha* Blume	BG-U	HEID	S	1.0	5	gl
ULM	*Ulmus carpinifolia* Gled.	BG-HEID	HEID	Pc	0.7		
ULM	*Ulmus fulva* Michx.	BG-HEID	HEID	Pc	0.7		gl
ULM	*Ulmus glabra* Huds.	BG-HEID		Pc	0.9		
ULM	*Zelkova carpinifolia* (Pall.) K. Koch	BG-K		Pc	0.9		
ULM	*Zelkova cretica* Spach.	BG-BONN	HEID	Pc	0.8		
ULM	*Zelkova serrata* (Thunb.) Mak.	BG-SIENA		Pc	0.9		
URT	*Boehmeria biloba* Miq.	BG-HEID		So	1.0		gl
URT	*Elatostema sessilis* Forst.	BG-BONN		S	0.7	3	
URT	*Laportea moroides* Wedd.	BG-BONN/HEID	HEID	So	1.1		gl
URT	*Myriocarpa longipes* Liebm.	BG-HEID		S	1.1	1	
URT	*Parietaria judaica* L. (=*ramiflora* Moench.)	HDB721605		S	1.0	1	
URT	*Parietaria officinalis* L.	BG-HEID	HEID	S	0.8	1	
URT	*Pellionia pulchra* N. E. Br.	BG-HEID		So	0.8		
URT	*Pilea repens* Liebm.	BG-BONN		So	0.8		
URT	*Urtica dioica* L.	BG-HEID/BONN	HEID	So	0.8		

cs, compound spherical; gl, globular; so, spindle shaped; tu, tubular; c, protein crystals; f, protein filaments; o, neither protein crystals nor filaments.

subtypes, and forms (Behnke 1981*b*). The two *types* are defined by presence (P-type) or absence (S-type) of protein accumulations, while each *form* is defined by a specific combination of three characters; starch grains (s), protein crystals (c), and protein filaments (f), or by none of these (o). The eight forms represented among the flowering plants can be explained as the maximum number of forms resulting from a free combination of the characters and at the same time being qualitatively distinct (cf. Behnke 1988): i.e. the six forms which belong to the P-type (c, f, cf, cs, fs, cfs) and the two forms of the S-type (s, o). Four forms of sieve-element plastids (Pc, Pcs, S, So) were recorded in the Hamamelidae. The presence of distinct morphological patterns among the sieve-element plastids of different taxa may unify several forms into a *subtype* (e.g. cuneate crystals in the P2-subtype, cf. Behnke 1981*a*). Quantitative aspects (size of plastids and amount of starch *vs.* protein accumulations) may provide additional characters, if there is a broad intergradation between S-type and P-type plastids (e.g. in the Magnoliidae, Behnke 1988).

Non-dispersive phloem (P)-protein bodies: P-protein is a characteristic component of angiosperm sieve elements which in young cells aggregates into small protein bodies (Fig. 6.1) and during cell differentiation disperses throughout the lumen (Behnke 1989). In some families an additional non-dispersive P-protein is formed which persists as discrete bodies in the mature sieve element. Several shapes of non-dispersive P-protein bodies have been described and their presence recorded within several, basically woody dicotyledon families (Behnke 1981*b*).

Distribution of the sieve-element characters within families of Hamamelidae

The families covered in the following are those recognized among the Hamamelidae or putatively related taxa by either Cronquist (1981), Dahlgren (1989), Takhtajan (1987), Thorne (1983), or in this symposium. The order below is alphabetical and the grouping of families suggested by sieve-element characters is given in Table 6.2 (plant material of Didymelaceae, Rhoipteleaceae, and Stylocerataceae was not available).

Altingiaceae (ALT): The sieve-element plastids of *Liquidambar* are of S-type with up to ten globular starch grains. With a diameter of about 1.6 μm these plastids are larger than those of (other) Hamamelidaceae (see also Behnke 1973, fig. 1).

Balanopaceae (BLN; Fig. 6.4): Both species investigated of this monogeneric family contain S-type plastids showing normal dicoty-

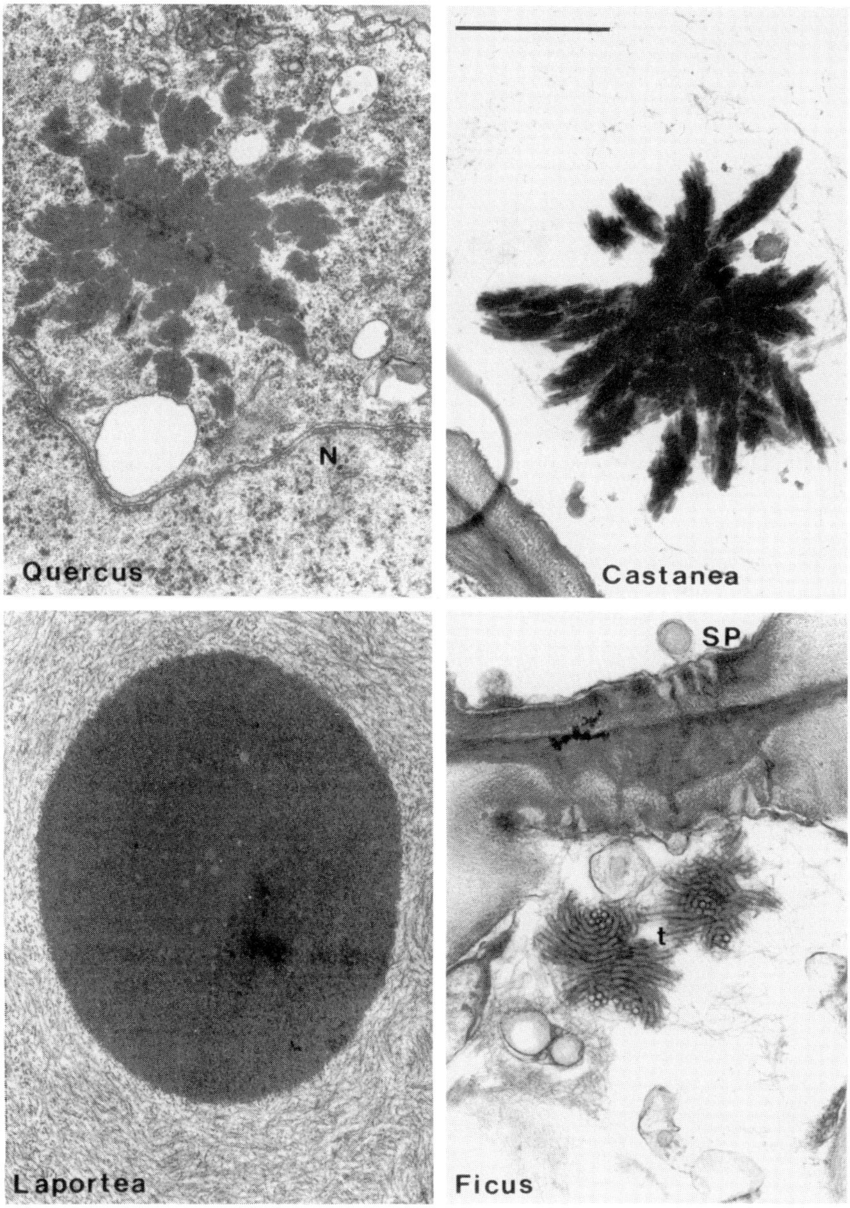

Fig. 6.1 Non-dispersive P-protein bodies in sieve elements: Compound spherical bodies in *Quercus robur* and *Castanea sativa*, globular body in *Laportea moroides*, and tubular bodies (t) in *Ficus benjaminae*. N, nucleus of young sieve element; SP, sieve pores; all ×20000. Scale bar = 1 μm.

Table 6.2 The families of the Hamamelidae (and allies) arranged according to their sieve-element characters. (Fam is the family acronym after Weber 1982, *n* is the number of species studied, and st the average number of starch grains in plastids.)

Fam	Taxon: Family name	Sieve-element plastids: Types	S-type *n*	S-type Diameter (μm)	S-type st	P type no.	P type diameter (μm)	Form So no.	Form So diameter (μm)	Non-dispersive P-protein
ECM	Eucommiaceae	S	1	1.7	10					
LTN	Leitneriaceae	S	1	1.1	5					
TTR	Tetracentraceae	S	1	1.5	5					+
TRC	Trochodendraceae	S	1	1.5	6					
CDP	Cercidiphyllaceae	S	1	1.6	10					
EPT	Eupteleaceae	S	1	1.3	10					
ALT	Altingiaceae	S	1	1.6	7					
HAM	Hamamelidaceae	S	6	1.3	6					
RHD	Rhodoleiaceae	S	1	1.2	10					
PLT	Platanaceae	S	1	1.7	10					
MTM	Myrothamnaceae	S	1		(5)					

BLN	Balanopaceae	S	2	1.3	8					
JUG	Juglandaceae	S	4	1.7	10					
MYR	Myricaceae	S	1	1.2	15					
DPH	Daphniphyllaceae	S	1	1.3	10					
NTH	Nothofagaceae	S	2	1.6	10					+
FAG	Fagaceae	S	7	1.1	4					+
BET	Betulaceae	S	4	1.2	5					+
CRL	Corylaceae	S	2	1.1	9					
CAS	Casuarinaceae	S	3	1.1	4					+
BRB	Barbeyaceae	S	1	1.2	10					
MOR	Moraceae	S,So	5	1.0	4			8	0.9	+
CAN	Cannabaceae	S	2	1.0	1					
CEC	Cecropiaceae	S,So	3	1.1	2			1	0.8	+
URT	Urticaceae	S,So	4	0.9	2			5	0.9	+
ULM	Ulmaceae	Pc(s),S,So	5	1.1	6	8	0.8	4	0.7	+
BUX	Buxaceae	Pc(s)				12	0.8			
SMM	Simmondsiaceae	S	1	1.0	8					

ledon features (diameter 1.3 µm; up to ten medium-sized, globular starch grains).

Barbeyaceae (BRB; Fig. 6.6): The monotypic *Barbeya oleoides* has S-type plastids (diameter 1.2 µm) with up to ten starch grains.

Betulaceae (BET; Fig. 6.4): Four species representing four of the five genera were investigated, all contain S-type plastids with an average diameter of about 1.2 µm and five globular starch grains. Globular non-dispersive protein bodies have been found in mature sieve elements of *Betula papyrifera* but have not been recorded from other species, so far.

Buxaceae (BUX; Fig. 6.4): Twelve species in three genera were investigated, all contain P-type plastids with rather peculiar characteristics (Pc or Pcs; diameter 0.8 µm; see also Behnke 1982).

Cannabaceae (CAN; Fig. 6.6): *Cannabis* and *Humulus* contain S-type plastids with almost identical characters: one large globular starch grain in a small plastid (diameter 1.0 µm). Other sieve-element features, such as endoplasmic reticulum (ER) membranes surrounding the plastids or aggregated to crystal-like complexes, make a distinction of the two genera on this basis impossible. See Behnke (1973) for a plastid of *Humulus*.

Casuarinaceae (CAS; Fig. 6.4): Three species in this monogeneric family were investigated, all contain small S-type plastids (diameter about 1.1 µm) with up to five starch grains. Sieve elements of the studied species also incorporate compound-spherical non-dispersive P-protein bodies.

Cecropiaceae (CEC; Fig. 6.6): Four species in three genera were investigated. All exhibit S-type plastids but one species has plastids of the So form. Plastid diameters range between 0.8 and 1.2 µm. Starch grains are few. A globular non-dispersive P-protein body was found in mature sieve elements of *Coussapoa*.

Cercidiphyllaceae (CDP; Fig. 6.2): *Cercidiphyllum japonicum* contains large S-type plastids (diameter 1.9 µm) which are often surrounded by ER membranes (see also Behnke 1971).

Corylaceae (CRL; Fig. 6.4): The two species investigated of this monogeneric family contain small S-type plastids (diameter 1.1 µm) with about ten starch grains.

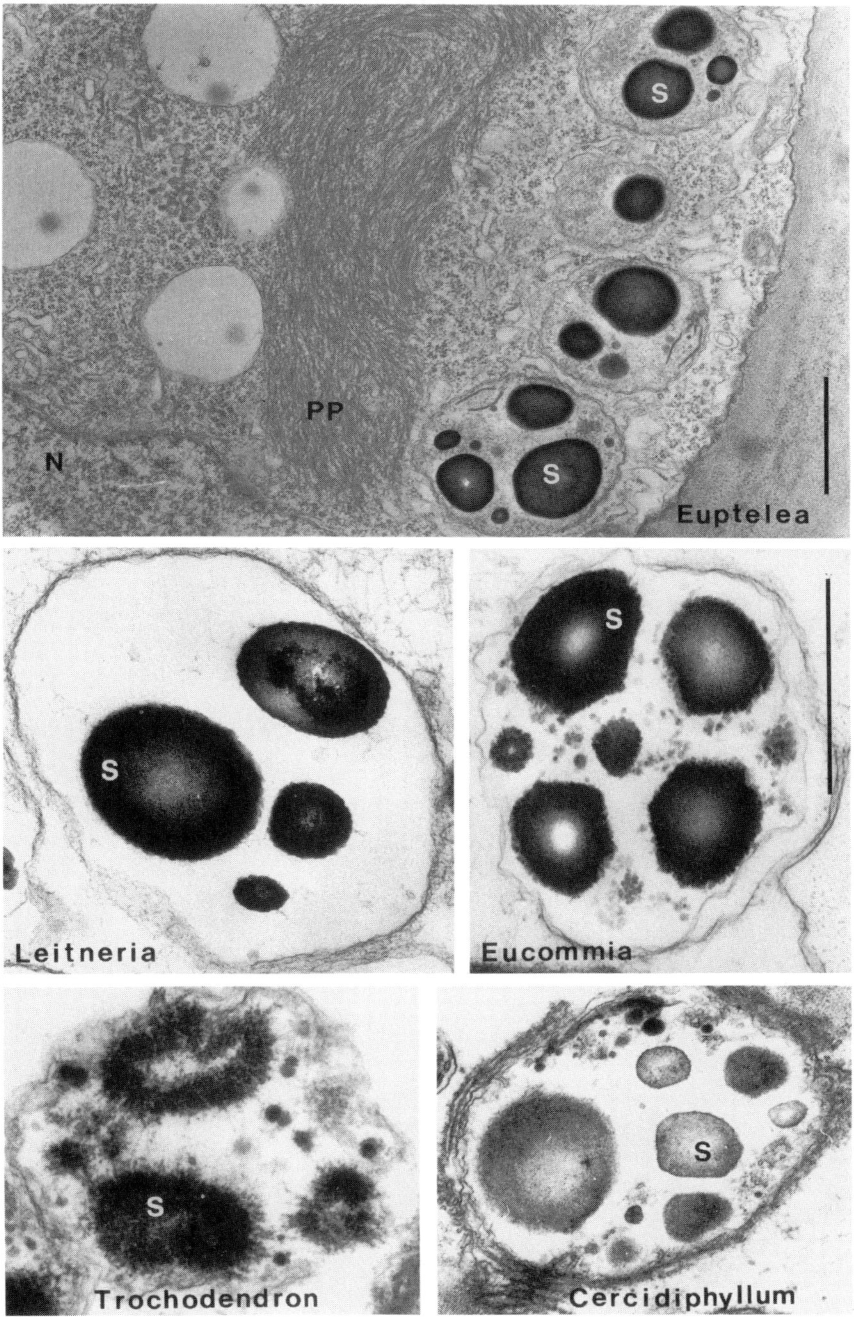

Fig. 6.2 Young sieve element of *Euptelea pleiosperma* (Eupteleaceae) with nucleus (N), P-protein (PP), and S-type plastids (S), ×16000. S-type sieve-element plastids of Leitneriaceae (*Leitneria floridana*), Eucommiaceae (*Eucommia ulmoides*), Trochodendraceae (*Trochodendron aralioides*), and Cercidiphyllaceae (*Cercidiphyllum japonicum*). S, starch grains; all ×30000. Scale bars = 1 μm.

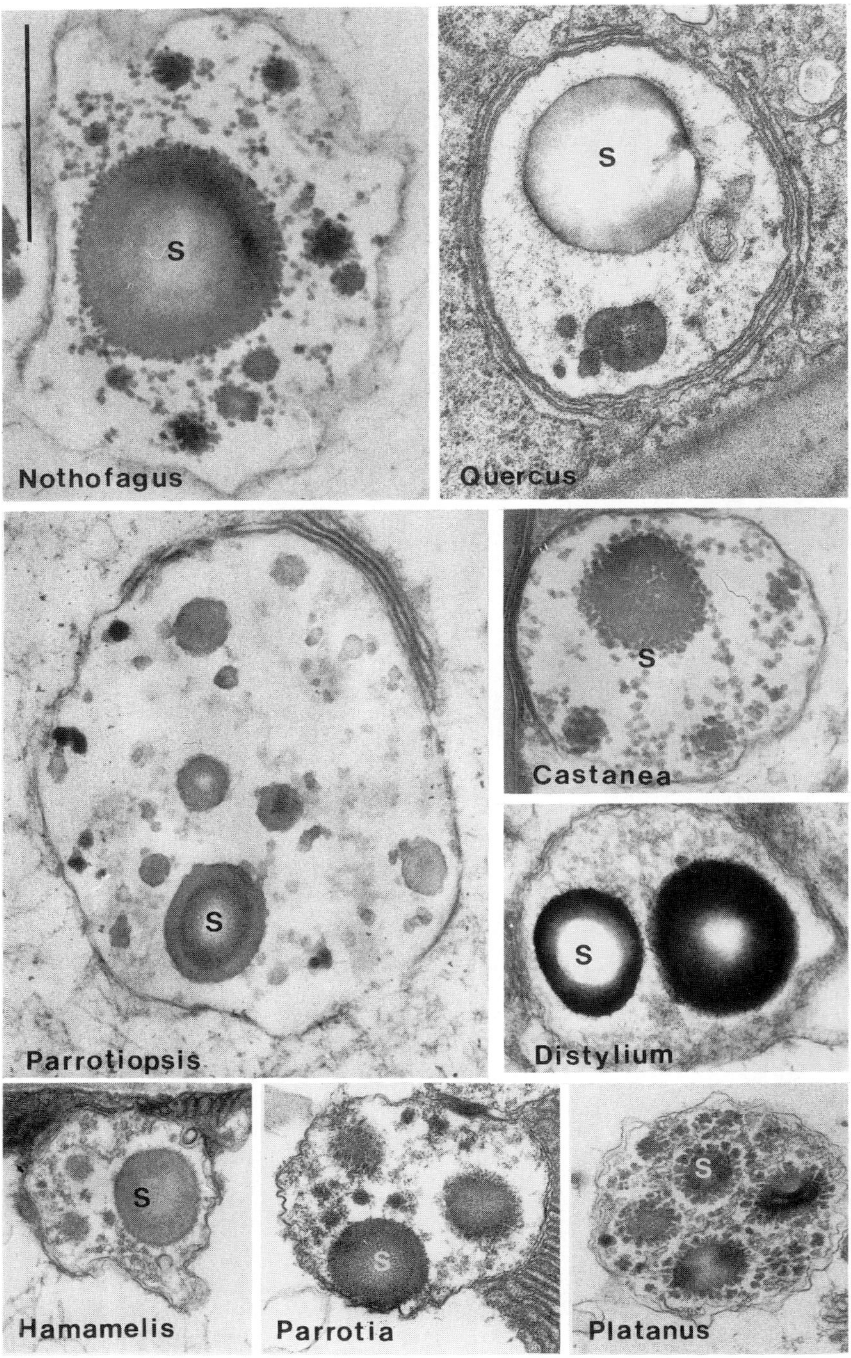

Fig. 6.3 S-type sieve-element plastids of Nothofagaceae (*Nothofagus obliqua*), Fagaceae (*Quercus robur, Castanea sativa*), Hamamelidaceae (*Parrotiopsis jaquemontiana, Distylium racemosum, Hamamelis mollis, Parrotia persica*) and Platanaceae (*Platanus occidentalis*). S, starch grains; all ×30000. Scale bar = 1 μm.

Daphniphyllaceae (DPH; Fig. 6.4): *Daphniphyllum macropodium* contains S-type plastids with about five globular starch grains and a diameter of 1.6 μm (see also Behnke 1982).

Eucommiaceae (ECM; Fig. 6.2): The monotypic *Eucommia ulmoides* contains the second largest sieve-element plastids Hamamelidae (S-type, 1.7 μm, about ten starch grains, see also Behnke 1973).

Eupteleaceae (EPT; Fig. 6.2): *Euptelea pleiosperma* contains S-type plastids with about ten globular starch grains and a diameter of 1.3 μm.

Fagaceae (FAG; Fig. 6.3): Seven species in five genera were investigated; all contain S-type plastids (see also Behnke 1973). About five globular starch grains are found in cross-sections. Plastid diameters range from 0.8 to 1.8 μm, but for *Fagus sylvatica* Schulz and Behnke (1987) were able to demonstrate that sieve-element plastids of secondary phloem in the stem were over twice as large as those of two-year-old branches. In other tree species investigated, comparable results have not been obtained, and there is no other evidence that in general plastid diameters are dependent on the part of the plant sampled. All studied Fagaceae were found to contain non-dispersive P-protein bodies (Fig. 6.1) which in most species are compound-spherical (Behnke 1981*b*, Parameswaran 1984). In *Fagus sylvatica* the P-protein body is spindle-shaped from the beginning (see Schulz and Behnke 1987).

Hamamelidaceae (HAM; Fig. 6.3): Six species from six genera were investigated; all contain S-type plastids. The plastid diameter is 1.3 μm on average, and five starch grains are commonly found.

Juglandaceae (JUG; Fig. 6.3): Four species in three genera were investigated; all contain S-type plastids (see also Behnke 1973). The average plastid diameter is 1.7 μm, and the presence of ten starch grains is not unusual.

Leitneriaceae (LTN; Fig. 6.2): The monotypic *Leitneria floridana* contains S-type plastids with a diameter of 1.1 μm and about five starch grains.

Moraceae (MOR; Fig. 6.6): Thirteen species in 11 genera were investigated, all contain S-type plastids of which eight are form So (see also Behnke 1973). The form S plastids have a diameter of 1.0 μm and contain an average of five starch grains, while form So plastids are considerably smaller (0.8 μm). Non-dispersive compound-spherical

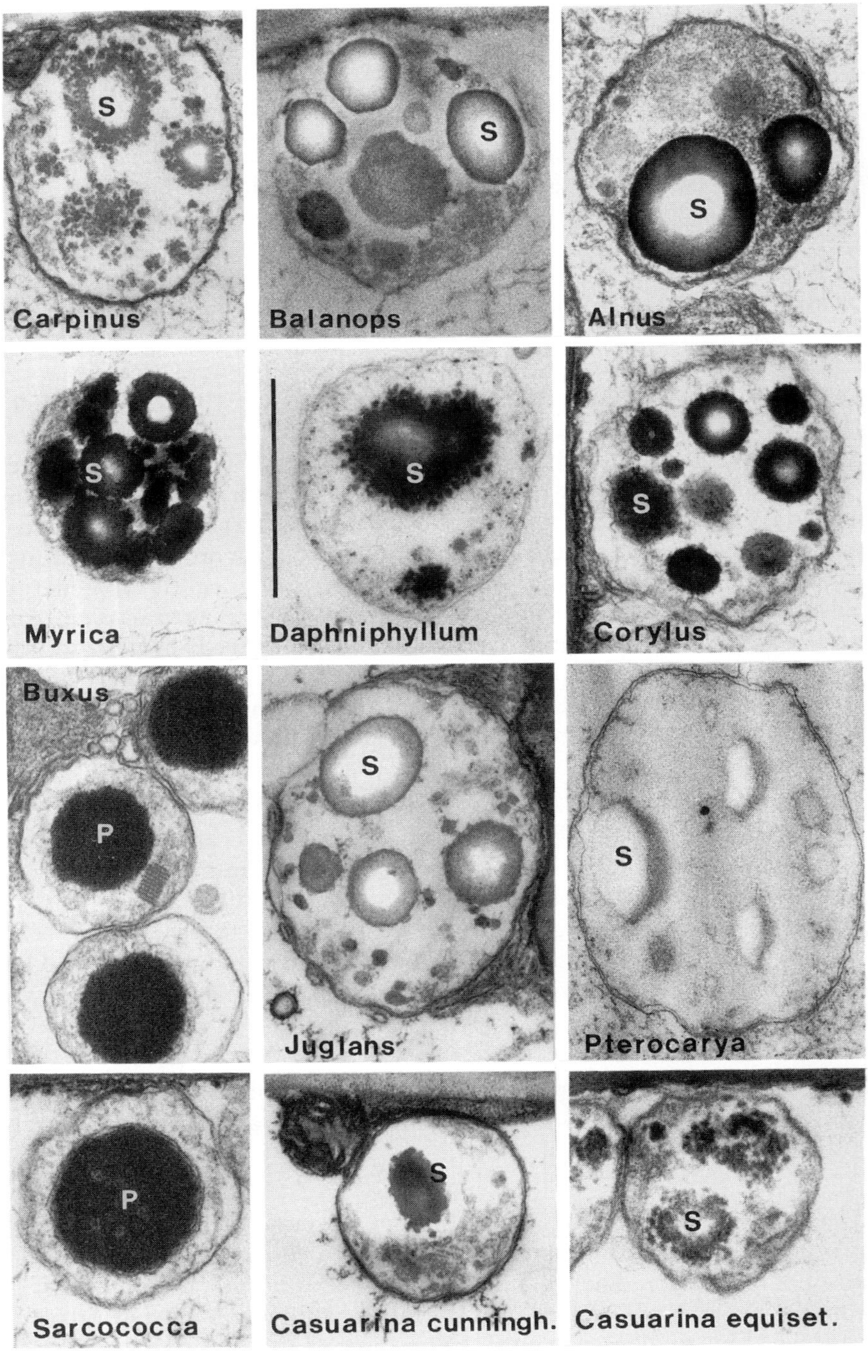
Carpinus
S
Balanops
S
Alnus
S
Myrica
S
Daphniphyllum
S
Corylus
S
Buxus
P
Juglans
S
Pterocarya
S
Sarcococca
P
Casuarina cunningh.
S
Casuarina equiset.
S

P-protein bodies are found in a few species (see Table 6.1), while in *Ficus* unique tubular bodies (of ER) are recorded in each mature sieve element (Fig. 6.1).

Myricaceae (MYR; Fig. 6.4): S-type plastids with a diameter of 1.2 μm and containing about 15 starch grains are recorded in *Myrica californica.*

Myrothamnaceae (MTM): The monotypic *Myrothamnus flabellifolius* certainly contains S-type plastids, but although three different fixations were studied, only starch grains released from plastids were seen. Measurements could not therefore be made.

Nothofagaceae (NTH; Fig. 6.3): The two *Nothofagus* species investigated contain S-type plastids with about ten starch grains and diameters of 1.8 and 1.3 μm respectively. Spindle-shaped P-protein bodies similar to those recorded for *Fagus* were found in *N. obliqua.*

Platanaceae (PLT; Fig. 6.3): S-type plastids with standard characteristics (ten starch grains, diameter 1.4 μm) were recorded for *Platanus occidentalis.*

Rhodoleiaceae (RHD): The monotypic *Rhodoleia championii* contains S-type plastids with about ten starch grains and a diameter of 1.2 μm.

Simmondsiaceae (SMM): The monotypic *Simmondsia chinensis* contains S-type plastids with a diameter of 1.0 μm and about eight starch grains (see Behnke 1982, fig. 5).

Tetracentraceae (TTR): *Tetracentron sinense* has been repeatedly investigated using material from three different sources. All fixations of the sieve elements were unsatisfactory, with only a few undamaged plastids. The sieve-element plastids are certainly of S-type and contain about five large starch grains. Their diameter probably is around 1.5 μm (mean diameters could not be calculated due to an insufficient number of intact plastids). Spindle-shaped non-dispersive P-protein bodies are found in mature sieve elements.

Fig. 6.4 S-type sieve-element plastids of Betulaceae (*Carpinus betulus, Alnus glutinosa*), Myricaceae (*Myrica californica*), Daphniphyllaceae (*Daphniphyllum macropodium*). Corylaceae (*Corylus colurna*), Juglandaceae (*Juglans cordiformis, Pterocarya fraxinifolia*), Casuarinaceae (*Casuarina cunninghamii, C. equisetifolia*), and P-type plastids of Buxaceae (*Buxus natalensis, Sarcococca saligna*). P, protein crystals; S, starch grains; all ×30000. Scale bar = 1 μm.

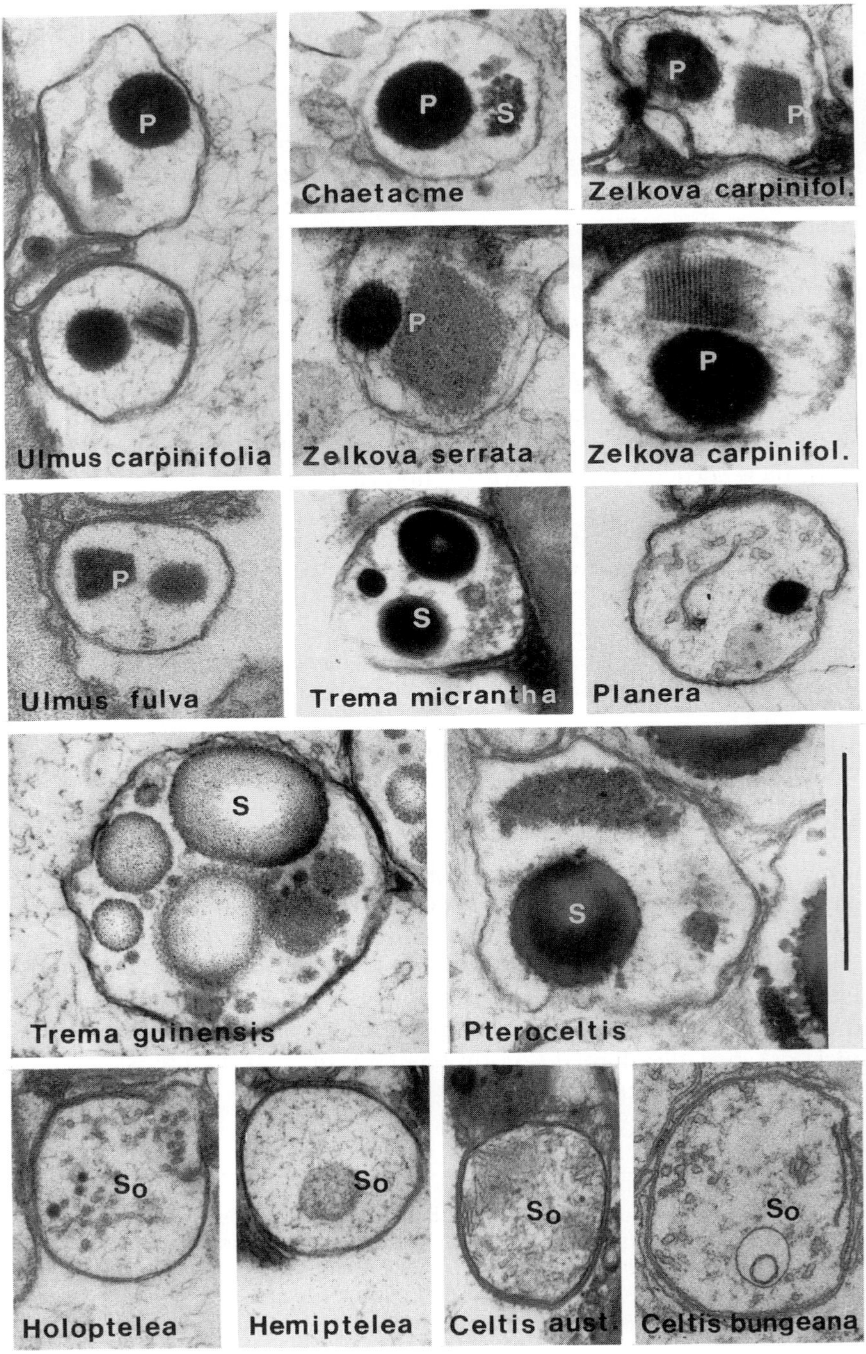
P
Ulmus carpinifolia
P
S
Chaetacme
P
P
Zelkova carpinifol.
P
Zelkova serrata
P
Zelkova carpinifol.
P
Ulmus fulva
S
Trema micrantha
Planera
S
Trema guinensis
S
Pteroceltis
So
Holoptelea
So
Hemiptelea
So
Celtis aust.
So
Celtis bungeana

Trochodendraceae (TRC; Fig. 6.2): The monotypic *Trochodendron aralioides* contains S-type plastids with about five large starch grains. The plastid diameter is 1.5 µm.

Ulmaceae (ULM; Fig. 6.5): Seventeen species from 11 genera were investigated; eight contain P-type and nine S-type plastids. With four forms from the two types the Ulmaceae show the greatest range in their sieve-element plastids among the Hamamelidae. P-type plastids are of form Pc in *Planera, Ulmus* and *Zelkova* and of form Pcs in *Chaetacme.* All contain a globular and another (often rhomboidal) protein crystal. These P-type plastids are 0.8 µm in diameter. Form So plastids are slightly smaller (diameter 0.7 µm), while form S plastids have a diameter of 1.1 µm and contain about five starch grains. P-type plastids in *Ulmus americana* were first depicted by Evert and Deshpande (1969); see also Behnke (1973) for other micrographs from Ulmaceae. In many species globular non-dispersive P-protein bodies are present (see Table 6.1).

Urticaceae (URT; Fig. 6.6): Nine species from eight genera were investigated. All contain S-type plastids of which five species are of form So (see also Behnke 1973). The form S plastids contain very few starch grains (one to three); the average diameter of form S and form So plastids is identical (0.9 µm). Globular non-dispersive P-protein bodies are found in *Laportea* (Fig. 6.1 and Behnke 1981*b*).

Arrangement of the hamamelid families according to sieve-element characters

Comparative analysis of the reported data supports previous conclusions (e.g. Thorne 1973) reflected in recent angiosperm systems (Thorne 1983; Takhtajan 1987; Dahlgren 1988) that the Hamamelidae do not represent a natural group. Sieve-element characters can be used to clearly separate three distinct core orders; viz. the Hamamelidales, Fagales, and Urticales, to which a few other (mainly monotypic) orders may be assigned.

The Hamamelidales with the families Hamamelidaceae (incl. Altingiaceae and Rhodoleiaceae), Platanaceae, Myrothamnaceae and the associate orders Trochodendrales, Cercidiphyllales, Eupteleales,

---

Fig. 6.5 Sieve-element plastids of Ulmaceae. P-type plastids in *Ulmus carpinifolia, U. fulva, Chaetacme aristata, Zelkova carpinifolia, Planera aquatica,* S-type plastids in *Trema micrantha, T. guinensis, Pteroceltis tartarinovii,* and form So plastids in *Holoptelea integrifolia, Hemiptelea davidii, Celtis australis* and *C. bungeana.* P, protein crystals; S, starch grains; So, form So plastids; all ×30000, except for *Zelkova carpinifolia* in second row which is ×50000. Scale bar = 1 µm.

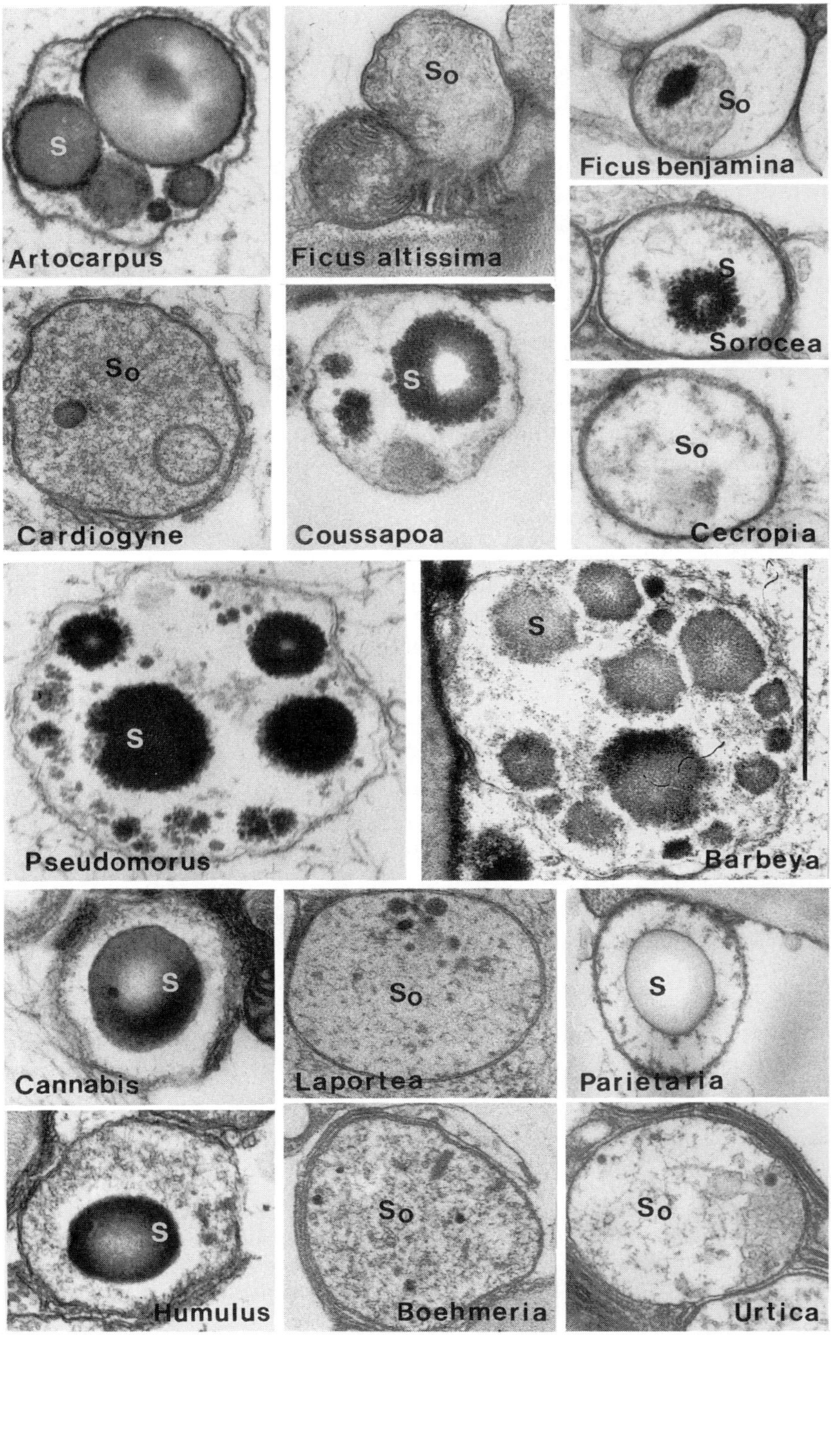
So
S
So
Ficus benjamina
Artocarpus
Ficus altissima
S
Sorocea
So
Cardiogyne
S
Coussapoa
So
Cecropia
S
S
Pseudomorus
S
Barbeya
S
Cannabis
So
Laportea
S
Parietaria
S
Humulus
So
Boehmeria
So
Urtica

and Eucommiales are characterized by comparatively large S-type sieve-element plastids and lack of non-dispersive P-protein bodies (only exception: *Tetracentron*). The average diameter of their plastids is about 1.5 μm which is higher than that of the other two groups and the average among all S-type angiosperms. The incorporation of Leitneriaceae into this alliance (see Cronquist 1981) would strengthen the concept of this group as being among the more primitive hamamelid taxa (i.e. with large-sized S-type plastids). Arguments for an addition of Balanopaceae, Daphniphyllaceae, Juglandaceae, and Myricaceae cannot be obtained from sieve elements plastids, but would on the other hand not be ruled out.

The Fagales with the core families Fagaceae (including Nothofagaceae) and Betulaceae (including Corylaceae) and the close associate Casuarinaceae are characterized by (mostly) compound-spherical non-dispersive P-protein bodies and S-type sieve-element plastids with a below-average diameter of 1.2 μm.

The Urticales families (Moraceae, Cannabaceae, Cecropiaceae, Urticaceae, Ulmaceae) are positively correlated by presence of both globular non-dispersive P-protein bodies and very small sieve-element plastids. P-protein bodies have not been found in Cannabaceae. Another common feature of this order is the presence of form-So plastids in all of its major families. The average diameter of all sieve-elements plastids investigated in this order is 0.9 μm, with 1.0 μm for S-type, and 0.8 μm for both P-type and form So plastids.

Ulmaceae is the only family in the entire Hamamelidae containing four different forms of sieve-element plastids and, except for the more peripheral Buxaceae, the only family with P-type plastids. Both the small diameter and the presence of the P-type suggest that Urticales and especially Ulmaceae are rather derived taxa within this alliance.

The form Pc plastids recorded in Ulmaceae are very unusual among angiosperms; the only other family to contain plastids with similar contents and identical diameters is the Buxaceae. The position of the Buxaceae is not yet settled (Cronquist 1981: in Euphorbiales; Thorne 1983: in Pittosporales, Rosiflorae; Takhtajan 1987: Buxales, Hamamelidanae; Dahlgren 1989: Buxales in vicinity to Fagales in Rosanae). Placement next to the Urticales as suggested by the sieve-element

---

Fig. 6.6 S-type sieve-element plastids of Moraceae (*Artocarpus integrifolia*, *Ficus altissima*, *Ficus benjaminae*, *Cardiogyne africana*, *Pseudomorus brunoniana*), Cecropiaceae (*Coussapoa schottii*, *Cecropia pachystachya*), Barbeyaceae (*Barbeya oleoides*), Cannabaceae (*Cannabis sativa*, *Humulus lupulus*), and Urticaceae (*Laportea moroides*, *Parietaria officinalis*, *Boehmeria biloba*, *Urtica dioica*). So, form So plastids; S, starch grains; all × 30000. Scale bar = 1 μm.

plastids adds another alternative. The Daphniphyllaceae would be excluded from buxacean alliance because of their larger, typical S-type plastids. On the other hand, Simmondsiaceae with small S-type plastids (diameter 1.0 μm) similar to those within Urticales may again be considered to be close to Buxaceae, if a placement of the latter in the vicinity of the Urticales is accepted (but see Behnke 1982).

Although an exact position cannot be given for some families based on sieve-element data (e.g. Barbeyaceae, Juglandaceae; see Thorne 1983 for alternatives) the relative position of all 27 families studied within the three higher taxa is listed in Table 6.2.

In conclusion, based on present results, sieve-element data cannot be used to definitely decide on the most likely position of the Urticales. An association with the Malvales (as proposed by Thorne 1983; Takhtajan 1987; Dahlgren 1989) is supported by the occurrence of non-dispersive P-protein bodies in both Urticales and Malvales. However, an average plastid diameter of 1.4 μm and purely S-type plastids in the Malvales does not support this interpretation. An alliance with the Fagales (and position within the Hamamelidae as proposed by Cronquist (1981) would find equivalent support in the P-protein bodies and the smaller plastid diameter (1.2 μm). Dahlgren (1989) treats the Hamamelidales (and allies) and Fagales as primitive taxa of his Rosiflorae. The type and relative diameter of the sieve-element plastids would conform with this interpretation, but the P-protein bodies of Fagales and the inclusion of Buxales into this alliance would remain anomalous.

Acknowledgements

This investigation was made possible only through the author's extensive collections of fresh material in botanical gardens (listed under 'origin' in Table 6.1) and made available by permission of the respective directors, and by the unselfish help of my colleagues who collected and sent fresh or fixed samples from various parts of the world: P. S. Ashton (Arnold Arboretum, USA), C. Bailey (Morris Arboretum, USA), T. Ebba (Dire Dewa, Ethiopia), D. M. C. Fourie (Pretoria, RSA), E. Köhler (Berlin, DDR), K. Kubitzki (Hamburg, FRG), W. C. Oechel (San Diego, USA) J. Popenoe (Miami, USA)—or helped the author during his collections at botanical gardens: L. B. Jørgensen (Copenhagen, Denmark), K. Kramer (Bonn, FRG), W. Morawetz (Wien, Austria), U. Sutisma (Bogor, Indonesia). Doris Laupp and Liliana Pop (Heidelberg, FRG) processed the material from fixation to the electron micrograph. The Deutsche Forschungsgemeinschaft gave continued travel and research support for these investigations. My sincere gratitude is extended to all the persons and institutions mentioned above.

References

Behnke, H.-D. (1971). Sieve-tube plastids of Magnoliidae and Ranunculidae in relation to systematics. *Taxon* **20,** 723–30.

Behnke, H.-D. (1973). Sieve-tube plastids of Hamamelididae. Electron microscopic investigations with special reference to Urticales. *Taxon* **22,** 205–10.

Behnke, H.-D. (1976). Ultrastructure of sieve-element plastids in Caryophyllales (Centrospermae), evidence for the delimitation and classification of the order. *Plant Systematics and Evolution* **126,** 31–54.

Behnke, H.-D. (1981*a*). Siebelement-Plastiden, Phloem-Protein und Evolution der Blütenpflanzen: II. Monokotyledonen. *Berichte der deutschen Botanischen Gesellschaft* **94,** 647–62.

Behnke, H.-D. (1981*b*). Sieve-element characters. *Nordic Journal of Botany* **1,** 381–400.

Behnke, H.-D. (1982). Sieve-element plastids, exine sculpturing and the systematic affinities of the Buxaceae. *Plant Systematics and Evolution* **39,** 257–66.

Behnke, H.-D. (1988). Sieve-element plastids, phloem protein and evolution of flowering plants: III. Magnoliidae. *Taxon* **37,** 699–732.

Behnke, H.-D. (1988*b*). Structure of the phloem. In *Transport of photoassimilates* (ed. D. A. Baker and J. A. Milburn), pp. 97–137 Longman, Harlow, Essex.

Behnke, H.-D. and Pop, L. (1981). Sieve-element plastids and crystalline P(hloem)-protein in Leguminosae: micromorphological characters as an aid to the circumscription of the family and subfamilies. In *Advances in legume systematics* (eds R. M. Polhill and P. H. Raven), pp. 707–15. Academic Press, London.

Cronquist, A. (1981). *An integrated system of classification of flowering plants.* Columbia University Press, New York.

Dahlgren, G. (1989). The last Dahlgrenogram—System of classification of the Dicotyledons. In *The Davis and Hedge Festschrift* (ed. K. Tan), 237–48. Edinburgh University Press.

Evert, R. F. and Deshpande, B. P. (1969). Electron microscope investigations of sieve-element ontogeny and structure in *Ulmus americana. Protoplasma* **68,** 403–32.

Parameswaran, N. (1984). Crystalline protein bodies in the sieve element of *Castanopsis borneensis. Beiträge zur Biologie der Pflanzen* **58,** 413–9.

Schulz, A. and Behnke, H.-D. (1987). *Feinbau und Differenzierung des Phloems von Buchen, Fichten und Tannen aus Waldschadensgebieten.* PEF-Berichte 16, Kernforschungszentrum, Karlsruhe.

Takhtajan, A. (1969). *Flowering plants: origin and dispersal.* Oliver and Boyd, Edinburgh.

Takhtajan, A. (1987). *Systema Magnoliophytorum.* Nauka, Leningrad (In Russian).

Thorne, R. F. (1973). The 'Amentiferae' or Hamamelidae as an artifical group: A summary statement. *Brittonia* **25,** 395–405.

Thorne, R. F. (1983). Proposed new realignments in the angiosperms. *Nordic Journal of Botany* **3,** 85–117.

Weber, W. A. (1982). Mnemonic three-letter acronymns for families of vascular plants: a device for more effective herbarium curation. *Taxon* **31,** 74–88.

7. Karyological patterns in the Hamamelidae

WILFRIED MORAWETZ AND
M. ROSABELLE A. SAMUEL

Institut für Botanik und Botanischer Garten der Universität Wien, A-1030 Wien, Austria

Abstract

Chromosome numbers and karyomorphological data on 14 families of the Hamamelidae including investigations by standard methods as well as Giemsa-C-banding and fluorochrome staining are presented, and their significance with respect to systematics and evolution is discussed. The numeric differentiation is thought to have occurred from an original base number $2n=16$ by orthoploidy, anorthoploidy, and dysploidy. Chromosomes are mostly small in all families, having an intermediate condensation behaviour and interphase nuclei of '*Hernandia* type'. Because of the relatively high degree of uniformity of karyomorphological characters in Hamamelidae they are interpreted to be a homogeneous (monophyletic) group, and the high karyotypic variation in Magnoliidae and Rosidae makes it difficult to see the much less variable Hamamelidae as an intermediate group. Karyotypic evolution patterns known from the Magnoliidae are repeated in Hamamelidae on a smaller scale. Banding patterns could be established for some Moraceae and Casuarinaceae but not in Hamamelidaceae, Juglandaceae, and Ulmaceae because of a different type of chromatin.

Introduction

The relationships of the Hamamelidae is one of the most crucial problems of angiosperm systematics (Ehrendorfer 1989, Vol. 40A, Ch.

Evolution, Systematics, and Fossil History of the Hamamelidae, Volume 1: Introduction and 'Lower' Hamamelidae (eds P. R. Crane and S. Blackmore), Systematics Association Special Volume No. 40A, pp. 129–154. Clarendon Press, Oxford, 1989.

1). Although it was originally described as a homogeneous group (see Stern 1973), doubts have been raised about its monophyly and systematic position (e.g. Thorne 1973). Chromosome counts of the Hamamelidae are relatively well known but contribute little to systematic interpretations. In contrast, karyomorphological characters such as interphase nuclei structure and condensation behaviour of chromosomes are highly valuable in macrosystematics (Okada 1975; Morawetz 1986*b*) and in this paper we present the first karyomorphological data for most of the critical families of Hamamelidae and discuss the results in regard to their systematics and evolution.

Materials and methods

The plants investigated come from wild sources and botanical gardens. Their origins, and as far as possible the available voucher numbers, are listed in Table 7.1. Standard chromosome preparations follow the HCl/Giemsa method of Guerra (1983); for Giemsa-C-banding, DAPI (4-6- diamidino-2-phenylindole), and CMA (chromomycin A3) see Morawetz and Waha (1985).

Descriptions of the chromosomes and interphase nuclei are based on studies of most of the Magnoliidae families (Morawetz 1986*a*,*b*; 1988). Prochromosomal nuclei or '*Gyrocarpus* type' have usually light-staining or non-staining euchromatin, and small to large, well-delimited, and isolated chromocentres often with a constriction. Their number is usually identical to that of the chromosomes. The '*Hedyosmum* type' has dark-staining chromatin all over the nucleus with numerous very small dot-like chromocentres. Sometimes parts of the chromatin fuse to form homogeneously-staining dark cloud-like aggregates. In the '*Hernandia* type', euchromatin is medium-staining with a reticulate, thread-like consistency, and is sometimes united to denser aggregations. A few to numerous, mainly small, chromocentres are visible. The '*Doryphora* type' consists of few large chromocentres which fuse with the surrounding euchromatin. In the '*Tetrameranthus* type', large chromocentres fuse to irregularly shaped dark-staining chromocentres which are embedded in a granular medium- to light-staining euchromatin.

The condensation behaviour of chromosomes is divided into three types: *continuous condensation* means that during prometaphase there is little or no differentiation between early and late condensing segments. This kind of condensation is usually linked with nuclei of the *Hedyosmum*, *Doryphora*, or *Tetrameranthus* type. *Intermediate condensation* is found in chromosomes where, during the prometaphase, the chromosomes consist of a number of early and late condensing segments including intercalary ones. This is mostly linked with nuclei of the *Hernandia* type.

Table 7.1 Chromosome counts in Hamamelidae and possibly related taxa.

Family/genera/species	Present count 2*n*	Locality/source herbarium voucher	Earlier reports 2*n*
Betulaceae			
Betula			
B. pendula Roth.	28	Lower Austria WM 11-25882	28 Woodworth (1929)
Balanopaceae			
Balanops			
B. australiana F. Muell.	40	Australia Ehrendorfer 260826–3524	42 Goldblatt (1979)
B. vieillardi Boill.	40	New Caledonia Mac Kee 38214	
Casuarinaceae			
Allocasuarina			
A. littoralis (Salisb.) L. Johnson	44	Australia	
A. topulosa L. Johnson	22	no voucher	
Casuarina			
C. equisetifolia Forst & Forst	18	Australia	18 Barlow (1959)
C. glauca Sieb.	18	no voucher	18 Barlow (1959)
C. stricta Miq.	18		26 Barlow (1959)
C. tortulosa. Miq.	18		24 Barlow (1959)
Cercidiphyllaceae			
Cercidiphyllum			
C. japonicum Sieb. & Zucc.	38	HBV WM 11-24688	38 Nakajima (1942) 38 Ratter and Milne (1976) 38 Whitaker (1933)
Eucommiaceae			
Eucommia			
E. ulmoides Oliver	34	HBV	34 Janaki Ammal (1945)

Table 7.1—*continued*

Family/genera/species	Present count 2*n*	Locality/source herbarium voucher	Earlier reports 2*n*
Eupteleaceae			
Euptelea			
E. polyandra Sieb. & Zucc.	28	Zürich	28 Sugiura (1936) 28 Whitaker (1933)
Fagaceae			
Castanopsis			
C. acuminatissima A.D.C.	24	Papua New Guinea WM 11-21187	
Cyclobalanopsis			
C. multinervus	24	HBV, FAG-27	
Fagus			
F. sylvatica L.	24	Vienna WM- 21-21583	22 Delay (1947) 22 Wetzel (1928, 1929)
Nothofagus			
N. sp.	26	HBV	
Pasania			
P. sp.	24	Papua New Guinea Ehrendorfer 04306	
Quercus			
Q. cerris L.	24	Lower Austria WM 11-25886	
Hamamelidaceae			
Disanthus			
D. cercidifolius Maxim.	16	Zürich	16 Goldblatt and Endress (1977)
Distylium			
D. myricoides Hemsl.	24	HBV, HA 12	—
D. racemosum Sieb. & Zucc.	24	HBV, HA 10	24 Sugiura (1936)

Fortunearia			
F. sinensis Rehder & Wilson	24	Zürich	
Hamamelis			
H. intermedia Rehder	24	Zürich	—
H. mollis Oliver	24	Zürich	—
H. vernalis Sargent	24	Zürich	24 Anderson and Sax (1935)
			24 Ernst (1963)
H. virginiana L.	24	Zürich	24 Ernst (1963)
Liquidambar			
L. styraciflua L.	32	Zürich	30 Anderson and Sax (1935)
			30 32 Ernst (1963)
			32 Pizzolongo (1958)
Molinadendron			
M. sinoloense (Stand. & Gentry) Endress	24	Zürich	
Noahdendron			
N. nicholasii (Hyland & Tracey) Endress	24	Australia leg. by Tracey	
Ostrearia			
O. australiana Rehd.	24	Zürich Braun 23.1.86	
Parrotiopsis			
P. jacquemontiana (Decne.) Rehd.	24	Zürich	24 Anderson and Sax (1935)
Parrotia			
P. persica C.A.	24	HBV	
Sinowilsonia			
S. henryi Hemsl.	24	Zürich	24 Anderson and Sax (1935)
			48 Ratter and Milne (1976)
Sycoparrotia			
S. semidecidua Endress	24	Zürich	
Sycopsis			
S. sinensis Oliver	36	Zürich	
Trichocladus			
T. ellipticus Eckl. & Zeyh.	24	Zürich	

Table 7.1—*continued*

Family/genera/species	Present count $2n$	Locality/source herbarium voucher	Earlier reports $2n$
Juglandaceae			
Platycarya			
P. strobilacea Sieb. & Zuc.	28	HBV, JUG 3	—
Juglans			
J. regia L.	32	HBV	32 Delay (1947)
			32 Mehra (1976)
			32 Mehra and Hans (1972)
			32 Woodworth (1930)
Moraceae			
Cudrania			
C. tricuspidata Bureau ex Lavalle	50	HBV	—
Dorstenia			
D. sp.	24	HBV	—
Ficus			
F. benjamina L.	26	HBV	26 Hans (1972)
			26 Mehra (1976)
F. elastica Roxb.	26	HBV	26 Condit (1933)
			39 Condit (1964)
			26 Hans (1972)
			26 Krause (1930)
			26 Mehra (1976)
			26 Sugiura (1936)

Myricaceae			
Myrica			
M. gale L.	16	HBV	96 Taylor and Mulligan (1968) 48 Oginuma *et al.* (1988)
Platanaceae			
Platanus			
P. acerifolia (Anton) Willd.	42	HBV	16 Brouwer (1924) 20 22 Bretzler (1924) 42 Sax (1933)
Ulmaceae			
Celtis			
C. africana Burm.	20	HBV, UL 44	—
C. spinosa Ruiz.	20	HBV, UL 5	—
C. yunnanensis Schneider.	30	HBV, UL 22	—
Chaetacme			
C. aristata Planch.	60	HBV, UL 10, UL 12	—
Pteroceltis			
P. tatarinowii Maxim.	20	HBV, UL 31	20 Oginuma *et al.* (1988)
Trema			
T. guinensis Priemer.	30	HBV, UL 8	—
Urticaceae			
Pellionia			
P. repentes W. T. Wang	24	HBV	—
Pilea			
P. cardierei Gagnep.	48	HBV	48 Sharma (1970)
P. crassifolia Blume	24	HBV	—
P. spruceana Wedd.	48	HBV	—

Finally, chromosomes with *proximally early and distally late condensing chromatin* occur predominantly in the prochromosomal nuclei (*Gyrocarpus* type).

Results

Altogether 14 families, more than 40 genera, and about 50 species have been investigated by standard methods (HC1/Giemsa) and for many of them banding patterns (Giemsa-C-banding and fluorochrome banding) have been established. Special attention has been drawn to one of the core families, the Hamamelidaceae. In general, all families are karyomorphologically similar and mainly characterized by small chromosomes with intermediate to continuous condensation behaviour and nuclei of the *Hernandia* type. Most of the taxa investigated are resistant to the Giemsa-C-banding procedure. Details of results for individual families are given below.

Balanopaceae (Fig. 7.3A): Two species of the genus *Balanops* were counted with $2n=40$. Chromosome size varies from 0.57–1.14 μm, and from metacentric to submetacentric. Interphase nuclei of *Balanops* show tendencies towards prochromosomal nuclei or *Gyrocarpus* type, having light- or non-staining euchromatin with small well-delimited chromocentres.

Betulaceae: *Betula pendula* ($2n=28$) has submetacentric chromosomes between 0.6 μm and 1.3 μm, and an intermediate condensation behaviour with tendencies towards proximally early condensing chromatin. Nuclei are light-staining and of the *Hernandia* type with a few well-delimited medium-sized chromocentres.

Casuarinaceae (Fig. 7.4A–D, 7.6A–B): Four species of *Casuarina* ($2n=18$) and two species of *Allocasuarina* ($2n=22$, 44) have been investigated. There is a remarkable differentiation in chromosome size. The largest chromosomes are found in *A. littoralis* (0.9–2.6 μm) and *A. topulosa* (1.3–2.2 μm). Chromosomes in *C.glauca* (0.9–2.2 μm) and *C. equisetifolia* (0.6–1.7 μm) have smaller ones. Usually there is one pair of NO-chromosomes (nucleolus organizer) but *A. littoralis* ($2n=44$) has two. Most of the chromosomes are meta-to submetacentric. Condensation behaviour is continuous to intermediate, sometimes with more proximal than intercalary early condensing chromatin. Interphase nuclei are of the *Hernandia* type with several medium-sized well-staining chromocentres. Giemsa-C-banding showed mostly proximal bands in all chromosomes which were DAPI negative and CMA neutral to positive (*Fig. 7.6A–B*). The NORs (nucleolus organizer regions) showed hot spots in CMA and negative fluorescence in DAPI staining. The total amount of heterochromatin lies within the normal range of woody plants.

Cercidiphyllaceae (Fig. 7.3B–C): *Cercidiphyllum japonicum* ($2n=38$) exhibits meta- to submetacentric chromosomes between 0.6 μm and 1.7 μm. Intermediate condensation behaviour with most early condensing segments situated proximally. Interphase nuclei are of the *Hernandia* type, with a few larger and several very small chromocentres.

Eucommiaceae (Fig. 7.4E–F): *Eucommia ulmoides* has $2n=34$, identical with previous counts. Metaphase chromosomes vary between 1.3 μm and 1.7 μm and are mainly submetacentric. The interphase nuclei do not correspond completely with the *Hernandia* type. They are characterized by a granular, relatively dark-staining euchromatin and large irregularly shaped, sometimes thread-like chromocentres which fuse in some cases.

Eupteleaceae: *Euptelea polyandra* ($2n=28$) has submetacentric chromosomes with predominantly proximal early-condensing chromatin. Nuclei are of the *Hernandia* type with numerous well-delimited medium-sized chromocentres and a number of smaller ones.

Fagaceae (Fig. 7.3D–E): Four genera and five species (Table 7.1) of Fagaceae all show $2n=24$; *Nothofagus* differs in having $2n=26$. Chromosome size is between 0.8 μm and 2.2 μm with the smallest chromosomes in *Fagus sylvatica* (0.9–1.5 μm) and the largest in *Cyclobalanopsis* (1.3–2.2 μm), all meta- to submetacentric. The condensing behaviour is intermediate, as in Ulmaceae, and interphase nuclei are of the *Hernandia* type with several small chromocentres and having different staining intensity of the euchromatin.

Hamamelidaceae (Figs 7.1–7.2): 14 genera and 18 species have been investigated (Table 7.1). Most taxa fit well into a general pattern. Chromosome size varies between 0.4 μm and 2.6 μm; the chromosomes within a karyotype are of continuously decreasing length and are mostly submetacentric. Condensation is characterized by a large proximally early condensing zone consisting of one or several segments. The distally late condensing parts are usually small. Interphase nuclei are of the *Hernandia* type with some bigger and several small chromocentres. In some genera certain differences in their karyomorphology could be detected. *Disanthus* ($2n=16$) has the biggest chromosomes (especially

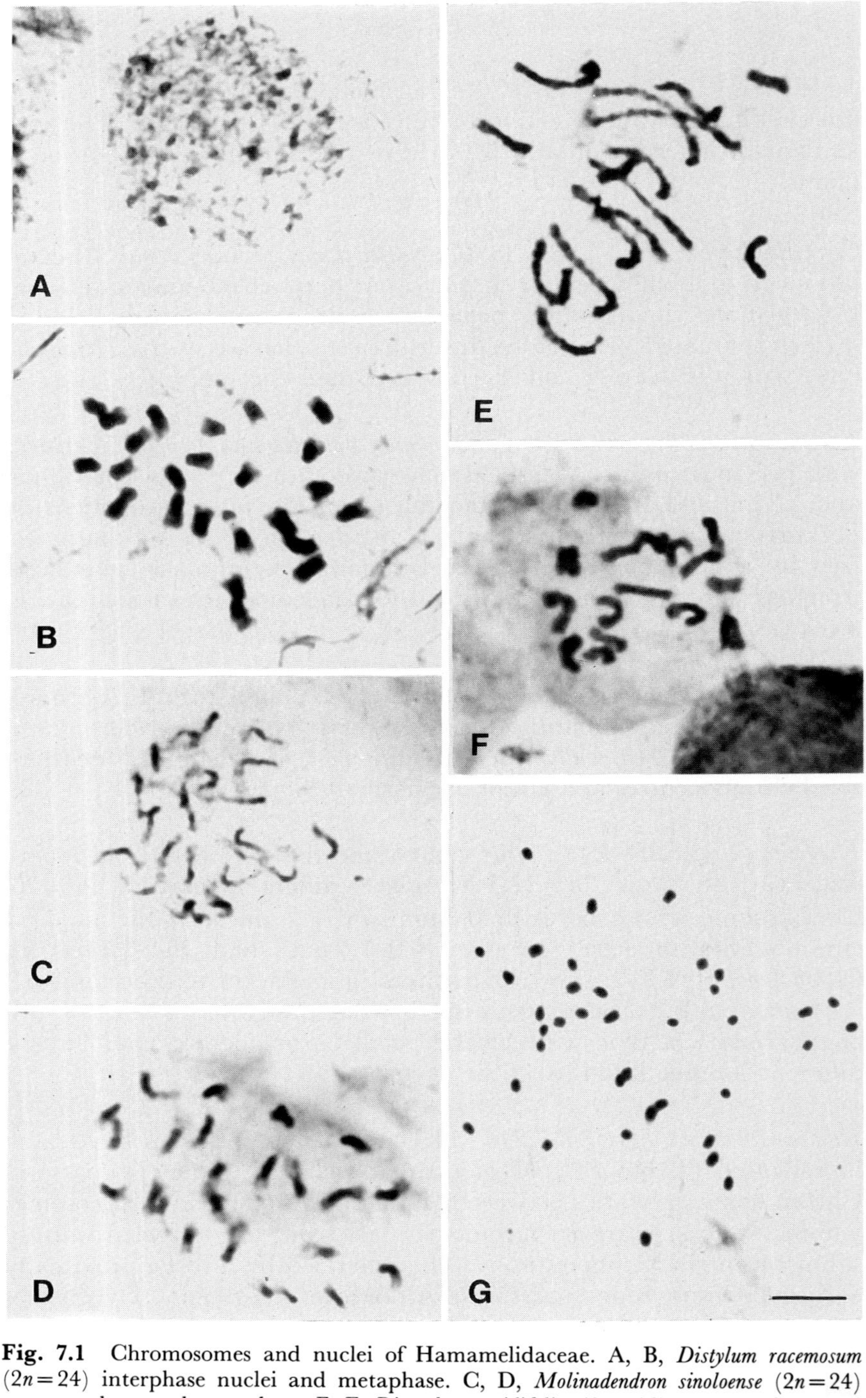

Fig. 7.1 Chromosomes and nuclei of Hamamelidaceae. A, B, *Distylum racemosum* ($2n = 24$) interphase nuclei and metaphase. C, D, *Molinadendron sinoloense* ($2n = 24$) prometaphase and metaphase. E, F, *Disanthus cercidifolius* ($2n = 16$) prometaphase and metaphase. G, *Sycopsis sinensis* ($2n = 36$). (Scale line 5 μm.)

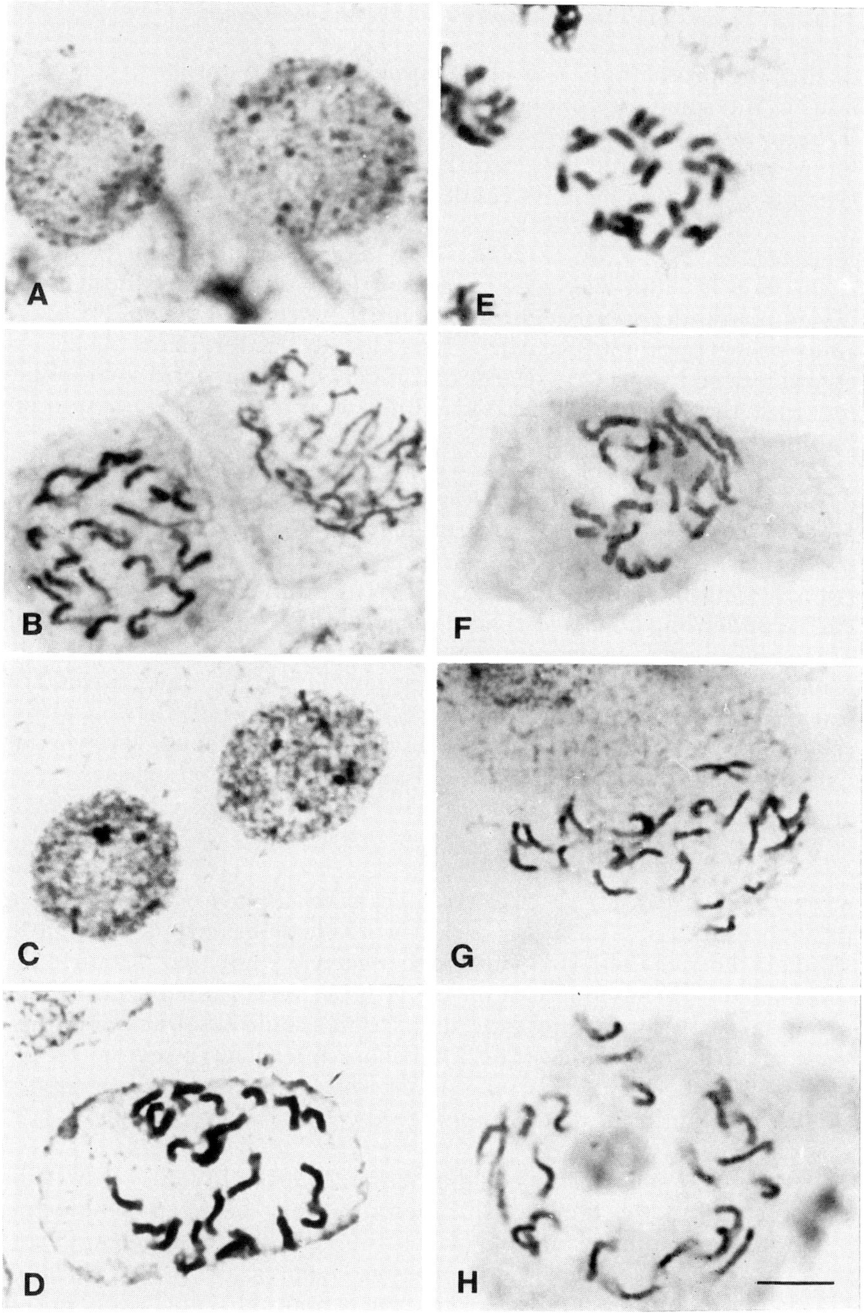

Fig. 7.2 Chromosomes and nuclei of Hamamelidaceae A, B, *Trichocladus ellipticus* ($2n=24$) interphase nuclei and prophase. C, D, *Ostrearia australiana* ($2n=24$), interphase nuclei and prometaphase. E, *Fortunearia sinensis* ($2n=24$) metaphase. F, *Hamamelis mollis* ($2n=24$) prometaphase. G, *Sinowilsonia henryi* ($2n=24$) prometaphase. H, *Noahdendron nicholasii* ($2n=24$) prometaphase. (Scale line 5 μm.)

during the prometaphase and metaphase, 1.5–2.6 μm) with a continuous to intermediate condensation behaviour and nuclei similar to the *Tetrameranthus* type. The smallest chromosomes (0.4–0.9 μm) were found in *Sycopsis* which also exhibit lighter-staining interphase nuclei. Giemsa-C-banding gave no results in several genera tested.

Juglandaceae (Fig. 7.3F, G, H): Two taxa were investigated, *Juglans regia* ($2n=32$) and *Platycarya strobilacea* ($2n=28$); both exhibit small meta- to submetacentric chromosomes, between 0.8 μm and 1.6 μm, condensing proximally earlier than distally. Interphase nuclei are characterized by one to two large chromocentres, in general corresponding with the *Hernandia* type, sometimes with a euchromatin staining lighter than is typical.

Moraceae (Figs 7.4G–H, 7.6 C–E): The chromosomes of *Ficus elastica* and *F. benjamina* ($2n=26$) are small (0.4–1.3 μm), those of *Cudrania tricuspidata* ($2n=50$) and *Dorstenia* sp. ($2n=24$) are bigger (1.3–3 μm). All are meta- to submetacentric, with *Ficus* showing a large proximal early-condensing block, whereas *Cudrania* and *Dorstenia* show intermediate to continuous condensation behaviour. Interphase nuclei are of the *Hernandia* type in *Ficus*, with several well-delimited chromocentres, and intermediate to the *Hedyosmum* type in *Cudrania* and *Dorstenia*, with relatively dense and dark-staining euchromatin but rarely with cloudlike aggregations. Giemsa-C-banding in *Dorstenia* showed very clear proximal bands which were DAPI positive and CMA negative (*Fig. 7.6C–E*).

Myricaceae: *Myrica gale* was counted with $2n=16$ (in contrast to previous reports, see Table 7.1). The chromosomes are meta to submetacentric, their size varying between 0.9 μm and 1.3 μm. The condensation behaviour is intermediate between the continuous and proximal early-condensing type, as far as this could be observed in these very small chromosomes. The interphase nuclei have several clearly visible chromocentres of varying shape, and the euchromatin consists of small well-staining threads, similar to the *Hernandia* type.

Platanaceae: *Platanus acerifolia* has $2n=42$, probably the only valid number within the family. Chromosomes vary between 1.8 μm and 1.6 μm at metaphase and are of continuously decreasing length and mostly submetacentric. Condensation behaviour is intermediate between a continously condensing and proximally early-condensing type. Interphase nuclei are of the *Hernandia* type, sometimes with more small dark-staining chromocentres than usual, reminiscent of the *Hedyosmum* type.

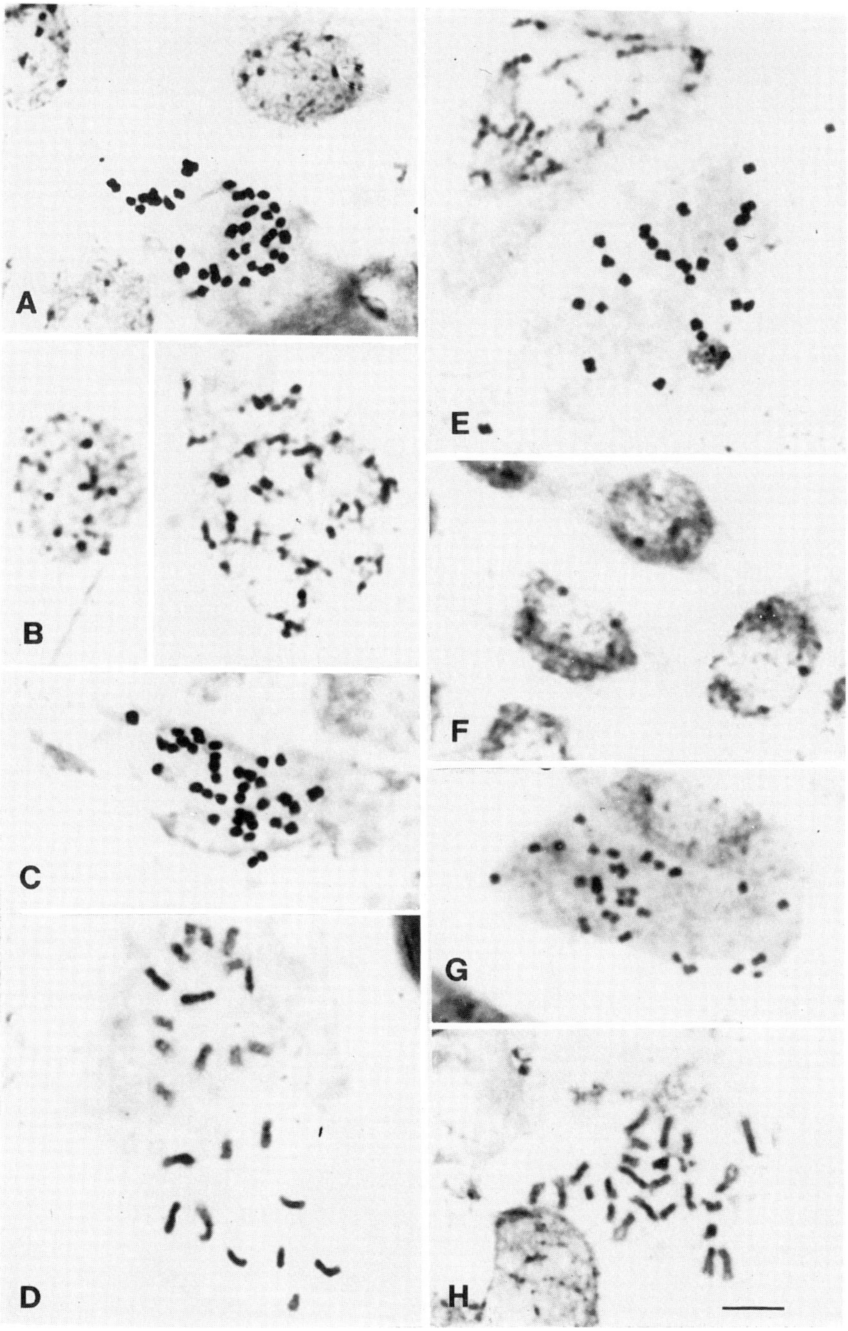

Fig. 7.3 Chromosomes and nuclei of Balanopaceae, Cercidiphyllaceae, Fagaceae, and Juglandaceae. A, *Balanops vieillardii* (2*n*=40) metaphase. B, C, *Cercidiphyllum japonicum* (2*n*=38) interphase, prophase, and metaphase. D, *Quercus cerris* (2*n*=24) metaphase. E, *Fagus sylvatica* (2*n*=24) prophase and metaphase. F, G, *Juglans regia* (2*n*=32) interphase and metaphase. H, *Platycarya strobilacea (2n*=28) metaphase chromosomes. (Scale line 5 μm.)

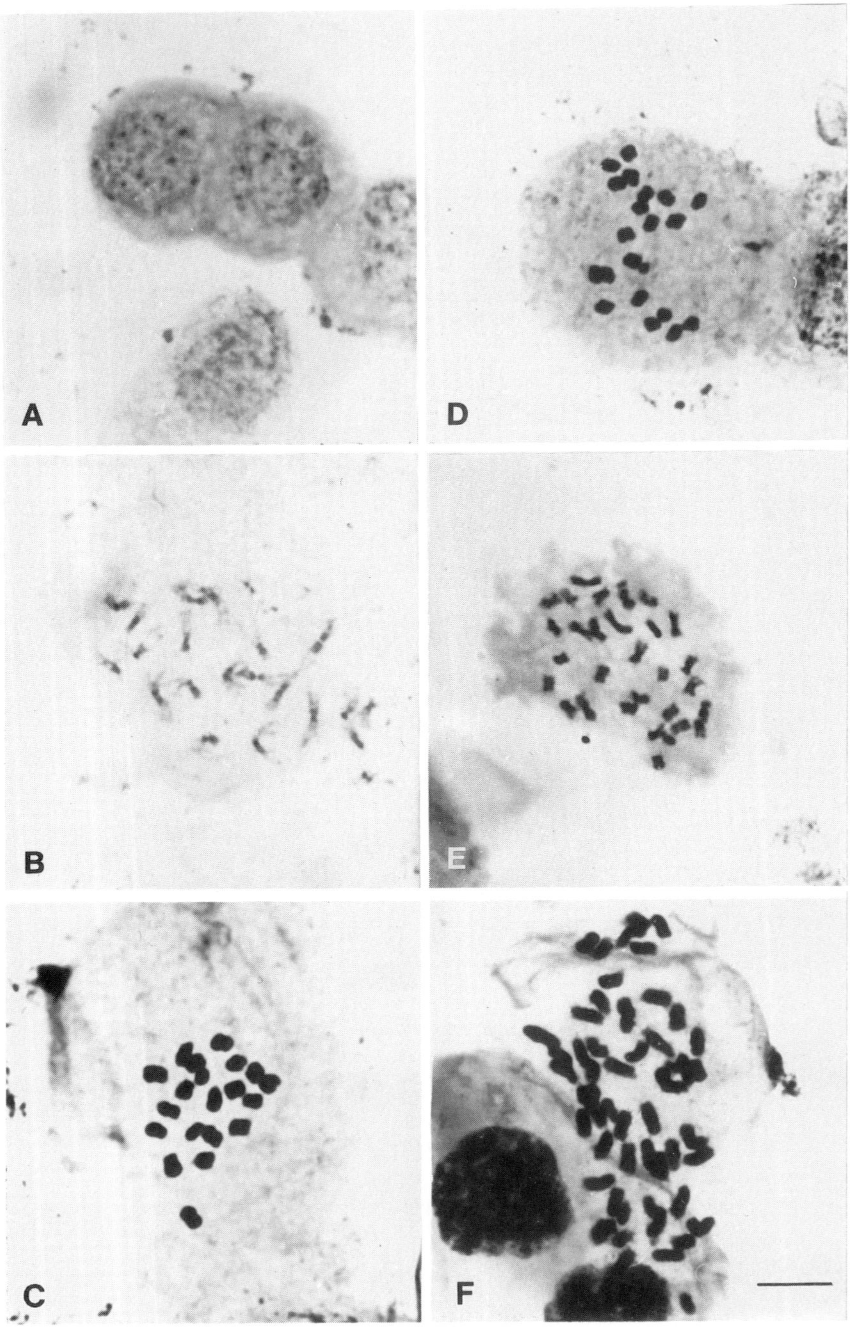

Fig. 7.5 Chromosomes and nuclei of Ulmaceae. A, B, C, *Celtis spinosa* ($2n = 20$) interphase, prometaphase, and metaphase. D, *Pteroceltis tatarinowii* ($2n = 20$) metaphase. E, *Celtis yunnanensis* ($2n = 30$) metaphase. F, *Chaetacme aristata* ($2n = 60$) metaphase. (Scale line 5 μm.)

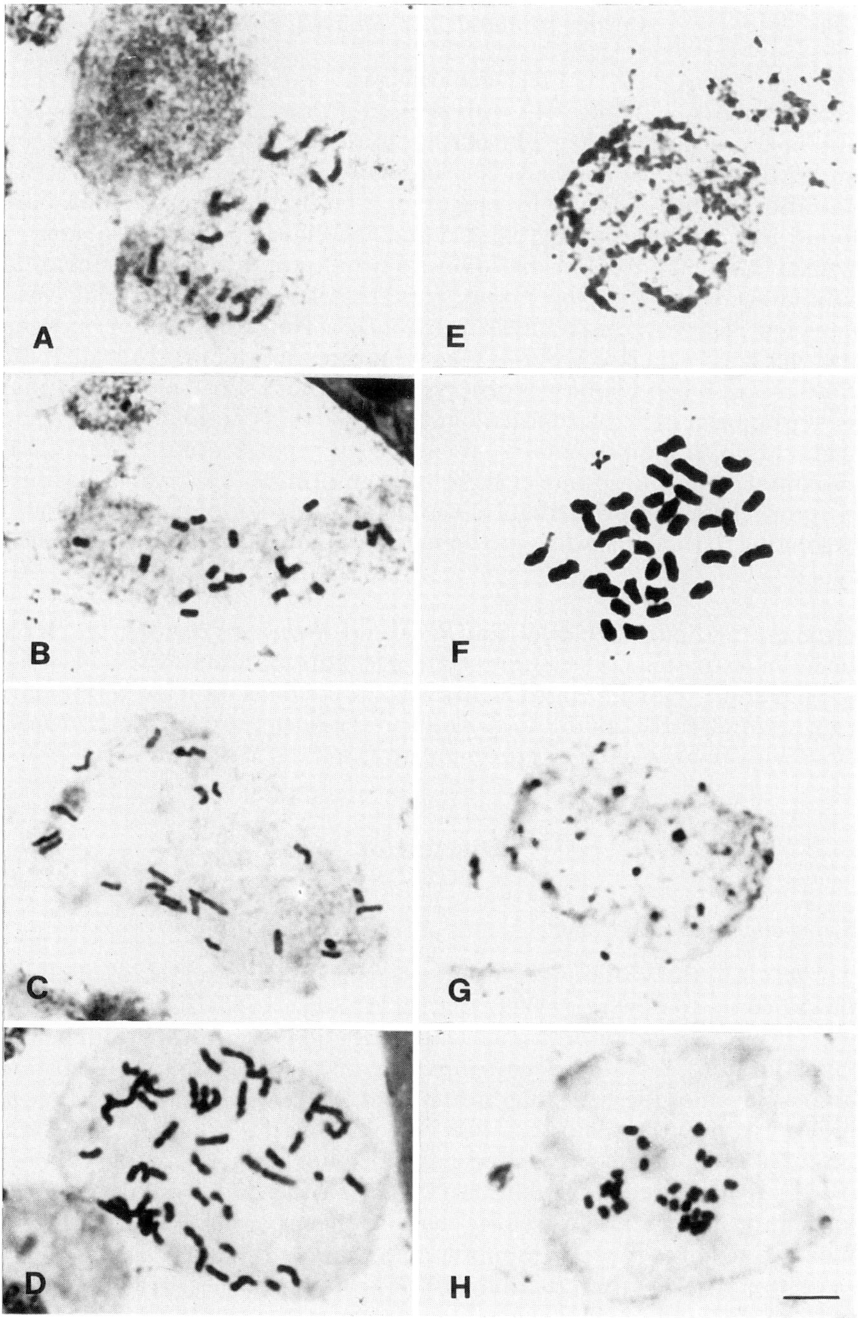

Fig. 7.4 Chromosomes and nuclei of Casuarinaceae, Eucommiaceae, and Moraceae, A, B, *Casuarina glauca* ($2n=18$) prophase and metaphase. C, *Allocasuarina topulosa* ($2n=22$) metaphase. D, *A. littoralis* ($2n=44$) early metaphase. E, F, *Eucommia ulmoides* ($2n=34$) interphase and metaphase. G. H, *Ficus elastica* ($2n=26$) interphase and metaphase. (Scale line 5 μm.)

Ulmaceae (Fig. 7.5): Five genera and seven species have been investigated (see Table 7.1) with diploids ($2n=20$), triploids ($2n=30$) and hexaploids ($2n=60$). The chromosome size ranges between 0.8 μm and 2.6 μm, although there is a certain inter- and infrageneric size variation within the taxa investigated. The smallest chromosomes are found in *Celtis yunnanensis* (0.8–1.5 μm), the largest occur in *Chaetacme aristata* (1.3–2.6 μm). The metaphasic chromosomes are mostly meta- to submetacentric, but in *Trema guinensis* one acrocentric pair was detected. Karyotype architecture reveals certain differences between the genera, which have not been studied in detail. Intermediate condensation behaviour was observed throughout. Pro- to prometaphasic chromosomes are composed of a series of early- and late-condensing segments, although the main part of the early-condensing chromatin is proximal. Interphase nuclei have a large number of small to minute chromocentres embedded in a reticulate, well-staining euchromatin belonging to the *Hernandia* type, sometimes reminiscent of the *Hedyosmum* type.

Urticaceae: Four species of the *Pilea* and *Pellionia* ($2n=24$, 48) have been investigated. The submetacentric chromosomes vary between 0.6 μm and 1.8 μm in length. Chromosome condensation is intermediate to proximal early condensing. Interphase nuclei are of the *Hernandia* type with relatively large chromocentres.

Discussion

Systematics of Hamamelidae

To test all the different systematic concepts of the Hamamelidae proposed up to now (e.g. Wettstein 1924; Stern 1973; Takhtajan 1973; Thorne 1973; Ehrendorfer and Dahlgren 1983) against karyomorphological data one would need to investigate a great part of all existing angiosperm families. Although many parallel investigations in Magnoliidae (Morawetz 1986*a*,*b*, 1988), Rosidae and others (Morawetz, unpublished) have been carried out, this paper focuses mainly on taxa listed in the system of Takhtajan (1973). Indeed the relatively great karyomorphological uniformity within the taxa investigated points more to a homogeneous monophyletic alliance in the classical sense (Takhtajan 1973) than to an artificial group (Thorne 1973). Based on our results and a thorough literature review the position of the Hamamelidae may be characterized as a sideline of the basic angiosperms with certain affinities to Magnoliidae and Rosidae, but the idea of an intermediate position of the Hamamelidae (Ehrendorfer, Vol.

40A, Ch. 1) is hardly supported by karyotypic characters. The high degree of karyomorphological diversification of woody Magnoliidae would have had to go through the evolutionary bottleneck of the karyomorphologically uniform Hamamelidae to increase again in the extremely variable Rosidae.

Systematic implications

The size variation of chromosomes in Fagaceae (Fig. 7.3D–E) is possibly correlated with the subfamily classification with the smallest chromosomes in Fagoideae (*Fagus*), the biggest in Quercoideae (*Quercus*, *Cyclobalanopsis*). *Nothofagus* is karyomorphologically very similar to *Fagus* and from this point of view it does not merit family status as has been suggested (Romero 1986; Jones 1986; Nixon Vol. 40B, Ch. 3) although it differs by its dysploid number $2n=26$ from the otherwise uniform family with $2n=24$.

Betulaceae *sensu lato* fits well with the Fagaceae (Endress 1977) and differs slightly by having somewhat lighter-staining interphase nuclei with tendencies towards prochromosomal nuclei. The chromosome number variation is larger with the basic number $2n=16$ and the (subtriploid?) $2n=22$ in the subfamily Coryleae, the records of $2n=28$ in *Corylus* seem less probable; Betuleae have $2n=28$ and are characterized by extensive polyploid series.

Balanopaceae were counted by us to have $2n=40$ (Fig. 7.3A, 2 spp., cultivated) in contrast to Goldblatt (1979, $2n=42$) although this makes little difference for interpretation. Karyotypic characters do not conflict with a position next to the Fagales (Takhtajan 1973; Cronquist 1981; Dahlgren 1983) but do not fit into the Pittosporales (Thorne 1976) as far as we know their karyomorphology.

Hamamelidaceae are divided into two groups by their base numbers: $x=8$ in the Disanthoideae, Exbucklandoideae, and Liquidambaroideae, $x=12$ in Hamamelidoideae and Rhodoleioideae (Goldblatt and Endress 1977). The general pattern of nuclei is closely related to Fagaceae, Betulaceae, and Platanaceae ($2n=42$), altogether obviously a natural group (Endress 1977), probably also including Myrothamnaceae ($2n=20$: Puff 1978). For chromosomal diversification patterns see below.

Trochodendraceae ($2n=40$), Tetracentraceae ($2n=$c.48, sometimes included into the Trochodendraceae), Cercidiphyllaceae, ($2n=38$), and Eupteleaceae ($2n=28$) are united in the Trochodendrales (Endress 1986). They exhibit typical nuclei of the *Hernandia* type, common in Hamamelidae, but are also comparable with for example Magnoliaceae both in chromosome numbers (*Trochodendron*, *Cercidiphyllum*) and chromatin structure (see Okada 1975; Morawetz 1981*a*, 1986*b*) which would support their proposed intermediate position between Magnoliidae and Hamamelidae (Endress 1986). Nevertheless this is also true for many other Hamamelidae and therefore these characters contribute little evidence on the systematic position of the Trochodendrales.

Different chromosome numbers in the Casuarianaceae are probably additional characters differentiating the genera *Casuarina* ($2n=18$) and *Allocasuarina* ($2n=20$, 22, 24, 26, 28), the latter also with polyploids (partly from Johnson 1982). Furthermore the chromosomes are larger in *Allocasuarina* than in *Casuarina*. Although standard staining with HCl/Giemsa shows nuclei similar to many other Hamamelidae, Giemsa-C-banding reveals unusual features. In contrast to Hamamelidaceae, Ulmaceae, and Juglandaceae clear proximal C-bands could be developed both in *Casuarina* and *Allocasuarina*, which also showed a certain differentiation after fluorochrome staining (Fig. 7.6A–B). This points to chemical differences of the nuclei as compared with most of the taxa investigated and underlines the isolated position of the family.

The systematic positions of Moraceae, Urticaceae, and Ulmaceae (Urticales) are especially controversial; they are placed either in the Hamamelidae (Takhtajan 1973; Cronquist 1981) or close to the Malvales (Thorne 1976; Berg 1977). Although the relatively great karyomorphological variability of all three families does not fit exactly into the general karyological pattern of the Hamamelidae this is not too surprising as they compose the largest part of the species of Hamamelidae and also exhibit a greater diversity of life forms than all other families.

In Moraceae the nuclei of *Ficus* (Fig. 7.4G–H) and *Cudrania* are related to the Hamamelidae, whereas the herbaceous *Dorstenia* (see below) has bigger chromosomes and well developed Giemsa-C and fluorochrome bands (Fig. 7.6C–E). *Humulus* (Cannabaceae, not listed here) is karyologically completely different with large chromosomes and nuclei of the *Hedyosmum* type. Urticaceae (*Pellonia*, *Pilea*) fits well into the Hamamelidae but one can expect greater variation within the family. In Ulmaceae we found the same variation in chromosome size and karyotype architecture reported by Oginuma *et al.* (1988, Fig. 5), but the differences in interphase nuclei structure reported by them probably reflect the normal variability of *Hernandia* type nuclei during the mitotic cycle. All three families are characterized by extensive polyploid series, with fewer different numbers in Ulmaceae.

Compared with the Malvales, which are karyomorphologically well circumscribed (Morawetz 1981*a*, 1986*b*), the Urticales do not differ too much. Chromosomes are in the same size range but have proximally early- and distally late-condensing chromatin in Malvales, whereas Urticales frequently show the intermediate type. Interphase nuclei are related but are frequently darker-staining in Malvales than Urticales.

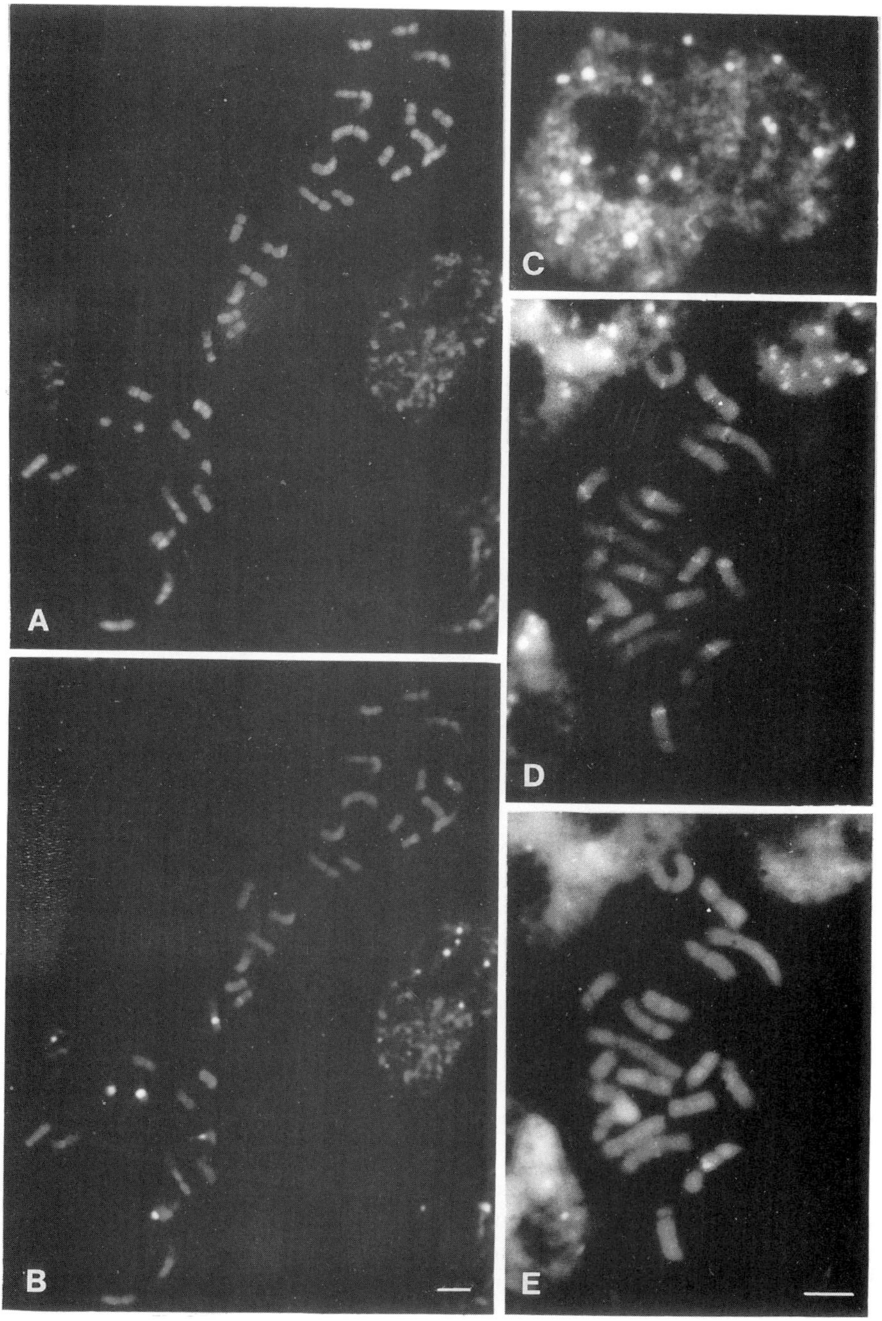

Fig. 7.6 Fluorochrome staining of chromosomes in Casuarinaceae (A, B) and Moraceae (C–E). A, B, DAPI/CMA staining of *Allocasuarina littoralis*. C, DAPI staining of *Dorstenia* sp. Interphase nucleus. D, E, DAPI/CMA staining of *Dorstenia* sp. (Scale line 5 μm.)

```
Moraceae                 ++++++++++++++++++++++++++++
Hamamelidaceae           +++++++++++++++++++++++
Casuarinaceae              +++++++++++++++++++++
Ulmaceae                    ++++++++++++++++++++
Fagaceae                     +++++++++++++++
Urticaceae                 +++++++++++++
Cercidiphyllaceae          ++++++++++++
Juglandaceae               ++++++++++
Betulaceae                 ++++++++
Balanopaceae               +++++++
Myricaceae                    +++++
Platanaceae                     +++++
Eucommiaceae                      +++++

                      *.........*.........*.........*
                                1µm       2µm       3µm
```

Fig. 7.7 Size range of chromosomes in Hamamelidae based on the original investigations of this study. Only fully contracted metaphasic chromosomes have been considered. For each species a mean was calculated from measurements of about five cells.

Thus there is no strong karyological contradiction in placing the Urticales next to the Malvales but, based on our karyological results, we would prefer to maintain them in the Hamamelidae.

In Myricaceae the base number is $2n=16$ and there are further infra- and interspecific polyploid series. The first report of a diploid *Myrica gale* extends its number range to $2n=16$, 48, 96. The karyomorphology and interphase nuclei structure fits well into the Hamamelidae.

Eucommia ulmoides with $2n=34$ and intermediate condensation behaviour (Fig. 7.4E, F; Tanaka and Oginuma 1983) does not differ significantly from the other taxa in Hamamelidae.

Base number

The original data presented here, and a thorough literature review (Morawetz and Rainer 1987), show that the most probable original base number is $2n=16$ and not $2n=14$ as previously suggested (e.g. Ehrendorfer 1976). Our arguments are based on the fact that the most frequent low diploid number in the more conservative woody groups is $2n=16$ (Hamamelidaceae, Betulaceae *sensu lato*, and Myricaceae). The diploid number $2n=14$ is present only in the karyologically more variable Moraceae *sensu lato* and Urticaceae, which is obviously a decreasing dysploid number from $2n=16$. In Leitneriaceae and Juglandaceae $2n=32$ is possibly also based on $x=8$. From other numbers without known lower diploids like $2n=42$ in Platanaceae, $2n=28$ in Eupteleaceae, and $2n=24$, 26 in Fagaceae an original base number can hardly be calculated: identical numbers may originate by various cytogenetic mechanisms on different base numbers (Morawetz 1986a). Finally the original base number $2n=16$ in Magnoliidae (Morawetz 1986*a*,*b*) suggests the same base number for the obviously related Hamamelidae.

Numeric differentiation

Polyploid series are the predominant evolutionary factor in the karyological differentiation in Hamamelidae (Table 7.2). For example, in Hamamelidaceae (Goldblatt and Endress 1977) there is a diploid ($2n=16$) in Disanthoideae, tetraploids ($2n=32$) in Liquidambaroideae, and octoploid ($2n=64$) in Exbucklandoideae. The typical number for Rhodoleioideae and Hamameloideae ($2n=24$) is probably of triploid origin on $x=8$ (Morawetz 1986a). For further numbers in Hamameloideae ($2n=36$, 48, 72) the ploidy level cannot be explained (Fig. 7.8). Typical polyploid series occur in Betulaceae (*Betula*), Myricaceae, and Ulmaceae. Dysploid differentiation is obviously characteristic mainly in Casuarinaceae but is also present in Moraceae and Urticaceae.

Karyotypic variation and evolution in Hamamelidae

Compared with woody Magnoliidae (Morawetz 1986*a*,*b*) and Rosidae (Morawetz, unpublished) the karyotypic variation in Hamamelidae is relatively low. There are mostly small metacentric chromosomes with intermediate condensation behaviour and interphase nuclei of the *Hernandia* type. A further character found in most of the taxa investigated is the negative response to Giemsa-C-banding which suggests a widely distributed special chromatin type possibly consisting of a mixture of hetero- and euchromatin.

If we compare the karyotypic evolution as suggested in Magnoliidae

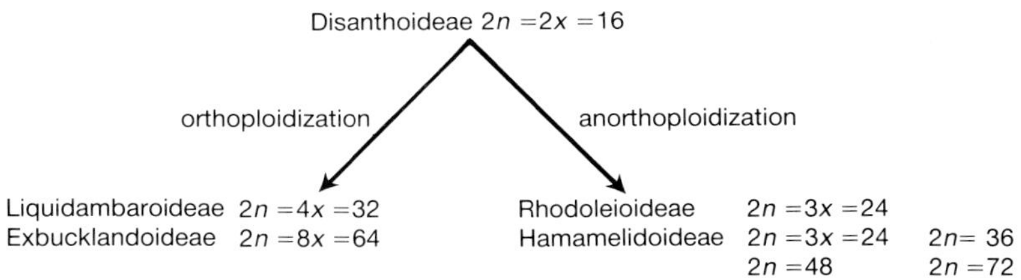

Fig. 7.8 Possible evolution of chromosome numbers within the Hamamelidaceae. Within the numbers $2n=36$, 48, and 72 the ploidy level is uncertain and could be explained in several ways.

Table 7.2 Chromosome numbers in Hamamelidae and possibly related taxa. The data are based on a bibliographic study (Morawetz and Rainer 1988) including most of the literature available. Obviously erroneous numbers have been excluded.

Family	2*n* numbers
Trochodendraceae	(38) 40
Tetracentraceae	(38) ca. 48
Cercidiphyllaceae	38
Eupteleaceae	28
Casuarinaceae	18 20 22 24 26 28 33 44 55 56
Hamamelidaceae	16 24 32 36 48 72
Platanaceae	42
Myrothamnaceae	20
Betulaceae (incl. Corylaceae)	16 22 28 42 56 62 64 70 84
Fagaceae	24 26 48 84
Balanopaceae	40 (−42)
Leitneriaceae	32
Juglandaceae	20 28 32 64
Myricaceae	16 32 48 96
Moraceae (incl. Cannabaceae and Cecropiaceae)	14 16 20 24 26 28 32 39 40 42 50 52 56 64 308
Urticaceae	14 16 20 22 24 26 28 32 42 48 50 52
Ulmaceae	20 (26) 28 30 40 56 60 80 84
Eucommiaceae	34

(Morawetz 1986*a*,*b*) with that of Hamamelidae we can see a certain parallelism, although the diversification is much lower and less well developed in the latter group. As a basic type of nucleus we would consider that of the diploid genus *Disanthus* (Hamamelidaceae) with $2n = 16$, which is thought to have also many primitive morphological characters (Endress 1977). Here we find somewhat larger chromosomes than usual, a continuous condensation behaviour and nuclei reminiscent of that of the *Tetrameranthus* type, all characters which are interpreted to be primitive in Magnoliidae and in woody plants in general (Morawetz 1986*a*,*b*). From that type we may derive the general and frequently occurring nuclei of the *Hernandia* type in Hamamelidae which is found in some more derived families of the Magnoliidae too, in both groups mostly linked to polyploid numbers. Finally, the most advanced nuclei are the prochromosomal ones (*Gyrocarpus* type) found to a certain extent in Balanopaceae or in some Betulaceae, in Magnoliidae frequent in Annonaceae and typical for many advanced woody plants like Melastomataceae, Myrtaceae, or some Lecythidaceae. Herbaceous members of the Hamamelidae may exhibit strongly differing karyological features compared to woody taxa. This is quite a frequent pattern among angiosperms, for example, comparing woody (small chromosomes) and herbaceous (big chromosomes) Moraceae we find the same differences as occur between Myristicaceae and Annonaceae or between Ranunculaceae and Nympheaceae.

References

Anderson, E. and Sax, K. (1935). Chromosome numbers in the Hamamelidaceae and their phylogenetic significance. *Journal of the Arnold Arboretum* **16,** 210–5.

Barlow, B. A. (1959). Chromosome numbers in the Casuarinaceae. *Australian Journal of Botany* **7,** 231–7.

Berg, C. C. (1977). Urticales, their differentiation and systematic position. *Plant Systematics and Evolution, Supplement* **1,** 349–74.

Brouwer, J. (1924). Studies in Platanaceae. *Recueil des Travaux Botaniques Néerlandaise* **21,** 369–82.

Bretzler, E. (1924). Beiträge zur Kenntnis der Gattung *Platanus*. *Botanisches Archiv* **7,** 388–417.

Condit, I. J. (1933). Cytological and morphological studies in the genus *Ficus*. II. Chromosome numbers and morphology in thirty-one species. *University of California Publications in Botany* **17,** 61–74.

Condit, I. J. (1964). Cytological studies in the genus *Ficus* III. Chromosome numbers in sixty-two species. *Madroño* **17,** 153–5.

Cronquist, A. (1981). *An integrated system of classification of flowering plants.* Columbia University Press, New York.

Dahlgren, R. (1983). General aspects of angiosperm evolution and macrosystematics. *Nordic Journal of Botany* **3,** 119–49.

Delay, C. (1947). Recherches sur la structure des noyaux quiescents chez les Phanerogrames. *Revue de Cytologie et de Cytophysiologie Végetales* **9,** 169–222; **10,** 103–229.

Ehrendorfer, F. (1976). Evolutionary significance of chromosomal differentiation patterns in gymnosperms and primitive Angiosperms. In *Origin and early evolution of Angiosperms* (ed. C. B. Beck) pp 220–40. Columbia University Press, New York.

Ehrendorfer, F. and Dahlgren, R. (eds) (1983). New evidence of relationships and modern systems of classification of the angiosperms. *Nordic Journal of Botany* **3,** 1–153.

Endress, P. K. (1977). Evolutionary trends in the Hamamelidales Fagales-group. *Plant Systematics and Evolution, Supplement* **1,** 321–47.

Endress, P. K. (1986). Floral structure, systematics, and phylogeny in Trochodendrales. *Annals of the Missouri Botanical Garden* **73,** 297–324.

Ernst, W. R. (1963). The genera of Hamamelidaceae and Platanaceae in the south-eastern United States. *Journal of the Arnold Arboretum* **44,** 193–210.

Goldblatt, P. (1979). Miscellaneous chromosome counts in angiosperms, II. Including new family and generic records. *Annals of the Missouri Botanical Garden* **66,** 856–61.

Goldblatt, P. and Endress, P. K. (1977). Cytology and evolution in Hamamelidaceae. *Journal of the Arnold Arboretum* **58,** 67–71.

Guerra, M. (1983): O uso de Giemsa na citogenetica vegetal—comparação entre a coloração simples e o bandeamento. *Ciencia e Cultura* **35,** 190–3.

Hans, A. S. (1972). Cytomorphology in arborescent Moraceae. *Journal of the Arnold Arboretum* **53,** 216–25.

Janaki Ammal, E. K. (1945). Cited in Darlington, C. D. and Janaki Ammal, E. K. (ed.) *Chromosome atlas of cultivated plants*, London.

Jones, J. H. (1986). Evolution of the Fagaceae: The implications of foliar features. *Annals of the Missouri Botanical Garden* **73,** 228–75.

Johnson, L. A. S. (1982). Notes on Casuarinaceae II. *Journal of the Adelaide Botanical Garden* **6,** 73–87.

Krause, O. (1930). Zytologische Studien bei den Urticales. *Berichte der deutschen botanischen Gesellschaft* **48,** 9–13.

Mehra, P. N. (1976). *Cytology of himalayan hardwoods.* Sree Saraswaty Press, Calcutta.

Mehra, P. N. and Hans, A. S. (1972). Comparison of diploid and natural polyploid hardwoods. *Journal of the Indian Botanical Society* **51,** 319–25.

Morawetz, W. (1981*a*). C-banding in *Liriodendron tulipifera* (Magnoliaceae): Some karyological and systematic implications. *Plant Systematics and Evolution* **138,** 209–16.

Morawetz, W. (1981*b*). Zur systematischen Stellung der Gattung Prockia: Karyologie und Epidermisskulptur im Vergleich zu *Flacourtia* (Flacourtiaceae), *Grewia* (Tiliaceae) und verwandten Gattungen. *Plant Systematics and Evolution* **139,** 57–76.

Morawetz, W. (1986*a*). Remarks on karyological differentiation patterns in tropical woody plants. *Plant Systematics and Evolution* **152,** 49–100.

Morawetz, W. (1986*b*). Systematics and karyoevolution in Magnoliidae: *Tetrameranthus* as compared with other Annonaceae genera of the same chromosome number. *Plant Systematics and Evolution* **154,** 147–73.

Morawetz, W, (1988). Karyosystematics and evolution of Australian Annonaceae as compared with Eupomatiaceae, Himantandraceae and Austrobaileyaceae. *Plant Systematics and Evolution* **159,** 49–79.

Morawetz, W. and Rainer, H. (1987) Die Chromosomenzahlen der Hamamelidae. *Sitzungsberichte der Österreichischen Akademie der Wissenschaften, Mathematisch-Naturwissenschaftliche Klasse* **197,** 157–72.

Morawetz, W. and Waha, M. (1985). A new pollen type, C-banded and fluorescent counterstained chromosomes and evolution in *Guatteria* and related genera (Annonaceae). *Plant Systematics and Evolution* **150,** 119–41.

Nakajima, G. (1942). Cytological studies in some flowering dioecious plants, with special reference to the sex chromosomes. *Cytologia* **12,** 262–70.

Okada, H. (1975). Karyomorphological studies in woody Polycarpicae. *Journal of Sciences of the Hiroshima University, Ser. B, Div. 2, Botany* **15,** 115–200.

Oginuma, K., Raven, P. H. and Tobe, H. (1988). Karyomorphology and Relationship of Celtidaceae and Ulmaceae (Urticales). *Plant Systematics and Evolution,* (In press).

Pizzolongo, P. (1958). Ricerche cariotassonomiche su alcune Hamamelidales. *Annali di botanica* **26,** 1: 1–18.

Puff, C. (1978). Zur Biologie von *Myrothamnus flabellifolia* Welw. (Myrothamnaceae). *Dinteria* **14,** 1–20.

Ratter, J. A. and Milne, C. (1973). Chromosome counts in Primitive angiosperms. *Notes of the Royal Botanical Garden Edinburgh* **32,** 423–28.

Ratter, J. A. and Milne, C. (1976). Chromosome counts in primitive Angiosperms 2. *Notes of the Royal Botanical Garden Edinburgh* **35,** 143–5.

Romero, E. J. (1986). Fossil evidence regarding the evolution of *Nothofagus* Blume. *Annals of the Missouri Botanical Garden* **73,** 276–83.

Sax, K. (1933). Species hybrids in *Platanus* and *Campsis*. *Journal of the Arnold Arboretum* **4,** 274–8.

Sharma, A. K. (1970). Annual report, 1967–68 *Research Bulletin, University of Calcutta (Cytogenetics Laboratory)* **2,** 1–50.

Stern, W. L. (1973). Development of the amentiferous concept. *Brittonia* **25,** 316–33.

Sugiura, T. (1936). Studies on the chromosome numbers in higher plants with special reference to the cytokinesis 1. *Cytologia* **7,** 544–95.

Takhtajan, A. L. (1973). *Ausbreitung und Evolution der Blütenpflanzen*. G. Fischer, Stuttgart.

Tanaka, R. and Oginuma, K. (1983). Karyomorphological study on *Eucommia ulmoides*. *Chromosome Information Service* **35,** 21–2.

Taylor, R. L. and Mulligan, G. A. (1968). Cytological aspects of the vascular plants. In *Flora of the Queen Charlotte Islands*, Part 2. Queen's Printer, Ottawa.

Thorne, R. F. (1973). The 'Amentiferae' or Hamamelidae as an artificial group: A summary statement. *Brittonia* **25,** 395–405.

Thorne, R. F. (1976). A phylogenetic classification of the Angiospermae. *Evolutionary Biology* **9,** 35–106.

Wettstein, R. (1924). *Handbuch der Systematischen Botanik*. F. Deuticke, Leipzig.

Wetzel, G. (1928). Chromosomenstudien bei den Fagales. *Berichte der Deutschen Botanischen Gesellschaft* **46,** 212–14.

Wetzel, G. (1929). Chromosomenstudien bei den Fagales. *Botanisches Archiv* **25,** 257–83.

Whitaker, T. W. (1933). Chromosome numbers and relationship in the Magnoliales. *Journal of the Arnold Arboretum* **14,** 376–85.

Woodworth, R. H. (1929). Cytological studies in the Betulaceae. I. *Betula. Botanical Gazette* **87,** 331–63.

Woodworth, R. H. (1930). Meiosis of microsporogenesis in the Juglandaceae. *American Journal of Botany* **17,** 863–9.

8. Reproductive structures of Cretaceous Hamamelidae

ELSE MARIE FRIIS

Department of Palaeobotany, Swedish Museum of Natural History, S-104 05 Stockholm, Sweden

PETER R. CRANE

Department of Geology, Field Museum of Natural History, Chicago, IL 60605, USA

Abstract

The earliest unequivocal fossil records of the subclass Hamamelid(id)ae are from the late Albian (Early Cretaceous) where the group is represented by leaves and reproductive structures of the Platanaceae, as well as putative trochodendralean foliage. In the Late Cretaceous the Hamamelidae are more extensively represented by fossil reproductive structures assignable to the Trochodendrales, Platanaceae, Hamamelidaceae, Juglandales/Myricales and possibly Fagales. Fossil flowers of juglandalean/myricalean affinity are especially diverse: more than 30 taxa are known, many of which produced triporate pollen of the Normapolles type.

Introduction

Palaeobotanical investigations of fossil pollen and leaves have clearly documented a major diversification of angiosperms through the mid-Cretaceous (Brenner 1963; Doyle 1969; Muller 1970; Doyle and Hickey 1976; Lidgard and Crane 1988) and studies of these remains have yielded important insights into angiosperm differentiation (Doyle and

Evolution, Systematics, and Fossil History of the Hamamelidae, Volume 1: Introduction and 'Lower' Hamamelidae (eds P. R. Crane and S. Blackmore), Systematics Association Special Volume No. 40A, pp. 155–74. Clarendon Press, Oxford, 1989.

Hickey 1976; Hickey and Doyle 1977). Frequently, however, fossil pollen and leaves have afforded only limited systematic resolution and knowledge of the relationships of Cretaceous flowering plants has been greatly enhanced by recent information on well-preserved fossil flowers (Dilcher 1979; Friis 1983, 1984, 1985; Crane 1989; Friis *et al.* 1988; Crane *et al.* in press). In this paper we assess the Cretaceous fossil history of the Hamamelidae mainly based on reproductive structures. We review Cretaceous evidence of Trochodendrales, Hamamelidales, Fagales, and Juglandales/Myricales, and consider the implications for phylogenetic interpretations of the Hamamelidae. The fossil history of the Urticales is outlined elsewhere (Collinson Vol. 40B, Ch. 18; Manchester Vol. 40B, Ch. 12), and the Casuarinales, Eucommiales, Daphniphyllales, Didymelales, Leitneriales, Buxales, and Simmondsiales have no well-documented Cretaceous fossil record.

Trochodendrales

The extant Trochodendrales *sensu* Endress (1986) include the families Trochodendraceae (*Trochodendron*, *Tetracentron*), Cercidiphyllaceae (*Cercidiphyllum*) and Eupteleaceae (*Euptelea*). We also treat Myrothamnaceae with this group, based on the probable close relationship between *Myrothamnus* and *Cercidiphyllum* (Endress Vol. 40A, Ch. 12; Hufford and Crane Vol. 40A, Ch. 9). The Trochodendrales are usually included with the Hamamelidae on the basis of several characteristics of which perhaps the most significant is the possession of triaperturate pollen. However, the Trochodendrales also exhibit significant similarities to certain Magnoliidae and may retain more primitive characters than other 'higher' dicotyledons. For this reason the Trochodendrales are usually treated as the basal group of the Hamamelidae (Cronquist 1981; Endress 1986) and it is perhaps surprising that they are not better represented in the mid-Cretaceous fossil record. Fossil pollen tetrads with strong morphological and ultrastructural similarities to those of extant *Myrothamnus* are present in the mid–late Albian (Ward and Doyle, personal communication) and fruiting structures (*Caspiocarpus paniculiger*), attached to leaves with actinodromous venation and a lobed margin are known from the middle Albian of Kazakhstan (Vakhrameev and Krassilov 1979). Similarities of both the fruiting structures and the leaves to late Cretaceous and early Tertiary Cercidiphyllaceae suggest that *Caspiocarpus* may be an early representative of the Trochodendrales. Other evidence of possible trochodendralean reproductive structures in the mid-Cretaceous is currently lacking, but certain fossil leaves of this age may be related to the Trochodendrales. These leaves have actinodromous venation, frequently cordate bases, marginal crenations and chloranthoid teeth (e.g. '*Populus*' *potomacensis*, *Populophyllum reni-*

forme, and *Menispermites potomacencis* from the mid–late Albian of the Potomac Group, eastern North America) but detailed comparative analyses of these leaves have not yet been completed (Crabtree 1987).

During the Late Cretaceous and early Tertiary unequivocal remains of the Trochodendrales are more common, and two distinct groups the *Joffrea–Nyssidium* complex and *Nordenskioldia*, may be recognized based on reproductive structures. The *Joffrea–Nyssidium* complex is one of the most frequent constituents of latest Cretaceous and early Tertiary floras from middle and high latitudes in the northern hemisphere. Some of the plants in this group are known from seedlings, shoots, leaves, pistillate flowers and inflorescences, infructescences, seeds, and possibly staminate inflorescences (Stockey and Crane 1983; Crane and Stockey 1985, 1986). The detailed information available clearly demonstrates that the *Joffrea–Nyssidium* complex is more closely related to *Cercidiphyllum* than to any other extant genus (Crane and Stockey 1986). Pollen grains similar to those of modern *Cercidiphyllum* have been described from Campanian strata of Canada (Jarzen 1978).

Nordenskioldia borealis Heer (1870) is based on infructescences bearing sessile schizocarp-like fruits that are frequently associated with leaves very similar to those of the *Joffrea–Nyssidium* complex. Remains of the *Nordenskioldia* plant are also common and widespread in the latest Cretaceous and early Tertiary of the northern hemisphere (Brown 1962; Chandrasekharam 1974; Krassilov 1976; Crane 1989; Crane *et al.* in press). Recent studies have demonstrated that at least some *Nordenskioldia* flowers were hermaphroditic with about 15 carpels and numerous stamens, and that the wood of both infructescence axes and vegetative shoots lacked vessels. The morphological and anatomical details currently available indicate that *Nordenskioldia* is probably more closely related to *Trochodendron* than any other extant genus (Crane 1989, Crane *et al.* in press)

Hamamelidales

The extant Hamamelidales *sensu* Takhtajan (1969, excluding Myrothamnaceae) comprises the families Platanaceae (*Platanus*) and Hamamelidaceae (with about 30 modern genera). The group is extensively represented in the mid-Late Cretaceous fossil record and although leaves and reproductive structures of Platanaceae are especially common the Hamamelidaceae are also present.

Platanaceae: In the mid–late Albian and Cenomanian platanoid leaves are one of the most abundant and widespread components of fossil floras in the northern hemisphere. Several platanoid reproductive structures are also known (Table 8.1) and these all consist of elongated

Table 8.1 Reproductive structures of platanaceous affinity described from Cretaceous strata (in stratigraphic order).

Platanus ex gr. *newberryana* (Maastr.–Dan)	Krassilov (1976)
Tricolpopollianthus burejensis (Maastr.–Dan.)	Krassilov (1973)
Platanus cf. *laevis* (Maastr.)	Knobloch & Mai (1986)
Platanocarpus sp. (Santon.–Camp.)	Friis *et al.* (1988)
Unassigned infructescence (Santon.–Camp.)	Friis *et al.* (1988)
Platanocarpus carolinensis (Santon.–Camp.)	Friis *et al.* (1988)
Platananthus scanicus (Santon.–Camp.)	Friis *et al.* (1988)
Unassigned stamens (Santon.–Camp.)	Friis *et al.* (1988)
Platananthus hueberi (Santon.–Camp.)	Friis *et al.* (1988)
Platanus quedlinburgensis (Santon.)	Pacltová (1982)
Platanus richteri (Santon.)	Knobloch and Mai (1986)
Platanus laevis (Cenoman.)	Velenovský (1881, 1889)
Platanus primaeva (Cenoman.)	Lesquereux (1892), Dilcher (1979)
'*Sparganium*' *aspensis* (Alb.)	Brown (1933)
Platanocarpus marylandensis (Alb.)	Friis *et al.* (1988)
Platananthus potomacensis (Alb.)	Friis *et al.* (1988)

inflorescences or infructescences bearing sessile balls of numerous staminate or pistillate flowers (Fig. 8.1A). The most completely understood of these early platanoid reproductive structures are *Platanocarpus marylandensis* (pistillate) and *Platananthus potomacensis* (staminate) both from the mid–late Albian Potomac Group of Maryland (Crane *et al.* 1986, Friis *et al.* 1988).

In *Platanocarpus marylandensis* the numerous flowers of each ball-like unit of the inflorescence have a prominent perianth consisting of short broad outer parts and long narrow inner parts. The flowers are unisexual with an apocarpous gynoecium of five carpels. The carpels are smooth, lack trichomes, have a clear ventral slit and a very poorly developed style. In *Platananthus potomacensis* each ball is composed of unisexual flowers with five stamens and pronounced perianth parts. Each stamen has a short filament, a poorly differentiated apical extension of the connective, and a long tetrasporangiate anther. The connective between the pollen sacs is well developed and dehiscence of

Fig. 8.1 Stamen structure in Cretaceous Platanaceae (A–E, H) and Hamamelidaceae (F–G). A, Inflorescence axis ('*Sparganium*' *aspensis*) with sessile balls, Cenomanian of Wyoming, ×20. B–D, Stamens (*Platananthus potomacensis*) from the Albian of Maryland, lateral view (B, ×70), dorsal view? (C, ×70) and transverse section (D, ×100). E, Stamen (unassigned) from the Santonian/Campanian of southern Sweden with distinct peltate expansion of connective, ×65. F–G, Stamens (unnamed Hamamelidaceae) from the Santonian/Campanian of southern Sweden, transverse section (F), dorsal view (G), ×75. H, Stamen (*Platananthus scanicus*) from the Santonian/Campanian of southern Sweden showing valvate dehiscence and elongated expansion of connective, ×40.

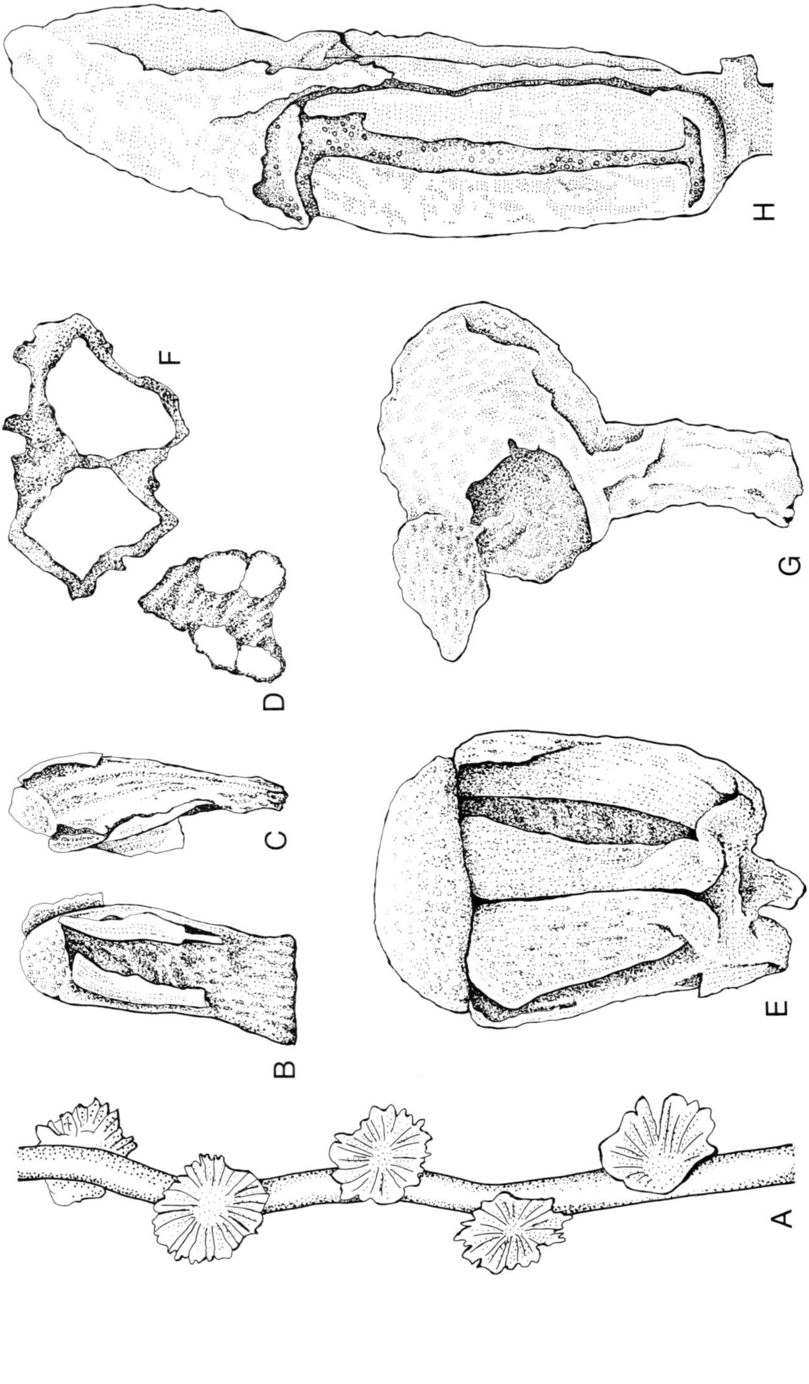
A
B
C
D
E
F
G
H

the anthers was apparently valvate (Figs. 8.1B–D). Pollen grains (*Tricolpites minutus*) are small (about 10 μm long), tricolpate, and reticulate. The occurrence of *Platanocarpus marylandensis* and *Platananthus potomacensis* at the same locality combined with the presence of *Tricolpites minutus* pollen on the *Platanocarpus marylandensis* flowers suggests that these different organs were produced by the same species.

Although Late Cretaceous platanoid flowers show considerable variability in characters such as the form of perianth, morphology of stamens (Figs 8.1E, H), and shape of the carpels they are consistently unisexual with either five carpels or five stamens surrounded by well-developed perianth parts (Friis *et al.* 1988). The same basic organization also occurs in early Tertiary platanoids (e.g. the *Macginitiea* plant, Manchester 1986), and it is possible that the irregular and much less fixed arrangement in extant *Platanus* has come about through a reduction of the perianth which no longer constrains the early development of carpels and stamens. In addition extant *Platanus* differs from all Cretaceous platanoid reproductive structures in which the relevant characters are known, in having larger pollen and obovoid carpels with abundant trichomes, rather than ellipsoidal carpels that lack hairs. Other possible chronological trends within the group may have been towards fewer and bigger heads per inflorescence and larger numbers of carpels per head.

Several of the differences between extant *Platanus* and the Cretaceous forms may be associated with changes in pollination and dispersal biology. While extant *Platanus* is wind pollinated, the very small pollen, the well-developed perianth and the fixed number of parts in the Cretaceous forms suggests that the ancestral condition in this clade was probably pollination by insects. These features are especially pronounced in the mid-Cretaceous taxa which have a more differentiated perianth, more open flowers, and smaller pollen than the later Cretaceous species. The possibility of insect pollination is also supported by the presence of pollenkitt and valvate dehiscence in some fossil taxa. Both these features are common among insect-pollinated 'lower' Hamamelidae (Endress 1986, 1989).

Hamamelidaceae: While the Early Tertiary record of the Hamamelidaceae is extensive and based on a large variety of leaves, inflorescences, fruits, seeds, and pollen, the Cretaceous record is poor. No unequivocal leaf fossils are known from the Late Cretaceous (Wolfe 1973) and the earliest unquestioned dispersed pollen is known from the Paleocene (Muller 1981). Knobloch and Mai (1986) have described 20 different species of seeds from the Late Cretaceous of middle Europe that they assigned to the Hamamelidaceae. These include two species referred to the modern genus *Disanthus* and one species referred to the modern genus *Rhodoleia*, all from the uppermost Cretaceous. Seventeen species were included in a new extinct genus *Klikovispermum* that ranges in age from the Cenomanian to the Maastrichtian. While the species assigned to *Disanthus* and *Rhodoleia* may well represent the family, those of *Klikovispermum* are much more uncertain. They constitute a very heterogenous assemblage of taxa that are probably related to different genera or even different families. Some of the species (e.g. *Klikovispermum rugulatum*) have characters that may suggest an affinity to the Magnoliidae.

Recently, the discovery of well-preserved flowers from southern Sweden (Santonian/Campanian) closely related to modern *Hamamelis* subfamily Hamamelidoideae, (Endress and Friis, unpublished) has documented that the Hamamelidaceae were present in the Late Cretaceous and probably had a long Cretaceous history. The flowers are bisexual with six or seven stamens and three carpels. The anthers have distinct valvate dehiscence and thin sections suggest that they were bisporangiate (Figs 8.1F–G), a feature that is considered advanced within the group (Endress 1989). The pollen grains are tricolpate, reticulate, and have granular aperture membranes.

One other subfamily, the Liquidambariodeae, was apparently also present in the Late Cretaceous, represented by spherical infructescences from the Maastrichtian–Danian of the Amur area (Krassilov 1976). The infructescences were assigned to the extinct genus *Steinhauera* that is sometimes included in the modern genus *Liquidambar* (Kirchheimer 1957). The extant genera of the Liquidambariodeae (*Altingia* and *Liquidambar*) have characteristic periporate pollen that may be traced back to the Paleocene (Muller 1981, see also Ferguson Vol. 40A, Ch. 13).

Fagales

The fossil history of the Fagales (Fagaceae *sensu lato*, Betulaceae) has received considerable recent attention (Crane Vol. 40B, Ch. 6; Crepet Vol. 40B, Ch. 4). Although there are no unequivocal pre-Tertiary records of either family based on macrofossils, the evidence from dispersed pollen grains suggests that the Fagales began to differentiate during the Late Cretaceous. Probable castaneoid pollen has been reported from the Maastrichtian of California (see Crepet Vol. 40B, Ch. 4) and the diagnostic pollen grains of *Nothofagus* are known from the Santonian, Campanian, and Maastrichtian of the southern hemisphere (Romero 1986; Tanai 1986). Pollen grains exhibiting synapomorphies of extant *Alnus* (Betulaceae) are known from the Campanian and Maastrichtian of North America and Japan (See Crane Vol. 40B, Ch. 6).

Two assemblages of fossil material from the northern hemisphere may provide support for the palynological evidence of Late Cretaceous Fagales. Distinctive palmately compound leaves with a pedate arrangement of leaflets from the Late Cretaceous and earliest Tertiary of the northern hemisphere have been assigned to the fossil genera *Debeya* and *Dewalquea*. Based on cuticular and leaf-architectural features, several authors (e.g. Kvaček 1983; Rüffle and Knappe 1977; Rüffle 1978, 1980) have pointed out the similarities of this leaf complex both to the Platanaceae and the Fagaceae. More details of these plants are needed to clarify their phylogenetic position and assess their relevance to the origin of the Fagales.

Only a single type of reproductive organ of possible fagalean affinity is known from the Cretaceous (Tiffney and Friis, unpublished). This material comprises small well-preserved flowers from the Santonian/Campanian of Massachusetts, eastern North America. The flowers are epigynous with six tepals in a single whorl, six stamens, and a trimerous gynoecium. The ovary is three-loculed and there are apparently two anatropous and apical ovules in each locule. The fossils remain to be studied in detail, but the information currently available strongly suggest a relationship to the Fagales.

Myricales and Juglandales

The Cretaceous record of Myricales and Juglandales is treated together because some of the fossils appear to have characters of both groups, and clear differentiation of the extant taxa may not have been established until the latest Cretaceous or earliest Tertiary. The Myricales consists of one family, the Myricaceae, that includes three modern genera *Myrica*, *Comptonia*, and *Canacomyrica* (MacDonald Vol. 40B, Ch. 8). The Juglandales includes two families, the Juglandaceae with eight modern genera (Stone Vol. 40B, Ch. 7) and the Rhoipteleaceae with a single modern genus, *Rhoiptelea*.

Although the presence of all three families in the Cretaceous have been suggested based on the record of dispersed pollen (Wolfe 1973; Muller 1981) this has not been confirmed on the basis of floral structures. In an extensive review of the fossil history of the Juglandaceae, Manchester (1987) established that the major diversification of the family occurred during the Paleocene (see also Stone Vol. 40B, Ch. 7).

It has long been suggested that the ancestors of the Juglandales and Myricales should be sought in the Late Cretaceous Normapolles complex (e.g. Krutzsch 1966). Pollen grains assigned to this complex are triporate, frequently with elaborate and often strongly protruding apertures. They constitute one of the most conspicuous elements in Late Cretaceous middle latitude palynofloras from Europe and eastern

North America with more than 70 genera and numerous species (see Batten Vol. 40B, Ch. 2; Kedves Vol. 40B, Ch. 1). The group first appeared in the mid-Cenomanian of Europe and is considered to have become extinct by the end of the Eocene (Pacltová 1966, 1981; Góczán *et al.* 1967; Batten 1981). The maximum diversification of the group was in the Turonian to Maastrichtian (Muller 1970).

Based on pollen morphology it has been suggested that at least some of the Normapolles genera were related to the Juglandaceae and the Rhoipteleaceae (Krutzsch 1966; Góczán *et al.* 1967; Skarby 1968; Wolfe 1973). The first evidence on the floral structure of Normapolles producing plants were provided by well-preserved flowers and fruits from the Santonian/Campanian of Scania, southern Sweden (Friis 1983). A large number of related forms have subsequently been reported from the Late Cretaceous of middle Europe (Knobloch and Mai 1986). More than 30 species can be included in the group with some confidence ranging from the Cenomanian to the Maastrichtian (Table 8.2), but probably the number of species is much higher (see below). The different species are often extremely abundant in fossil assemblages, and in the flora from Scania several thousand specimens were recovered suggesting that Normapolles plants constituted an important element in the vegetation.

The species enumerated in Table 8.2 appear to be grouped in three main types represented by the three genera *Manningia*, *Antiquocarya*, and *Caryanthus*. They are all small, bisexual, and epigynous with a single whorl of perianth parts and a unilocular ovary containing a single orthotropous ovule. The three types are distinguished mainly in number of parts and floral symmetry. Pollen grains included in the flowers are also different for the three types. *Manningia* and *Antiquocarya* are radially symmetrical with three carpels. The number of tepals and stamens is five in *Manningia* and six in *Antiquocarya*. *Caryanthus* is bisymmetrical with four tepals, six stamens, and two carpels.

The *Antiquocarya* type has so far been recorded only from the Late Cretaceous of Sweden, and the *Caryanthus* type (Fig. 8.2) is by far the most common of the currently known Normapolles flowers. Three species of *Caryanthus* have been identified from southern Sweden (Friis 1983) and eight species have been established from the Late Cretaceous floras of middle Europe (Knobloch and Mai 1986). An additional undescribed specimen of *Caryanthus* has also been recognized from the Campanian of Aachen. The Aachen flower shows the same general structure as those from southern Sweden, and Normapolles grains of the *Plicapollis* type have been recovered in both the Scanian *Caryanthus* as well as the flower from Aachen (Figs 8.2C, F). Several other species from the Late Cretaceous of middle Europe referred to the genera *Budvaecarpus* and *Walbeckia* (Knobloch and Mai 1986) show similar

Table 8.2 Reproductive structures of juglandalean/myricalean affinity described from Cretaceous strata (in alphabetical order)

Taxon	Reference
Antiquocarya nuda (Santon.–Camp.)	Friis (1983)
Antiquocarya verruculosa (Santon.–Camp.)	Friis (1983)
Budvaecarpus serialis (Cenoman.–Santon.)	Knobloch and Mai (1986)
Calathiocarpus calyciferus (Maastr.)	Knobloch and Mai (1986)
Calathiocarpus hollandicus (Camp.)	Knobloch and Mai (1986)
Calathiocarpus minimus (Turon.–Santon.)	Knobloch and Mai (1986)
Calathiocarpus octocostatus (Cenoman.–Paleoc.)	Knobloch and Mai (1986)
Carpolithus costatus (Turon.–Santon.)	Knobloch and Mai (1986)
Caryanthus communis (Turon.–Santon.)	Knobloch and Mai (1986)
Caryanthus deltoides (Turon.–Maastr.)	Knobloch and Mai (1986)
Caryanthus elongatus (Santon.–Camp.)	Friis (1983)
Caryanthus hamrensis (Cenoman.–Maastr.)	Knobloch and Mai (1986)
Caryanthus knoblochii (Santon.–Camp.)	Friis (1983)
Caryanthus microtriasseris (Turon.–Santon.)	Knobloch and Mai (1986)
Caryanthus multiaseris (Cenoman.–Maastr.)	Knobloch and Mai (1986)
Caryanthus pseudooctocostatus (Maastr.)	Knobloch and Mai (1986)
Caryanthus trebecensis (Turon.–Camp.)	Knobloch and Mai (1986)
Caryanthus triasseris (Cenoman.–Maastr.)	Knogloch and Mai (1986)
Caryanthus sp. (Camp.)	present work
Unnamed (*Caryanthus* sp.?) (Santon.–Camp.)	Friis (1983)
Costaecarpus pterocarpus (Turon.–Santon.)	Knobloch and Mai (1986)
Manningia crassa (Turon.–Camp.)	Friis (1983); Knobloch and Mai (1986)
Rhoiptelea pontwallensis (Camp.)	Vangerow (1954); Knobloch and Mai (1986)
Walbeckia ajkoensis (Santon.–Camp.)	Knobloch and Mai (1986)
Walbeckia aquisgranensis (Camp.)	Knobloch and Mai (1986)
Walbeckia costata (Maastr.)	Knobloch and Mai (1986)
Walbeckia fricii (Turon.–Santon.)	Knobloch and Mai (1986)
Walbeckia guttaeformis (Cenoman.–Maastr.)	Knobloch and Mai (1986)
Walbeckia hlubocensis (Turon.–Santon.)	Knobloch and Mai (1986)
Walbeckia libusae (Turon.–Santon.)	Knobloch and Mai (1986)
Walbeckia orthocostata (Maastr.)	Knobloch and Mai (1986)
Walbeckia scutata (Maastr.)	Knobloch and Mai (1986)

bisymmetrical structure, and remains of epigynous perianth parts indicate that these are probably closely related to *Caryanthus*. Two species assigned by Knobloch and Mai (1986) to the Urticaceae (*Debregeasia walbeckensis*, *Urticoidea minutissima*) also show some resemblance to *Caryanthus* species, but their affinity is less certain.

The *Manningia* type (Fig. 8.3) is represented by a single species from

Fig. 8.2 Cretaceous flowers with Normapolles pollen (*Plicapollis*). A–C, *Caryanthus knoblochii* Friis from the Santonian/Campanian of southern Sweden. D–F, *Caryanthus* sp. from the Campanian of Aachen, Germany. A and D, Lateral view of flowers showing epigynous position of perianth (A, ×40; D, ×45). B and E, Apical view of flowers showing bisymmetrical arrangement of perianth (B, ×60; E, ×45). C and F, Pollen grains from flowers (C, ×2300; F, ×2000).

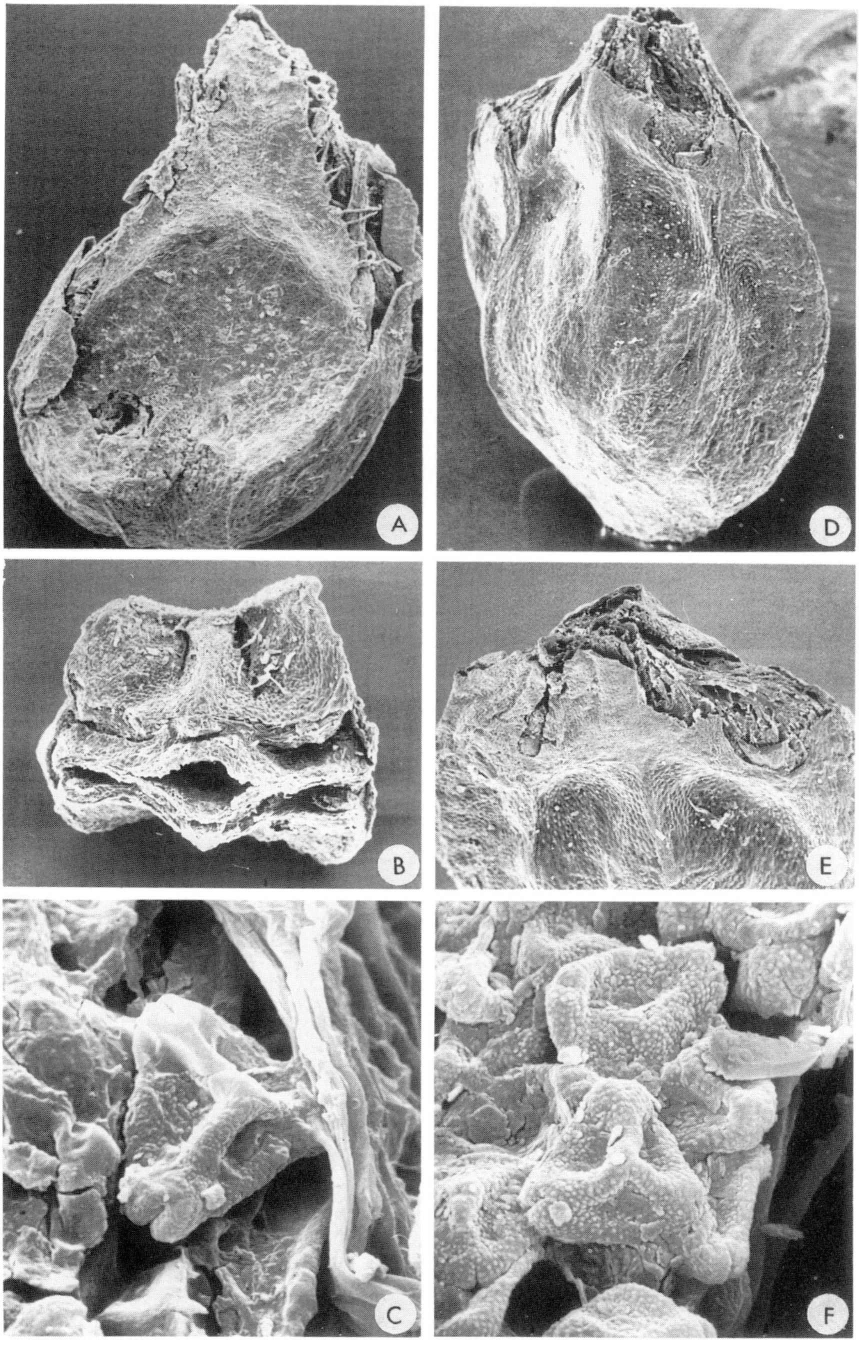
A
D
B
E
C
F

the Santonian/Campanian of southern Sweden (Friis 1983) and four species from the Late Cretaceous of middle Europe assigned to the genus *Calathiocarpus* (Knobloch and Mai 1986). Species of the two genera are very uniform in their major features and pollen grains of the *Trudopollis* type have been found in *Manningia* as well as in *Calathiocarpus* (Figs 8.3C, F). It is therefore questionable whether the later taxon should be maintained as a distinct genus.

Another group of Cretaceous fossil plants that probably also produced Normapolles pollen had small hypogynous flowers in contrast to the epigynous forms described above. All of the hypogynous forms so far recovered are also from the Santonian/Campanian of Scania. Each flower has a single whorl of five perianth parts and a trimerous, unilocular gynoecium with a single orthotropous ovule (Fig. 8.4). No stamens have so far been seen on any of the specimens, but large numbers of one type of Normapolles pollen have been observed at stigmatic surfaces and inside the perianth (Figs 8.4C–D). The four species of *Vangerowia* and *Wietersdorfia trigona* described by Knobloch and Mai (1986) from the Late Cretaceous of middle Europe have a very similar structure and may be closely related to the Scanian material.

The characters of the fossil Normapolles-producing flowers and fruits indicate a close relationship with the Myricales and the Juglandales. In particular members of the Juglandaceae show strong similarities, but the fossils cannot be included in any modern family with certainty. They differ from the Juglandaceae mainly in the very small size of the fruits, the bisexual flowers, and the undivided locule.

The three types of flowers possess features generally associated with wind pollination, but the hermaphroditic condition may indicate that wind pollination was not fully developed in these Cretaceous plants. The ancestral condition may well have been pollination by insects. Based on the foveolate to reticulate pollen-wall sculpture in some early Normapolles genera (e.g. *Atlantopollis*), Batten (1986) has also suggested that certain ancestral types in Normapolles lineages were probably insect pollinated.

Other Cretaceous hamamelids

Recent palaeobotanical investigations of mid–late Albian plant fossils from the Potomac Group of Maryland, eastern North America, have

---

Fig. 8.3 Cretaceous flowers with Normapolles pollen (*Trudopollis*). A–C, *Manningia crassa* Friis from the Santonian/Campanian of southern Sweden. D–F, *Calathiocarpus minimus* Knobloch and Mai from the Turonian–Santonian of Czechoslovakia. A and D, Lateral view of flowers showing epigynous position of perianth (A, × 45; D, × 90). B and E, Apical view of flowers showing pentamerous arrangement of perianth (B, × 55; E, × 115). C and F, Pollen grains from flowers (× 2000).

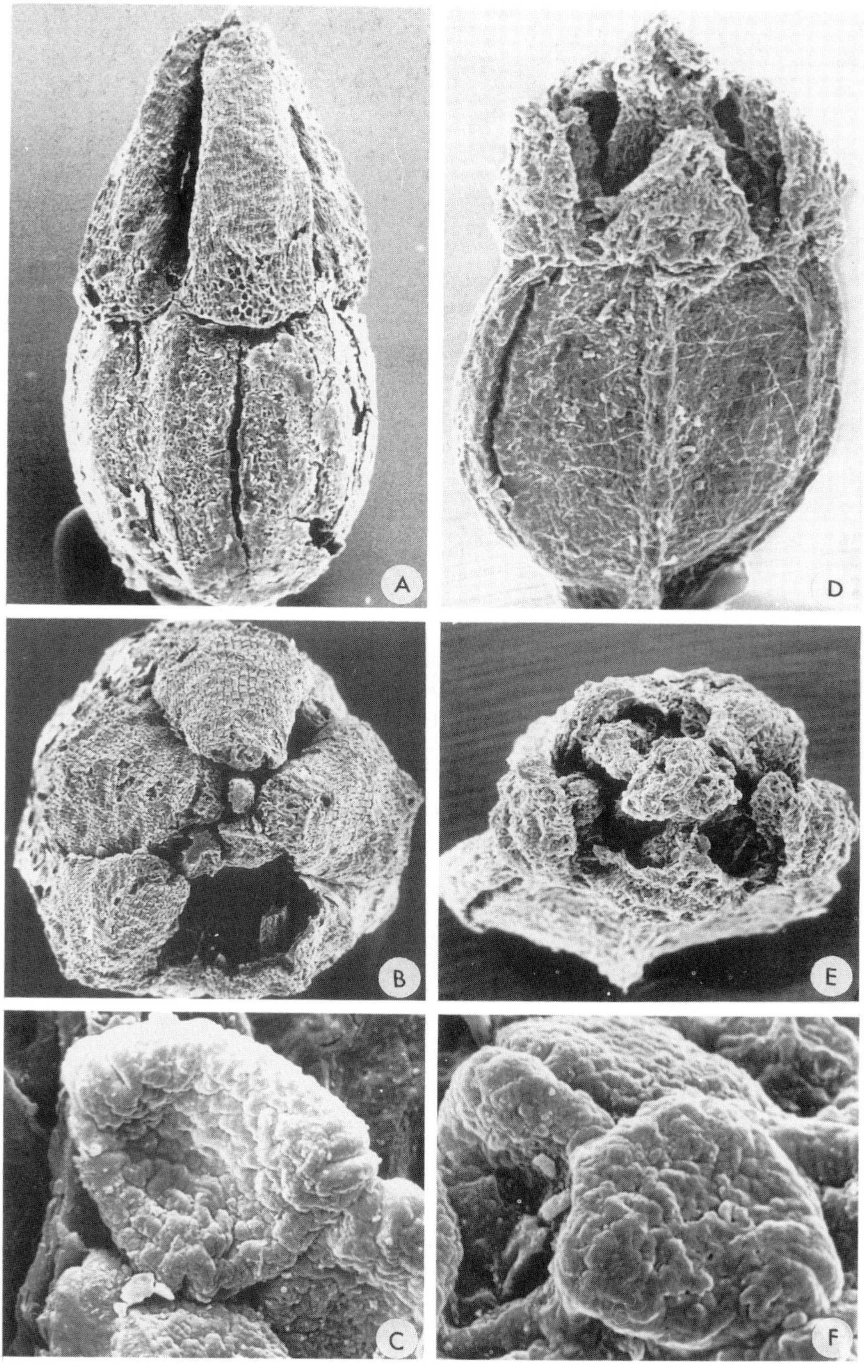
A
D
B
E
C
F

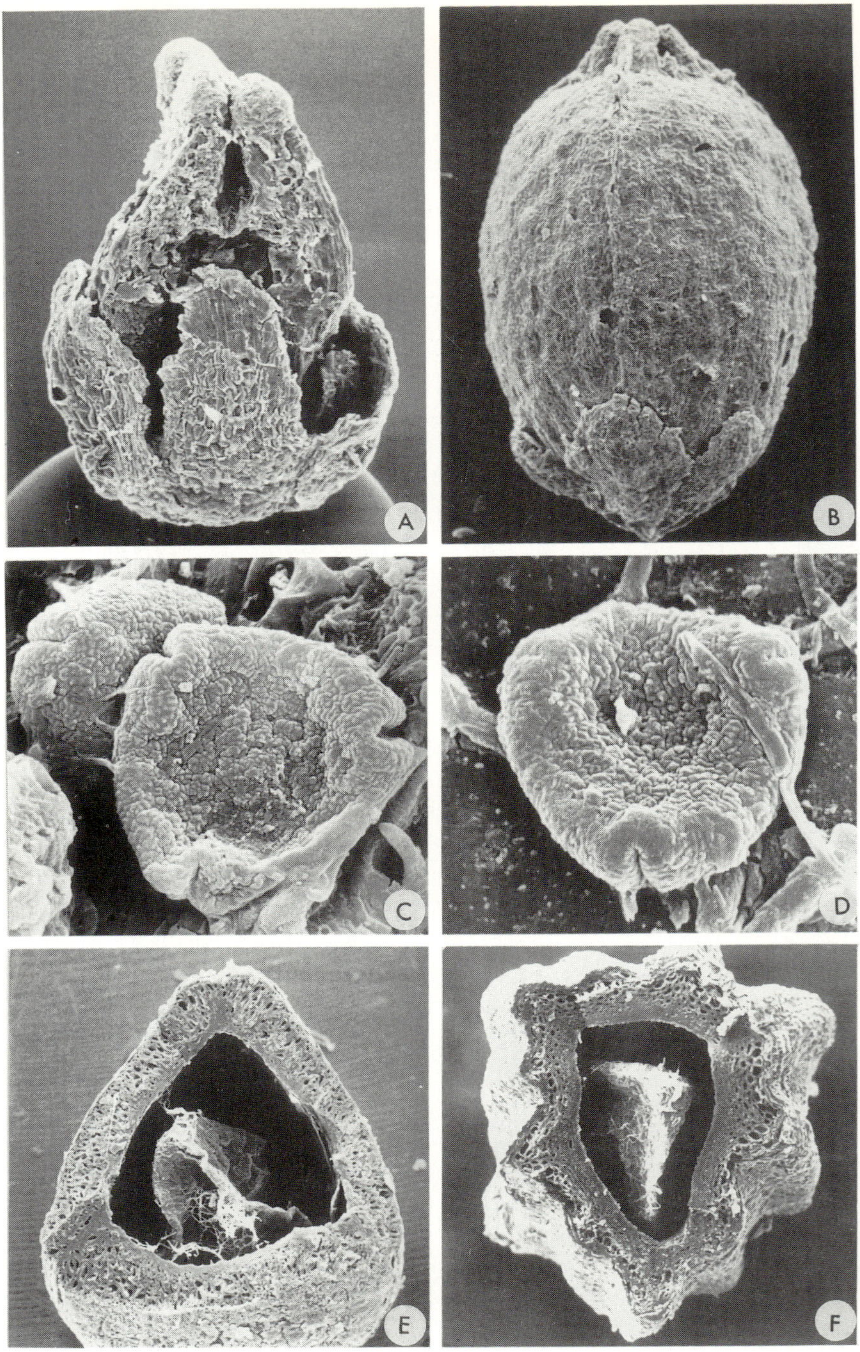

Fig. 8.4 A–E, Hypogynous flowers with Normapolles pollen from the Santonian/Campanian of southern Sweden. A–B, Lateral view of flowers showing hypogynous position of perianth (A, × 120; B, × 70). C, Pollen from perianth area of two different flowers similar to that in B (× 2000). E, Transverse section of gynoecium (× 65). F, *Manningia crassa* from the Santonian/Campanian of southern Sweden, transverse section of gynoecium (× 65).

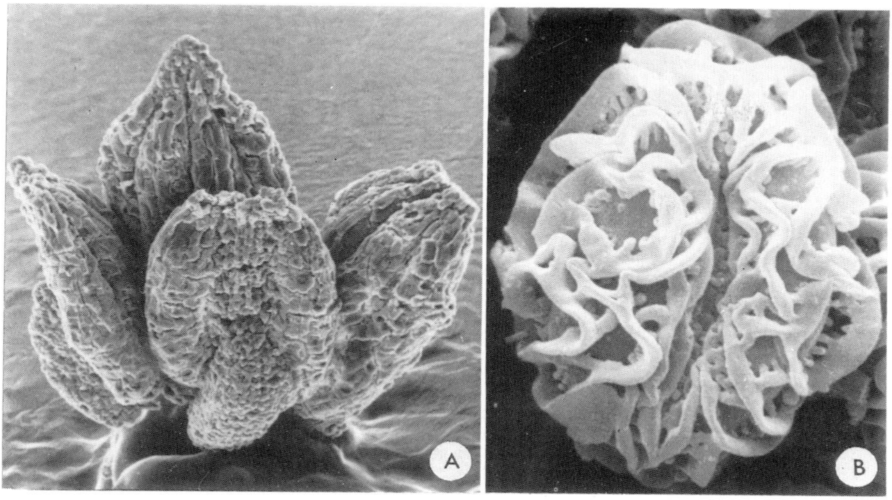

Fig. 8.5 Undetermined reproductive structure from the mid–late Albian of Maryland, eastern North America. A, Staminate flower of four stamens, × 120. B, Pollen grains from isolated stamen similar to those in A, × 5000.

yielded a variety of early angiosperm reproductive structures of which several may be referable to the Hamamelidae. One of these is a small androecium about 0.5 mm in diameter consisting of four free stamens (Fig. 8.5). Each stamen has a short stout filament and a large dorsifixed, tetrasporangiate anther with a pronounced distal extension of the connective. The anthers contain tricolpate pollen with an open reticulum and striate–rugulate organization of the muri. Similar pollen grains are attached to the surface of follicles resembling those of modern Winteraceae in having a long decurrent stigmatic area. Pollen grains are more similar to those of *Trochodendron* and *Tetracentron*. Detailed analyses of this and other Albian material are still in progress.

Conclusion

Although very few Cretaceous Hamamelidae are currently understood in detail, the fossils available suggest several patterns of probable evolutionary significance. By the late Albian the Hamamelidae were represented by unequivocal Platanaceae and putative Trochodendrales. Early platanaceous flowers were unisexual with an apocarpous pentamerous gynoecium or five free stamens producing tricolpate pollen. By the mid-Cenomanian a second major type of hamamelid floral structure, of probable juglandalean/myricalean relationship, is present, and information from better preserved, but slightly younger material suggests that these were characterized by a syncarpous ovary with two or three styles, bisexual flowers and triporate pollen of the Normapolles type. The structural characteristics of both types of fossil hamamelid flowers, as well as Cretaceous Hamamelidaceae and putative Fagales, suggest that insect pollination and unspecialized fruit or seed dispersal are primitive characteristics in many hamamelid lineages. This contrasts markedly with the biology of early Tertiary and later representatives of the Platanaceae, Juglandales/Myricales, and many Fagales which are largely anemophilous and exhibit obvious morphological specialization in association with animal or wind dispersal of propagules (Tiffney 1986; Manchester 1987, Vol. 40B, Ch. 12; Crane Vol. 40B, Ch. 6; Stone Vol. 40B, Ch. 7).

The palaeobotanical evidence currently available also raises several interesting phylogenetic questions, particularly with respect to the precise relationship between the Hamamelidae, Magnoliidae, and Rosidae. Derived similarities between mid-Cretaceous platanoid and *Sapindopsis* plants in leaf architecture, cuticular anatomy, inflorescences, and floral structure strongly suggest a close phylogenetic relationship between the platanaceous lineage and at least one lineage of early rosids (Friis *et al.* 1988; Crane 1989, and work in progress). However, conclusions with respect to the position of the Fagales are much less certain and the leaf-architectural similarities between platanoids, the *Debeya–Dewalquea* group and the Fagaceae (Rüffle 1978, 1980; Crane 1989), conflict with the fagalean–rosid relationship suggested by palynological evidence (Crepet Vol. 40B, Ch. 4; Nixon Vol. 40B, Ch. 3). Evidence from fossil flowers indicates a great diversity of the rosid group during the Late Cretaceous (e.g. Figs 8.6K–P) and resolution of this issue will depend on a more detailed understanding of these taxa, as well as rosids and putative Fagales (*Debeya–Dewalquea*), from the late Albian and Cenomanian.

With respect to the phylogenetic position of the Hamamelidae (see Donoghue and Doyle Vol. 40A, Ch. 3) tricolpate pollen diagnostic of the higher dicotyledous occurs very early in the angiosperm radiation (Barremian–Aptian boundary) particularly at low palaeolatitudes (Brenner 1976; Hickey and Doyle 1977). This palynological evidence indicates that fossils relevant to assessing the relationship between magnoliids and hamamelids may be preserved in pre-Albian rocks. The early occurrence of tricolpate pollen also suggests that some hamamelid lineages may have diverged while angiosperm diversity was still low with respect to pteridophytes and gymnosperms (Lidgard and Crane 1988), and probably prior to any extensive angiosperm radiation at the magnoliid grade.

Acknowledgements

We are grateful to Drs K. Raunsgaard Pedersen and P. K. Endress for valuable discussions, and to Drs A. Drinnan and B. H. Tiffney for their collaboration in some aspects of this study. Dr E. Knobloch kindly provided fossil material from the middle European Upper Cretaceous for comparative studies. We also thank Y. Arremo, K. Khullar, and I Svensson for assistance in preparing the manuscript and illustrations. This work was supported in part by the Swedish Natural Science Research Council grants R-RA 9381-301 and G-GU 9381-300 (E.M.F.) and U.S. National Science Foundation grant BSR-8708460 (P.R.C.).

References

Basinger, J. F. and Dilcher, D. L. (1984). Ancient bisexual flowers. *Science* **224,** 511–3.

Batten, D. J. (1981). Stratigraphic, palaeogeographic and evolutionary significance of Late Cretaceous and early Tertiary Normapolles pollen. *Review of Palaeobotany and Palynology* **35,** 175–208.

Batten, D. J. (1986). Possible functional implications of exine sculpture and architecture in some Late Cretaceous Normapolles Group. In *Pollen and Spores—Form and Function* (eds. S. Blackmore and I. K. Ferguson), pp. 209–32. Academic Press, London.

Brenner, G. J. (1963). The spores and pollen of the Potomac Group of Maryland. *State of Maryland Department of Geology, Mines and Water Resources Bulletin* **27,** 1–215.

Brenner, G. J. (1976). Middle Cretaceous floral provinces and early migrations of angiosperms. In *Origin and early evolution of angiosperms* (ed. C. B. Beck), pp. 23–47. Columbia University Press, New York.

Brown, R. W. (1933). Fossil plants from the Aspen shale of south-western Wyoming. *Proceedings of the United States National Museum* **82,** Art. 12, 1–10.

Brown, R. W. (1962). Paleocene flora of the Rocky Mountains and Great Plains. *United States Geological Survey Professional Paper* **375,** 1–119.

Chandrasekharam, A. (1974). Megafossil flora from the Genesee locality, Alberta, Canada. *Palaeontographica B* **147**, 1–41.

Crabtree, D. R. (1987). Angiosperms of the northern Rocky Mountains: Albian to Campanian (Cretaceous) megafossil floras. *Annals of the Missouri Botanical Garden* **74,** 707–47.

Crane, P. R. (1989). Palaeobotanical evidence on the early radiation of nonmagnoliid dicotyledons. *Plant Systematics and Evolution* **162,** 177–91.

Crane, P. R., Friis, E. M., and Pedersen, K. R. (1986). Angiosperm flowers from the Lower Cretaceous: Fossil evidence on the early radiation of the dicotyledons. *Science* **232,** 852–54.

Crane, P. R., Friis, E. M. and Pedersen, K. R. (in press). Reproductive structure and function in Cretaceous Chloranthaceae. *Plant Systematics and Evolution* **162,** 177–91.

Crane, P. R., Manchester, S. R. and Dilcher, D. L. (in press). Leaves and well preserved reproductive structures from the Fort Union Formation (Paleocene) near Almont, North Dakota, U.S.A. *Fieldiana, Geology.*

Crane, P. R. and Stockey, R. A. (1985). Growth and reproductive biology of *Joffrea speirsii* gen. et sp. nov., a *Cercidiphyllum*-like plant from the Late Paleocene of Alberta, Canada. *Canadian Journal of Botany* **63,** 340–64.

Crane, P. R. and Stockey, R. A. (1986). Morphology and development of pistillate inflorescences in extant and fossil Cercidiphyllaceae. *Annals of the Missouri Botanical Garden* **73,** 382–93.

Cronquist, A. (1981). *An integrated system of classification of flowering plants.* Columbia University Press, New York.

Dilcher, D. L. (1979). Early angiosperm reproduction: An introductory report. *Review of Palaeobotany and Palynology* **27,** 291–328.

Doyle, J. A. (1969). Cretaceous angiosperm pollen of the Atlantic coastal plain and its evolutionary significance. *Journal of the Arnold Arboretum* **50,** 1–35.

Doyle, J. A. and Hickey, L. J. (1976). Pollen and leaves from the mid-Cretaceous Potomac Group and their bearing on early angiosperm evolution. In *Origin and early evolution of angiosperms* (ed. C. B. Beck), pp. 139–206. Columbia University Press, New York.

Endress, P. K. (1986). Floral structure, systematics and phylogeny in Trochodendrales. *Annals of the Missouri Botanical Garden* **73,** 297–324.

Endress, P. K. (1989). Aspects of Evolutionary Differentiation of the Hamamelidaceae and the Lower Hamamelidae. *Plant Systematics and Evolution* **162,** 193–211.

Friis, E. M. (1983). Upper Cretaceous (Senonian) floral structures of juglandalean affinity containing Normapolles pollen. *Review of Palaeobotany and Palynology* **39,** 161–88.

Friis, E. M. (1984). Preliminary report of Upper Cretaceous angiosperm reproductive organs from Sweden and their level of organization. *Annals of the Missouri Botanical Garden* **71,** 403–18.

Friis, E. M. (1985) Structure and function in Late Cretaceous angiosperm flowers. *Biologiske Skrifter Danske Videnskabernes Selskab* **25,** 1–37.

Friis, E. M. and Skarby, A. (1982). *Scandianthus* gen. nov., angiosperm flowers of Saxifragalean affinity from the Upper Cretaceous of southern Sweden. *Annals of Botany* **50,** 569–83.

Friis, E. M., Crane, P. R. and Pedersen, K. R. (1988). Reproductive structures of Cretaceous Platanaceae. *Biologiske Skrifter Danske Videnskabernes Selskab* **31,** 1–56.

Góczán, F., Groot, J. J., Krutzsch, W. and Pacltová, B. (1967). Die Gattungen des 'Stemma Normapolles Pflug 1953b' (Angiospermae). Neubeschreibungen und Revision europäischer Formen (Oberkreide bis Eozän). *Paläontologische Abhandlungen B II* **3,** 427–633.

Heer, O. (1870). Die Miocene Flora und Fauna Spitzbergens. *Kongliga Svenska Vetenskapsakademiens Handlingar* **8,** 1–98.

Hickey, L. J. and Doyle, J. A. (1977). Early Cretaceous fossil evidence for angiosperm evolution. *Botanical Review* **43,** 3–104.

Jarzen, D. M. (1978). Some Maestrichtian palynomorphs and their phytogeographical and paleoecological implications. *Palynology* **2,** 29–38.
Kirchheimer, F. (1957). *Die Laubgewäche der Braunkohlenzeit.* Veb Wilhelm Knapp Verlag, Halle (Saale).
Knobloch, E. and Mai, D. H. (1986). Monographie der Früchte und Samen in der Kreide von Mitteleuropa. *Rozpravy ústredního ústavu geologického, Praha* **47,** 1–219.
Krassilov, V. A. (1973). Upper Cretaceous staminate heads with pollen grains. *Palaeontology* **16,** 41–4.
Krassilov, V. A. (1976). *Tsargayan flora of the Amur region* (in Russian). Nauka, Moscow.
Krutzsch, W. (1966). Die Sporenstratigraphische Gliederung der Oberkreide im Nordlichen Mitteleuropa—Methodische Grundlagen und Gegenwärtiger Stand der Untersuchungen. *Abhandlungen des Zentrales Geologischen Instituts, Paläontologische Abhandlungen* **8,** 79–111.
Kvaček, Z. (1983). Cuticular studies in angiosperms of the Bohemian Cenomanian. *Acta Palaeontologica Polonica* **28,** 159–70.
Lesquereux, L. (1892). The flora of the Dakota group. *United States Geological Survey Monograph* **17,** 1–400.
Lidgard, S. and Crane, P. R. (1988). Quantitative analyses of the early angiosperm radiation. *Nature* **331,** 344–6.
Manchester, S. R. (1986). Vegetative and reproductive morphology of an extinct plane tree (Platanaceae) from the Eocene of western North America. *Botanical Gazette* **147,** 200–26.
Manchester, S. R. (1987). The fossil history of the juglandaceae. *Monographs in Systematic Botany from the Missouri Botanical Garden* **21,** 1–137.
Muller, J. (1970). Palynological evidence on early differentiation of angiosperms. *Biological Reviews* **45,** 417–50.
Muller, J. (1981). Fossil pollen records of extant angiosperms. *Botanical Review* **47,** 1–142.
Pacltová, B. (1966). Pollen grains of angiosperms in the Cenomanian Peruc Formation in Bohemia *Palaeobotanist,* **15,** 52–4.
Pacltová, B. (1981). The evolution and distribution of Normapolles pollen during the Cenophytic. *Review of Palaeobotany and Palynology* **35,** 175–208.
Pacltová, B. (1982). Some pollen of recent and fossil species of the genus *Platanus* L. *Acta Universitatis Carolinae—Geologica, Pokorný* **4,** 367–91.
Romero, E. J. (1986). Fossil evidence regarding the evolution of *Nothofagus* Blume. *Annals of the Missouri Botanical Garden* **73,** 276–83.
Rüffle, L. (1978). Evolutionary and ecological trends in Cretaceous floras particularly in some Fagaceae. *Courier Forschungsinstitut Senckenberg* **30,** 77–83.
Rüffle, L. (1980). Wachstums-Modus und Blatt-Morphologie bei altertümlichen *Fagales* und *Hamamelidales* der Kreide und der Gegenwart. *100 Jahre Arboretum* (1879–1979), 329–41.
Rüffle, L. and Knappe, H. (1977). Entwicklungsgeschichtliche und ökologische Aspekte zur Oberkreide Flora, besonders einiger Fagaceae (Hamamelidaceae). *Zeitschrift Geologische Wissenschaften, Berlin* **5,** 269–303.

Skarby, A. (1968). *Extratriporopollenites* (Pflug) emend. from the Upper Cretaceous of Scania, Sweden. *Stockholm Contributions in Geology* **16,** 1–60.

Stockey, R. A. and Crane, P. R. (1983). *In situ Cercidiphyllum*-like seedlings from the Paleocene of Alberta, Canada. *American Journal of Botany* **70,** 1564–68.

Takhtajan, A. (1969). *Flowering plants. Origin and dispersal.* Oliver & Boyd, Edinburgh.

Tanai, T. (1986). Phytogeographic and phylogenetic history of the genus *Nothofagus* Bl. (Fagaceae) in the southern hemisphere. *Journal of the Faculty of Science, Hokkaido University, Series IV* **21,** 505–82.

Tiffney, B. H. (1986). Fruit and seed dispersal and the evolution of the Hamamelidae. *Annals of the Missouri Botanical Garden* **73,** 394–416.

Vakhrameev, V. A. and Krassilov, V. A. (1979). Reproductive organs of flowering plants from the Albian of Kazakhstan (in Russian). *Paleontologicheskiy Zhurnal* **1,** 121–8.

Vangerow, E. F. (1954). Megasporen und andere pflanzliche Mikrofossilien aus der Aachener Kreide. *Palaeontographica B* **96,** 24–38.

Velenovský, J. (1881). Die Flora der böhmischen Kreideformation. *Beiträge zur Palaeontologie von Oesterreich-Ungarn* **2,** 1–32.

Velenovský, J. (1889). Květena Českého Cenomanu. *Rozpravy Královské Česke Společnosti Nauk VII* **3,** 1–75.

Wolfe, J. A. (1973). Fossil forms of Amentiferae. *Brittonia* **25,** 334–55.

9. A preliminary phylogenetic analysis of the 'lower' Hamamelidae

LARRY D. HUFFORD*

Department of Botany, University of Kansas, Lawrence, KS 66045, USA

PETER R. CRANE

Department of Geology, Field Museum of Natural History, Chicago, IL 60605, USA

Abstract

We present a cladistic analysis of 'lower' Hamamelidae as an attempt to clarify phylogenetic relationships among several basal clades of 'higher' dicotyledons. Our results show a well-supported sister-group relationship for *Trochodendron* and *Tetracentron*, and these two genera form the sister group to other 'lower' hamamelids. *Cercidiphyllum* and *Myrothamnus* are resolved as sister taxa and are themselves placed as the sister taxa of *Euptelea* and *Eucommia*. *Cercidiphyllum*, *Myrothamnus*, *Euptelea*, and *Eucommia* form the sister group to *Platanus* and Hamamelidaceae. Hamamelidaceae are shown to be monophyletic and form the sister group to *Platanus*.

Introduction

The 'lower' Hamamelidae, comprising Trochodendrales and Hamamelidales as circumscribed by Cronquist (1981), have been suggested to be transitional between Magnoliidae and 'higher' Hamamelidae such as Betulaceae, Fagaceae, and perhaps the Juglandaceae and Myricaceae

Evolution, Systematics, and Fossil History of the Hamamelidae, Volume 1: Introduction and 'Lower' Hamamelidae (eds P. R. Crane and S. Blackmore), Systematics Association Special Volume No. 40A, pp. 175–92. Clarendon Press, Oxford, 1989.

*Present address: Department of Ecology and Evolutionary Biology, University of Arizona, Tucson, AZ 85721, USA.

(Takhtajan 1969; Endress 1986). The group has an extensive fossil record extending back to the Lower Cretaceous (Crane 1989; Friis *et al.* 1988; Friis and Crane Vol. 40A, Ch. 8), and it is clear that 'lower' hamamelids diverged from magnoliid dicotyledons at a very early stage in angiosperm evolution. Extant and fossil taxa at this level have also been significant in clarifying the relationships of other taxa currently included in the Rosidae (Hickey and Wolfe 1975; Hickey and Doyle 1977; Crane 1989; Wolfe Vol. 40A, Ch. 5) and Dilleniidae (Walker and Walker 1984; Wolfe Vol. 40A, Ch. 5). Because of the likely central position of the group in angiosperm phylogeny it is important to begin to develop explicit hypotheses of their interrelationships. To assist in developing such hypotheses, we present the results of a preliminary cladistic analysis.

Methods

We focus on the patterns of relationship among twelve taxa (Table 9.1). These include the genera of Cronquist's (1981) Trochodendrales and Hamamelidales, except that we represent the Hamamelidaceae with five taxa by using a single genus from each of the traditionally recognized subfamilies. *Eucommia* has also been included because it has been suggested to be closely related to both *Cercidiphyllum* and *Euptelea* (e.g. Harms 1897; Solereder 1899), although it is not usually placed with 'lower' Hamamelidae in modern systems of classification.

It is doubtful whether these 12 taxa constitute a monophyletic group, but it is clearly impractical to include representatives of all dicotyledonous taxa that could conceivably be nested among our 'lower' hamamelid study group. One means of circumventing this difficulty would be to use hypothesized patterns of character evolution in higher

Table 9.1 Taxa included in the cladistic analysis. Abbreviations indicate those that are used in the data matrix to specify the taxa.

Trochodendron (TR)	*Tetracentron* (TE)
Euptelea (EU)	*Cercidiphyllum* (CE)
Myrothamnus (MY)	*Eucommia* (EC)
Platanus (PL)	
Hamamelidaceae:	
Altingioideae (represented by *Altingia*; AL)	
Disanthoideae (represented by *Disanthus*; DI)	
Exbucklandoideae (represented by *Exbucklandia*; EX)	
Hamamelidoideae (represented by *Corylopsis*; HA)	
Rhodoleioideae (represented by *Rhodoleia*; RH)	
Outgroups:	
Degeneria (DE)	Winteraceae (WI)

dicotyledons to 'reconstruct' primitive rosids, dilleniids, and caryophyllids that may then be included in the analysis to represent higher taxa. In this case, however, we feel that this is undesirable because of very uncertain patterns of character evolution and relationship in the rosid, dilleniid, and caryophyllid assemblages. Rather than introduce too many assumptions at this stage, these groups are not included in our analysis.

Selection of outgroups is problematic because 'lower' Hamamelidae have not been shown to be closely related to any other group. We have chosen to use Winteraceae as one outgroup and *Degeneria* as a second outgroup to polarize character states (using the ROOT-ANCESTOR option in PAUP with *Degeneria* designated as HYPANC). Although *Degeneria* undoubtedly has certain autapomorphies, it has been widely considered to be a good representative for many primitive angiospermous characteristics. Similarities between Winteraceae and Hamamelidaceae have been shown (Hallier 1903), and cladistic analyses (Donoghue and Doyle Vol. 40A, Ch. 3) of basal angiosperms have shown that Winteraceae are part of a clade that is the sister group to other angiosperms that includes 'lower' Hamamelidae as one of the plesiomorphic subclades (making it likely that Winteraceae and some 'lower' Hamamelidae will share numerous primitive features that could be valuable for establishing character state polarity). Recent cladistic analyses (Donoghue and Doyle Vol. 40A, Ch. 3) have suggested that Ranunculidae with *Nelumbo* may be sister taxa to the Hamamelidae, but because character evolution and relationships in the Ranunculidae are currently uncertain they are difficult to use as an outgroup. In addition, because ranunculids were incorporated in the analysis of basal angiosperms as a single reconstructed taxon (Donoghue and Doyle Vol. 40A, Ch. 3), their monophyly and the possibility that they may be nested within the Hamamelidae has not yet been tested. In the Donoghue and Doyle analysis, within two steps of the most parsimonious tree (178 steps) there are several tree topologies with different secondary outgroups to the Hamamelidae and Ranunculidae. These include paleoherbs (i.e. Lactoridaceae, Aristolochiaceae, Piperaceae, Saururaceae, Nymphaeaceae, Cabombaceae, Liliopsida), Schizandraceae and Illiciaceae, the Laurales *sensu lato* (Monimiaceae, Amborellaceae, Trimeniaceae, Chloranthaceae, Austrobaileyaceae, Calycanthaceae) and all Magnoliidae and Liliopsida (minus Nymphaeaceae, Cabombaceae, and *Nelumbo*).

Character data (see the Appendix) were analysed using the computer algorithm Phylogenetic Analysis Using Parsimony (PAUP; version 2.4.1 written by D. L. Swofford) to attempt to identify most parsimonious character state trees. No *a priori* assessments of character state transitions were made for multistate characters (listed as

UNORDERED in PAUP). Most parsimonious trees were sought using global branch swapping (SWAP=GLOBAL in conjunction with the MULPARS option in PAUP). More complete descriptions of characters (than those in the Appendix) will be provided in a subsequent publication.

Results and discussion

Parsimony analysis of the data results in one cladogram (Fig. 9.1) with 223 character state changes (consistency index=0.51). *Trochodendron* and *Tetracentron* form a monophyletic subclade that is the sister group to the rest of the 'lower' hamamelid study group. *Trochodendron* and *Tetracentron* are unified by a number of features, of which several seem particularly significant. The two genera share seed-coat characteristics (flattened, elongate cells in the outer layers of the seed coat and a persistent tegmen; see Nast and Bailey 1945; Melikian 1973; Mohana Rao 1974; Corner 1976) that are not present among other 'lower' Hamamelidae. In addition, both *Trochodendron* and *Tetracentron* have a partially inferior ovary (which recurs in most Hamamelidaceae). Nectarial tissue on the dorsal surface of the carpel (van Tieghem 1900; Endress 1986) is a further specialized feature which is unknown among other 'lower' Hamamelidae and is unusual at this level of organization among the angiosperms (see Smets 1986, 1987). The two genera have almost identical reticulate–striate tectal sculpturing on their pollen grains (Walker 1976; Endress 1986) which is unique for 'lower' Hamamelidae.

Cercidiphyllum and *Myrothamnus* were resolved as sister taxa, and this significant result supports the nearest-neighbour phenetic relationship for the two genera demonstrated by Barabé *et al.* (1987) as well as the conclusions of Endress (Vol. 40A, Ch. 12). Several previous investigators (Jäger-Zürn 1966; Hufford and Endress 1989) have called attention to similarities between these genera, but various possible relatives have been suggested for each. In recent macrosystematic treatments, *Myrothamnus* has been considered to be most closely related to Hamamelidaceae (Takhtajan 1980, but not Takhtajan 1987;

Fig. 9.1 Most parsimonious cladogram (223 character state changes, consistency index=0.51) based on the data given in the Appendix. Numbers on the left of branches are character numbers; numbers on the right of branches refer to the apomorphic state at that level (see Appendix for list of characters and states). Where parallelisms or reversals provide equally likely hypotheses of character evolution, characters are optimized on the cladogram such that the ratio of reversal to parallelism is maximized (Farris optimization in combination with ROOT=ANCESTOR in PAUP).

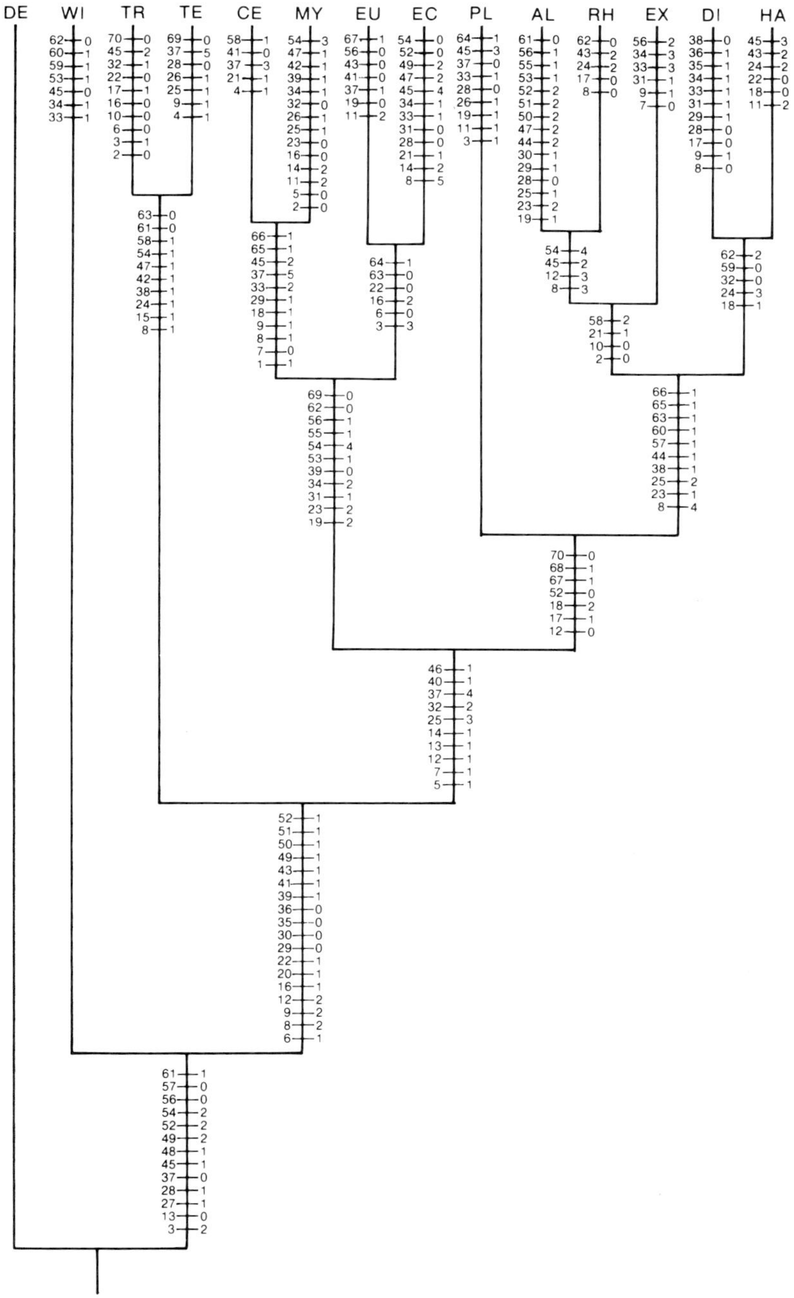
DE
WI
TR
TE
CE
MY
EU
EC
PL
AL
RH
EX
DI
HA

Cronquist 1981; Dahlgren 1983; Thorne 1983 considered *Myrothamnus* to be of dilleniid ancestry in being related to Hydrostachyaceae). Two features, opposite leaves and dioecy, in addition to a number of other characters that occur elsewhere in our cladogram (i.e. as homoplasies), are shown to support the sister-taxa relationship of *Cercidiphyllum* and *Myrothamnus*. In addition, the two taxa share anther stomial characteristics. Anthers with stomia that have both proximal and distal bifurcations, leading to valvate dehiscence, have been hypothesized to be plesiomorphic for Hamamelidae (Endress 1989; Hufford and Endress in press), and this hypothesis is supported by our cladogram. Both *Cercidiphyllum* and *Myrothamnus* share a modification of the plesiomorphic stomial condition because they lack the distal bifurcation but retain the proximal bifurcation (Hufford and Endress 1989). The same feature also occurs in *Euptelea*. In *Euptelea*, however, the more plesiomorphic equal bifurcation of the proximal end of the stomium is retained, whereas in *Cercidiphyllum* and *Myrothamnus* the branches of the proximal bifurcation are unequal.

Eucommia is placed as the sister taxon of *Euptelea* in our results but because other putatively closely related taxa (e.g. Urticales, Violales) are not included its position may be modified as more taxa are added. Our aim here is to explore the possible position of *Eucommia* in light of suggestions that it was derived from Hamamelidales (Takhtajan 1969; Cronquist 1981) and possibly related to *Cercidiphyllum* and *Euptelea* (Harms 1897; Solereder 1899). Features that support the alliance of *Eucommia* and *Euptelea* include the shared derivation of unilacunar from trilacunar nodes, a polytelic (intercalary) inflorescence, and indehiscent fruits (a potential parallelism with *Platanus*). In addition, this sister-taxon relationship of *Euptelea* and *Eucommia* is supported by several other features that occur as parallelisms in other parts of our cladogram; for example, the absence of stipules and the persistence of the tegmen are shared with *Trochodendron* and *Tetracentron*, and their pedicellate flowers are shared by *Trochodendron* and some Hamamelidoideae. Endress (1969) considered the similarities between *Euptelea* and *Eucommia* to be parallelisms, and, on the basis of leaf-architectureal data, Wolfe (Vol. 40A, Ch. 5) suggests that *Eucommia* had a violalean–thealean ancestry (particularly a potential relationship with Stachyuraceae). Although our results resolve *Euptelea* and *Eucommia* as sister taxa, *Eucommia* also shows divergences from what would otherwise be synapomorphies of *Euptelea*, *Cercidiphyllum*, and *Myrothamnus*, for example in pollen wall architecture and aperture shape and in stomial structure of anthers. More than any other genus in our analysis we expect that the relationships of *Eucommia* will change with more inclusive cladistic studies. Previous suggestions of a close relationship to Hamamelidales (e.g. Takhtajan 1969; Cronquist 1981) seem to have emphasized plesiomorphic characters and parallel simplifications of floral structures.

The position of *Platanus* as the sister taxon of Hamamelidaceae is consistent with placements in recent macrosystematic treatments (e.g. Cronquist 1968, 1981; Dahlgren 1980, 1983; Takhtajan 1980, 1987; Thorne 1981, 1983). The least equivocal features shared by *Platanus* and Hamamelidaceae are chemical. Both groups possess apigenin, myricetin (in parallel with *Euptelea*), and chalcones (in parallel with *Trochodendron*) (chemical data are from Giannassi 1986 in which data for Hamamelidaceae are generalized, and variation in the family cannot be ruled out). *Platanus* and Hamamelidaceae (except Altingioideae) also possess pollen colpi that are long and thin (a parallelism with *Eucommia*). Pollen apertures among most 'lower' Hamamelididae tend to be shorter and broader than they are among most Hamamelidaceae, especially Disanthoideae and Hamamelidoideae (see figures in Praglowski 1974; Walker 1976; Endress 1986; Zavada and Dilcher 1986). Other structural features that appear to unite *Platanus* and Hamamelidaceae are more equivocal.

In our analysis, Hamamelidaceae are monophyletic. They are defined by such characters as a perianth that is differentiated with a calyx and corolla, slight epigyny (a parallelism with *Trochodendron* and *Tetracentron*), multicellular stigmatic papillae (suggested by Endress 1989), presence of ellagic and gallic acid (both in parallel with *Cercidiphyllum* and *Myrothamnus*), and a large seed (Corner 1976) with the persistent epidermis of the inner integument remaining as the only element of tegmen (Mohana Rao 1974).

Curiously, the strictly bicarpellate gynoecium that is often considered to be a defining feature for Hamamelidaceae is shown to be a defining feature for all taxa above the derivation of *Trochodendron* and *Tetracentron*. This is interesting because it calls attention to the bicarpellate gynoecium that was identified in the *Cercidiphyllum*-like fossil *Joffrea speirsii* (Crane and Stockey 1985, 1986). If *Joffrea* is representative of a condition that is plesiomorphic for the clade that includes *Cercidiphyllum*, *Myrothamnus*, *Euptelea*, and *Eucommia*, then the origin for a bicarpellate gynoecium as indicated in our cladogram could be correct. The origin of the bicarpellate gynoecium as indicated in our cladogram may be largely influenced by the position of *Eucommia* (which is coded as being bicarpellate), and if it is inappropriately included with our study group then the origin for a bicarpellate gynoecium may change.

Diverse gynoecial patterns are present among such taxa as *Euptelea*, *Cercidiphyllum*, *Myrothamnus*, *Platanus*, and Hamamelidaceae. Among these taxa, *Myrothamnus* has a gynoecium that is most like that found in *Trochodendron* and *Tetracentron*, but *Platanus* and Hamamelidaceae share with them a basically similar carpellary morphology. Differences among these taxa with similar carpels are largely meristic. *Cercidiphyllum* and *Euptelea* are apocarpous and have notably distinct carpellary structures that probably indicate extreme paedomorphosis with developmental alterations instituted during early floral ontogeny.

Our preliminary results for Hamamelidaceae suggest that the family consists of two subclades, one formed by Exbucklandoideae, Rhodoleioideae, and Altingioideae (reflecting discussions by Bogle 1986, Vol. 40A, Ch. 11) and the other composed of Disanthoideae and Hamamelidoideae. These results do not support contentions (Niedenzu 1891, and implicit in Endress Vol. 40A, Ch. 10) that *Disanthus* is most closely related to Exbucklandoideae (or other groups along that subclade). Our results suggest that we should regard *Disanthus* and Hamamelidoideae as more closely related to each other than either is to the clade with Exbucklandoideae, Rhodoleioideae, and Altingioideae. The most unequivocal character states that support the sister-taxa relationship of *Disanthus* and Hamamelidoideae are in seed-coat structure (both have numerous cell layers in the seed coat and many, continuous layers are composed of thick-walled cells, Melikian 1973; see also Endress 1989 for other seed and fruit similarities shared by *Disanthus* and Hamamelidoideae).

Summary

We demonstrate that *Trochodendron* and *Tetracentron* share unique features among 'lower' Hamamelidae that emphasize the likelihood that they form a monophyletic subclade. *Cercidiphyllum* is shown to be more closely related to *Myrothamnus* than to any other member of our study group. *Cercidiphyllum* and *Myrothamnus* form a sister group to *Euptelea* and *Eucommia* on a clade that is characterized by reductions in inflorescence and floral morphology, but the possibility that *Eucommia* is not closely related to our study group seems likely. *Platanus* and Hamamelidaceae are shown to be sister taxa, although structural features that have previously been emphasized as indicative of a close relationship between them seem equivocal. Hamamelidaceae are shown to be monophyletic with the five traditionally recognized subfamilies forming two subclades. Exbucklandoideae, Rhodoleioideae, and Altingioideae form one subclade, and Disanthoideae and Hamamelidoideae form a coordinate subclade.

With this understanding of interrelationships within the 'lower' Hamamelididae, it should now be possible to 'collapse' certain well-supported monophyletic groups (e.g. *Trochodendron* and *Tetracentron*, *Cercidiphyllum* and *Myrothamnus*) and expand the study to incorporate basal ranunculids, rosids, dilleniids, and caryophyllids toward establishing relationships among major groups of higher dicotyledons. Detailed evaluation of the relationships of basal rosids will be of particular interest. While there is strong paleobotanical evidence to support a close evolutionary relationship between *Platanus* and some rosid taxa, the Hamamelidaceae also exhibit several significant rosid features such as differentiation of perianth into calyx and corolla and bicarpellate gynoecia. With respect to ultimately determining the relative phylogenetic positions of 'lower' and 'higher' hamamelids the details of relationships among basal rosids, *Platanus*, and Hamamelidaceae will be of critical importance.

Acknowledgements

We thank P. K. Endress and A. L. Bogle for important discussions and for sharing their unpublished data with us. This work was supported in part by NSF grant BSR-8708460 to P. R. Crane.

Appendix

Characters and Character States

References for the sources of our character definitions and character state assignments (see data matrix) are provided following either the general heading for groups of characters or with particular characters.

```
DE  0910009009031900090000000000099090011300000002900000000002102001209009999
WI  0920000009030000090000000001111001111000000000000120021200001110299909199
TR  0010000120020010100100010011000100000110111020111111010901200100000010
TE  0121010111020011000901011110000000000511019101011111101000120010000000 1
EU  0130109221211102000100203011001202001001000011011111141000201001001001
CE  1121119111011101012111203011101222003001001021011111141101201020110001
MY  1020010111211200012901001111101021005019111029111111131100201020110001
EC  0130101521011202002910293010000211004009101949212110101100999991000001
PL  0110111221101109121191003110000210990019901031011110020000201121001110
AL  0020111329031109121911201010110200004111191229211222141112210110111110
RH  0020111020031109020111122011000200004111102129011110040012211010111110
DI  0120119011091101010101132010101011114011191119011110020010011210119110
HA  0120111421291101100900122011000000004111192139011110020910011210111110
EX  0020110419001109120911102011001232994111191119011110020212211110111110
```

Vegetative Structure (Bailey and Smith 1984; Bailey and Nast 1945*a*; Cronquist 1981; Endress 1986, vol. 40A, Ch. 10; Howard 1979; Leroy 1988; Metcalfe and Chalk 1950; Nast and Bailey 1946; Puff 1978*a*,*b*; Swamy and Bailey 1949; Vink 1957).

1. Leaf arrangement: alternate (0), opposite (1).

2. Leaf retention: 'evergreen' (0), annually deciduous (1).
3. Nodes: variable (0), multilacunar (1), trilacunar (2), unilacunar (3).
4. Short shoots bearing a single leaf: absent (0), present (1).
5. Oil cells in appendages: present (0), absent (1).

Leaves (Baas 1969; Baranova 1972, 1983; Bogle vol. 40A, Ch. 11; Cronquist 1981; Endress 1978, Vol. 40A, Ch. 10; Exell 1933; Harms 1930*a*; Hickey 1973; Hickey and Wolfe 1975; Kaul and Kapil 1974; Metcalfe and Chalk 1950; Morley and Chao 1977; Upchurch 1984*a*,*b*; Vink; 1957; Weaver 1969; Wolfe 1973, Vol. 40A, Ch. 5).

6. Stipules: absent (0), present (1).

7. Foliar trichomes: absent (0), present (1).

8. Leaf margin tooth types: entire (0), chloranthoid (1), platanoid (2), fothergilloid (3), spinose (4), stachyuroid (5). Definition of characters follows Wolfe (1973, Vol. 40A, Ch. 5) and Hickey and Wolfe (1975).

9. Basic venation pattern: pinnate without basal crowding of secondary veins (0), actinodromous (1), pinnate with basal crowding of secondary veins (2).

10. Secondary vein junction with midrib: decurrent (0), non-decurrent (1).

11. Secondary venation near margin: festooned brochidodromous (0), brochidodromous (1), regularly craspedodromous (2).

12. Stomates: paracytic–hemiparacytic–laterocytic–cyclocytic (variable within a leaf) (0), anomocytic (1), laterocytic (2), paracytic (3). Variability characteristics are based on Upchurch (1984*a*,*b*).

Wood (Brush 1917; Carlquist 1976; Cronquist 1981; McLaughlin 1933; Metcalfe 1987; Metcalfe and Chalk 1950; Moseley 1973; Tippo 1938, 1940).

13. Vessel elements: absent (0), present (1).

14. Length of imperforate wood elements: >3500 μm (0), 1000–3500 μm (1), <1000 μm (2).

15. Imperforate element pitting: only circular (0), circular, scalariform and transitional (1).

Inflorescences (Bailey and Nast 1945*b*; Bailey and Smith 1942; Bogle 1986, Vol. 40A, Ch. 11; Boothroyd 1930; Endress 1969, 1971, 1977, 1986; Harms 1930*a*,*b*; van Heel 1987; Jäger-Zürn 1966; Swamy and Bailey 1949; Vink 1970).

16. Inflorescence termination: anthotelic (0), anauxotelic (1), polytelic (2). Definition of character states follows Briggs and Johnson (1979).

17. Inflorescence ramification: simple (0), compound (1). Inflorescences are simple if they consist solely of a single flower or if flowers are only terminal or in lateral positions along a single axis without having lateral inflorescence branches with more than one flower; they are compound if they possess lateral inflorescence branches that have more than one flower.

18. Inflorescence axis: elongated (0), non-elongated and cylindrical (1), non-elongated and laterally expanded (2). A non-elongated internode would exemplify a rosette morphology and includes capitate inflorescences.

Flowers (Bailey and Smith 1942; Bogle 1970, 1986, 1987, Vol 40A, Ch. 11; Boothroyd 1930; Cronquist 1981; Endress 1986, 1989, in press, Vol. 40A, Ch. 10; Harms 1930*a*,*b*; van Heel 1987; Jäger-Zürn 1966; Kaul and Kapil 1974; Morley and Chao 1977; Niedenzu and Engler 1930; Thien 1980; Vink 1956, 1970; Weaver 1969; Wisniewski and Bogle 1982).

19. Floral sex distribution: bisexual flowers (0), monoecious (1), dioecious (2).

20. Symmetry of floral primordium: radial (0), bilateral (1).

21. Floral axis prophylls: regularly present (even if only associated with the first formed flowers of the inflorescence) (0), absent (1). This character evaluates whether two sterile leaf-like appendages develop on the floral axis (i.e. that axis that undergoes organogenesis of an androecium and gynoecium).

22. Floral pedicels: present (pedicellate) (0), absent (sessile) (1).

23. Floral axis phyllomes (distal of prophylls, if present, and subjacent to the androecium/gynoecium): undifferentiated (tepals) (0), differentiated into calyx and corolla (1), absent (2). Our evaluation of this character follows Endress (1986, 1987) in noting that floral phyllomes must be considered in more of an inflorescence context rather than a simple evaluation of just perianth features, hence undifferentiated phyllomes above the prophylls must be considered in this category in which the state of the 'perianth' is evaluated.

24. Nectary: absent (0), gynoecial (1), androecial (2), perianth (3).

Androecium (Bailey and Nast 1943; Bogle Vol. 40A, Ch. 11; Eckardt 1963; Endress 1977, 1986, 1987; Endress and Hufford in press; Hufford and Endress 1989; Jäger-Zürn 1966; Kaul and Kapil 1974).

25. Androecial pattern: helical (0), haplostemony (1), cyclic polyandry (2), unordered in association with lack of perianth (3). The cyclic polyandry character state follows the definition of Ronse-Decraene and Smets (1987) and that for the unordered state (3) which follows the characterization of Endress (1987).

26. Stamen number: more than five (0), five or fewer (1). Stamen number includes staminodes.

27. Stamen morphology: not differentiated into regions of filament and anther (0), differentiated into anther and filament regions (1).

28. Anther theca length/filament length: more than one (0), less than one (1).

29. Proximal juncture of thecae with connective: level (0), below (1). This character evaluates whether the thecae overhang (below) their junction with the connective or are level with it.

30. Distal juncture of thecae with connective: level (0), above (1). This character evaluates whether the thecae project above their juncture with the connective or whether they are level with it.

31. Shape of connective protrusion apex: rounded (0), pointed (1).

32. Position of stomial termini: on thecal shoulders (distal and proximal) (0), on rim of thecal shoulder (1), lateral (side of theca) (2).

33. Proximal stomial terminus: equal bifurcation (0), unbifurcated straight (1), unequal bifurcation (2), unbifurcated arced (3). This character evaluates

whether the proximal end of the stomium bifurcates or not and describes its general appearance.

34. Distal stomial terminus: equal bifurcation (0), unbifurcated straight (1), unbifurcated arcs (2). This character evaluates whether the distal end of the stomium bifurcates or not and describes its general appearance.

35. Connective hypodermis with thick-walled cells on ventral side of anther: present (0), absent (1).

36. Connective hypodermis with thickened walls on dorsal side of anther: present (0), absent (1).

Gynoecium (Bailey and Smith 1942; Bogle 1986; Cronquist 1981; Eckardt 1963; Endress 1977, 1986, 1989; Harms 1930*a*; Jäger-Zürn 1966; Kaul and Kapil 1974; Leroy 1977; Nast and Bailey 1945, 1946; Puff 1978*b*; Sampson and Kaplan 1970; Sampson and Tucker 1978; Smith 1943; Swamy 1949; Swamy and Bailey 1949; Tucker and Sampson 1979; Vink 1957, 1978).

37. Carpel number: few (1–about 12) and variable (0), several (about 12 to about 50) and variable (1), many (more than 50) and variable (2), one (largely without variation) (3), two (largely without variation) (4), three–four (largely without variation) (5).

38. Gynoecial insertion: superior (0), partially inferior (1).

39. Carpel stipitation: stipitate (0), sessile (1).

40. Carpel peltation: peltate (0), epeltate (1).

41. Carpel fusion: free (0), synascidiate (1). Consult Leinfellner (1950) or Guédès (1979) for carpellary terminology.

42. Postgenital carpel fusion: absent (0), present (1).

43. Stigma morphology: protuberant slit or cap on surface of ovary (0), elevated and elongate along a long, distal protrusion (1), restricted to a small region at the apex of a well-defined style (2).

44. Stigmatic papillae: unicellular (0), multicellular (1), multicellular and unicellular (2).

45. Ovule number per carpel: few–many (1 to 50+) and variable (0), few (1–10) and variable (1), several (10–40) and variable (2), one (rarely variable) (3), two (rarely variable) (4).

46. Ovule maturity: mature at anthesis (0), immature at anthesis (1). This character used by Endress (1977, 1986) refers to the presence of a fully developed embryo sac at anthesis.

Pollen (Bailey and Nast 1943; Bogle and Philbrick 1980; Endress 1986; Praglowski 1974; Swamy 1949; Walker 1976; Walker and Walker 1984; Zavada and Dilcher (1986).

47. Size of pollen (equatorial diameter): 21–30 μm (0), <21 μm (1), >30 μm (2).

48. Shape and symmetry: boat-shaped and bilateral (0), prolate-oblate (including spheriodal) and radiosymmetry (1), prolate-oblate and isobilateral (*sensu* Walker and Doyle 1975) (2).

49. Amb: oval (0), circular (1), lobate (2).

50. Aperture number: one (0), three (1), four or more (2).

51. Aperture arrangement: distal (0), zono (1), peri (2).

52. Aperture length: width ratio: >4:1 (1), 2–3:1 (1), 1:1 (2).

53. Aperture termini: pointed (0), rounded (1). The importance of this character is taken from Praglowski (1974) and Endress (1986).

54. Primary sculpture of tectum: no primary sculpture (0), perforate–striate (1), reticulate (2), clavate (3), perforate–irregular, thick rugae (4). This character is meant to convey the major (gross) sculptural elements of the tectum, although strict distinctions between tectum sculpture and exine structure are difficult (see the discussion in Walker and Doyle 1975).

55. Secondary sculpture of tectum: smooth (0), papillate (1). This character is meant to convey minor sculptural elements that may or may not be found on the primary sculpture.

56. Endexine thickness: thicker near apertures (0), uniform thickness (1), no endexine (2).

Seeds (Corner 1976; Harms 1930*a*; Kaul and Kapil 1974; Melikian 1973; Mohana Rao 1974; Niedenzu and Engler 1930; Vink 1970).

57. Size: small (0), large (1).

58. Wings: unwinged (0), chalazal wing with looped vascular bundle (1), micropylar wing (2).

59. Seed-coat layers: more than ten (0), three (1), four to ten (2).

60. Outer layer of seed coat with thickened cell walls: absent (0), present (1).

61. Shape of cells in outer layers of seed coat: flattened, elongate (0), large, rectangular (1).

62. Inner layers of seed coat with thickened cell walls: absent (0), few, discontinuous layers (1), numerous, continuous layers (2).

63. Tegmen: persistent (0), only epidermis persistent (1), not persistent (2).

Fruits (Corner 1976; Cronquist 1981; Endress 1989).

64. Fruit dehiscence: ventral (0), indehiscent (1).

Chemistry (Giannasi 1986; Jay 1968).

65. Gallic acids: absent (0), present (1).

66. Ellagic acid: absent (0), present (1).

67. Myrecetin: absent (0), present (1).

68. Apigenin: absent (0), present (1).

69. Protoanthocyanins: absent (0), present (1).

70. Chalcones: absent (0), present (1).

References

Baas, P. (1969). Comparative anatomy of *Platanus kerrii* Gagnep. *Botanical Journal of the Linnean Society* **62,** 413–21.

Bailey, I. W. and Nast, C. G. (1943). The comparative morphology of the Winteraceae I. Pollen and stamens. *Journal of the Arnold Arboretum* **24,** 340–6.

Bailey, I. W. and Nast, C. G. (1945*a*). Morphology and relationships of

Trochodendron and *Tetracentron*, I. Stem, root, and leaf. *Journal of the Arnold Arboretum* **26,** 143–54.

Bailey, I. W. and Nast, C. G. (1945*b*). The comparative morphology of the Winteraceae. VII. Summary and conclusions. *Journal of the Arnold Arboretum* **26,** 37–47.

Bailey, I. W. and Smith, A. C. (1942). Degeneriaceae, a new family of flowering plants from Fiji. *Journal of the Arnold Arboretum* **23,** 356–65.

Barabé, D., Bergeron, Y, and Vincent, G. (1987). La répartition des caractères dans la classification des Hamamelididae (Angiospermae). *Canadian Journal of Botany* **65,** 1756–67.

Baranova, M. (1972). Systematic anatomy of the leaf epidermis in the Magnoliaceae and some related families. *Taxon* **21,** 447–69.

Baranova, M. (1983). On the laterocytic stomatotype in angiosperms. *Brittonia* **35,** 93–102.

Bogle, A. L. (1970). Floral morphology and vascular anatomy of the Hamamelidaceae: the apetalous genera of Hamamelidoideae. *Journal of the Arnold Arboretum* **51,** 310–66.

Bogle, A. L. (1986). The floral morphology and vascular anatomy of the Hamamelidaceae: subfamily Liquidambaroideae. *Annals of the Missouri Botanical Garden* **73,** 325–47.

Bogle, A. L. (1987). Inflorescence and flower ontogeny in the pseudanthium of *Rhodoleia* (Hamamelidaceae). *American Journal of Botany* **74,** 607–8.

Bogle, A. L. and Philbrick, C. T. (1980). A generic atlas of hamamelidaceous pollens. *Contributions from the Gray Herbarium* **210,** 29–103.

Boothroyd, L. E. (1930). The morphology and anatomy of the inflorescence and flower of the Platanaceae. *American Journal of Botany* **17,** 678–93.

Briggs, B. G. and Johnson, L. A. S. (1979). Evolution in the Myrtaceae—evidence from inflorescence structure. *Linnean Society of New South Wales* **102,** 157–256.

Brush, W. D. (1917). Distinguishing characters of North American sycamore woods. *Botanical Gazette* **64,** 480–96.

Carlquist, S. (1976). Wood anatomy of *Myrothamnus flabellifolia* (Myrothamnaceae) and the problem of multiperforate perforation plates. *Journal of the Arnold Arboretum* **57,** 119–26.

Corner, E. J. H. (1976). *The seeds of dicotyledons*. Cambridge University Press, Cambridge.

Crane, P. R. (1989). Palaeobotanical evidence on the radiation of non-magnoliid dicotyledons. *Plant Systematics and Evolution* **162,** 165–91.

Crane, P. R. and Stockey, R. A. (1985). Growth and reproductive biology of *Joffrea speirsii* gen. et sp. nov., a *Cercidiphyllum*-like plant from the Late Paleocene of Alberta, Canada. *Canadian Journal of Botany* **63,** 340–64.

Crane, P. R. and Stockey, R. A. (1986). Morphology and development of pistillate inflorescences in extant and fossil Cercidiphyllaceae. *Annals of the Missouri Botanical Garden* **73,** 382–93.

Cronquist, A. (1968). *The evolution and classification of flowering plants*. Houghton Mifflin, New York.

Cronquist, A. (1981). *An integrated system of classification of flowering plants*. Columbia University Press, New York.

Dahlgren, R. M. T. (1980). A revised system of classification of the angiosperms. *Botanical Journal of the Linnean Society* **80,** 91–124.

Dahlgren, R. M. T. (1983). General aspects of angiosperm evolution and macrosystematics. *Nordic Journal of Botany* **3,** 119–149.

Eckardt, T. (1963). Some observations on the morphology and embryology of *Eucommia ulmoides* Oliv. *Journal of the Indian Botanical Society* **42**A, 27–34.

Endress, P. K. (1969). Gesichtspunkte zur systematische Stellung der Eupteleaceen (Magnoliales). *Berichte der Schweizer Botanischen Gesellschaft* **79,** 229–78.

Endress, P. K. (1971) Blütenstände und morphologische Interpretation der Blüten bei apetalen Hamamelidaceen. *Berichte der Deutsch Botanischen Gesellschaft* **84,** 183–5.

Endress, P. K. (1977) Evolutionary trends in the Hamamelidales–Fagales-group. *Plant Systematics and Evolution, Supplement* **1,** 321–347.

Endress, P. K. (1978). Stipules in *Rhodoleia* (Hamamelidaceae). *Plant Systematics and Evolution* **130,** 157–60.

Endress, P. K. (1986). Floral structure, systematics, and phylogeny in Trochodendrales. *Annals of the Missouri Botanical Gardens* **73,** 297–324.

Endress, P. K. (1987). Floral phyllotaxis and floral evolution. *Botanische Jahrbücher für Systematik* **108,** 417–38.

Endress, P. K. (1989). Aspects of evolutionary differentiation of the Hamamelidaceae and the lower Hamamelididae. *Plant Systematics and Evolution* **162,** 193–211.

Endress, P. K. (in press). Evolution of reproductive structures and functions in primitive angiosperms. *Memoirs of the New York Botanical Garden.*

Endress, P. K. and Hufford, L. D. (1989). The diversity of stamen structures and dehiscence patterns among Magnoliidae. *Botanical Journal of the Linnean Society* **100,** 45–85.

Exell, A. W. (1933). A revision of the genus *Rhodoleia. Sunyatsenia* **1,** 95–102.

Friis, E. M., Crane, P. R., and Pendersen, K. R. (1988). Reproductive structures of Cretaceous Platanaceae. *Biologiske Skrifter Danske Videnskabernes Selskab* **31,** 1–56.

Giannasi, D. E. (1986). Phytochemical aspects of phylogeny in Hamamelidae. *Annals of the Missouri Botanical Garden* **73,** 417–37.

Guédès, M. (1979). *Morphology of seed plants.* Cramer, Vaduz.

Hallier, H. (1903). Über den Umfang, die Gliederung und die Verwandtschaft der Familie der Hamamelidaceen. *Beihefte zum Botanischen Centralblatt* **14,** 247–60.

Harms, H. (1897). Uber die Stellung der Gattung *Tetracentron* Oliv. *Berichte der Deutsch Botanischen Gesellschaft* **15,** 350–60.

Harms, H. (1930*a*). Eucommiaceae. In *Die natürlichen Pflanzenfamilien* (2nd edn) (eds. A. Engler and K. Prantl) **18a,** 348–52. Englemann, Leipzig.

Harms, H. (1930*b*). Hamamelidaceae. In *Die natürlichen Pflanzenfamilien* (2nd edn) (eds. A. Engler and K. Pratl), **18a,** 303–45, 487. Englemann, Leipzig.

Heel, W. A. van (1987). Note on the morphology of the male inflorescences in *Cercidiphyllum* (Cercidiphyllaceae). *Blumea* **32,** 303–9.

Hickey, L. J. (1973). Classification of the architecture of dicotyledonous leaves. *American Journal of Botany* **60,** 17–33.

Hickey, L. J. and Doyle J. A. (1977). Early Cretaceous fossil evidence for angiosperm evolution. *Botanical Review* **43,** 3–104.

Hickey, L. J. and Wolfe, J. A. (1975). The bases of angiosperm phylogeny: vegetative morphology. *Annals of the Missouri Botanical Garden* **62,** 538–89.

Howard, R. A. (1979). The stem-node-leaf continuum of the Dicotyledoneae. In *Anatomy of the dicotyledons* (2nd edn) (eds. C. R. Metcalfe and L. Chalk). Clarendon Press. Oxford.

Hufford, L. D. and Endress, P. K. (1989). The diversity of anther structures and dehiscence patterns among Hamamelidae. *Botanical Journal of the Linnean Society* **99,** 301–46.

Jäger-Zürn, I. (1966). Infloreszenz- und blütenmorphologische, sowie embryologische Untersuchungen an *Myrothamnus* Welw. *Beitrage zur Biologie der Pflanzen* **42,** 241–71.

Jay, M. (1968). Distribution des flavonoides chez les Hammélidacées et familles affines. *Taxon* **17,** 136–47.

Kaul, U. and Kapil, R. N. (1974). *Exbucklandia populnea*—from flower to fruit. *Phytomorphology* **24,** 217–28.

Leinfellner, W. (1950). Der Bauplan des Synkarpen Gynoeceums. *Österreiche Botanisches Zeitschrift* **97,** 403–36.

Leroy, J.-F. (1977). A compound gynoecium in Winteraceae. *Science* **196,** 997–8.

Leroy, J.-F. (1988). L'organisation fondamentale d'une Angiosperme archaïque: *Tetracentron* (Tetracentraceae). *Comptes rendus hebdomadaire des seances de l'Académie des Sciences, Paris,* **306,** Serie 3, 229–35.

McLaughlin, R. P. (1933). Systematic anatomy of the woods of the Magnoliales. *Tropical Woods* **34,** 3–39.

Melikian, A. P. (1973). Seed-coat types of Hamamelidaceae and allied families in relation to their systematics. *Botanicheskii Zhurnal* **58,** 350–9.

Metcalfe, C. R. (1987). *Anatomy of the dicotyledons Vol. III. Magnoliales, Illiciales, and Laurales,* (2nd edn). Clarendon Press, Oxford.

Metcalfe, C. R. and Chalk, L. (1950). *Anatomy of the dicotyledons.* Clarendon Press, Oxford.

Mohana Rao, P. R. (1974). Seed anatomy in some Hamamelidaceae and phylogeny. *Phytomorphology* **31,** 18–23.

Morley, B. and Chao, J.-M. (1977). A review of *Corylopsis* (Hamamelidaceae). *Journal of the Arnold Arboretum* **58,** 382–415.

Moseley, M. F. (1973). Vegetative anatomy and morphology of Amentiferae. *Brittonia* **25,** 356–70.

Nast, C. G. and Bailey, I. W. (1945). Morphology and relationships of *Trochodendron* and *Tetracentron* II. inflorescence, flower and fruit. *Journal of the Arnold Arboretum* **26,** 267–76.

Nast, C. G. and Bailey, I. W. (1946). Morphology of *Euptelea* and comparison with *Trochodendron*. *Journal of the Arnold Arboretum* **27,** 186–91.

Niedenzu, F. (1891). Hamamelidaceae. *Die natürlichen Pflanzenfamilien* (eds. A. Engler and K. Prantl) **III 2A,** 115–36, Englemann, Leipzig.

Niedenzu, F. and Engler, A. (1930). Myrothamnaceae. In *Die natürlichen Pflanzenfamilien* (eds. A. Engler and K. Prantl), **18a,** 262–5. Englemann, Leipzig.

Praglowski, J. (1974). The pollen morphology of the Trochodendraceae, Tetracentraceae, Cercidiphyllaceae, and Eupteleaceae, with reference to taxonomy. *Pollen et Spores* **16,** 449–67.

Puff, C. (1978*a*). The nodal anatomy of *Myrothamnus flabellifolius* (Myrothamnaceae): Another example of a 'split-lateral' condition. *Journal of the Arnold Arboretum* **59,** 192–6.

Puff, C. (1978*b*). Zur Biologie von *Myrothamnus flabellifolius* Welw. (Myrothamnaceae). *Dinteria* **14,** 1–20.

Ronse-Decreane, L.-P. and Smets, E. 1987. The distribution and systematic relevance of the androecial characters oligomery and polymery in the Magnoliophytina. *Nordic Journal of Botany* **7,** 239–53.

Sampson, F. B. and Kaplan, D. R. (1970). Origin and development of the terminal carpel in *Pseudowintera traversii*. *American Journal of Botany* **57,** 1185–96.

Sampson, F. G. and Tucker, S. C. (1978). Placentation in *Exospermum stipitatum* (Winteraceae). *Botanical Gazette* **139,** 215–22.

Smets, E. (1986). Localization and systematic importance of the floral nectaries in the Magnoliatae (Dicotyledons). *Bulletin du Jardin Botanique Nationale Belgique* **54,** 183–211.

Smets, E. (1987). Characters and character-states of floral nectaries: a contribution towards a better understanding of the relationships among the higher taxa of the Magnoliophytina (Angiospermae). *XIV International Botanical Congress (Berlin) Abstracts* p. 236.

Smith, A. C. (1943). Taxonomic notes on the Old World species of Winteraceae. *Journal of the Arnold Arboretum* **24,** 119–64.

Solereder, H. (1899). Zur Morphologie und Systematik der Gattung *Cercidiphyllum* Sieb. et Zucc., mit Berücksichtigung der Gattung *Eucommia* Oliv. *Berichte der Deutschen Botanischen Gesellschaft* **17,** 387–406.

Swamy, B. G. L. (1949). Further contributions to the morphology of the Degeneriaceae. *Journal of the Arnold Arboretum* **30,** 10–38.

Swamy, B. G. L. and Bailey, I. W. (1949). The morphology and relationships of *Cercidiphyllum*. *Journal of the Arnold Arboretum* **30,** 187–210.

Takhtajan, A. L. (1969). *Flowering plants: origin and dispersal.* Oliver and Boyd, Edinburgh.

Takhtajan, A. L. (1980). Outline of the classification of flowering plants (Magnoliophyta). *Botanical Review* **46,** 225–359.

Takhtajan, A. L. (1987). *Systema Magnoliophytorum.* Nauka, Leningrad.

Thien, L. B. (1980). Patterns of pollination in the primitive angiosperms. *Biotropica* **12,** 1–13.

Thorne, R. F. (1981). Phytochemistry and angiosperm phylogeny, a summary statement. In *Phytochemistry and angiosperm phylogeny* (eds. D. A. Young and D. S. Seigler). pp. 233–95. Praeger, New York.

Thorne, R. F. (1983). Proposed new realignments in the angiosperms. *Nordic Journal of Botany* **3,** 85–117.

Tieghem, P. van (1900). Sur les Dicotylédones du groupe des Homoxylées. *Journal de botanique* **14,** 259–97, 330–61.

Tippo, O. (1938). Comparative anatomy of the Moraceae and their presumed allies. *Botanical Gazette* **100,** 1–99.

Tippo, O. (1940). The comparative anatomy of the secondary xylem and the phylogeny of the Eucommiaceae. *American Journal of Botany* **27,** 832–38.

Tucker, S. C. and Sampson, F. B. (1979). The gynoecium [of] winteraceous plants. *Science* **203,** 920–21.

Upchurch, G. R. (1984*a*). Cuticle evolution in Early Cretaceous angiosperms from the Potomac Group of Virginia and Maryland. *Annals of the Missouri Botanical Garden* **71,** 522–50.

Upchurch, G. R. (1984*b*). Cuticular anatomy of angiosperm leaves from the Lower Cretaceous Potomac Group. I. Zone I leaves. *American Journal of Botany* **71,** 192–202.

Vink, W. (1957). Hamamelidaceae. *Flora Malesiana* **1,** 363–79.

Vink, W. (1970). The Winteraceae of the Old World. I. *Pseudowintera* and *Drimys*—morphology and taxonomy. *Blumea* **28,** 225–354.

Vink, W. (1978). The Winteracaeae of the Old World. III. Notes of the ovary of *Takhtajania*. *Blumea* **24,** 521–5.

Walker, J. W. (1976). Evolutionary significance of the exine in the pollen of primitive angiosperms. In *The evolutionary significance of the exine* (eds. I. K. Ferguson and J. Muller), pp. 251–308. Academic Press, London.

Walker, J. W. and Doyle, J. A. (1975). The bases of angiosperm phylogeny: palynology. *Annals of the Missouri Botanical Garden* **62,** 664–723.

Walker, J. W. and Walker, A. G. (1984). Ultrastructure of Lower Cretaceous angiosperm pollen and the origin and early evolution of flowering plants. *Annals of the Missouri Botanical Garden* **71,** 464–521.

Weaver, R. E. (1969). Studies in the North American genus *Fothergilla* (Hamamelidaceae). *Journal of the Arnold Arboretum* **50,** 599–619.

Wisniewski, M. and Bogle, A. L. (1982). The ontogeny of the inflorescence and flower of *Liquidambar styraciflua* L. (Hamamelidaceae). *American Journal of Botany* **69,** 1612–24.

Wolfe, J. A. (1973). Fossil forms of Amentiferae. *Brittonia* **25,** 334–55.

Zavada, M. S. and Dilcher, D. L. (1986). Comparative pollen morphology and its relationship to phylogeny of pollen in the Hamamelidae. *Annals of the Missouri Botanical Garden* **73,** 348–81.

10. The systematic position of the Myrothamnaceae

PETER K. ENDRESS

Botanischer Garten und Institut für Systematische Botanik der Universität Zürich, CH-8008 Zürich, Switzerland

Abstract

Evidence from various fields shows that the family Myrothamnaceae, although highly isolated, is best included in the Trochodendrales near the Cercidiphyllaceae.

Introduction

The Myrothamnaceae are a small, isolated family of one genus and two species in tropical and southern Africa and in Madagascar. They are perhaps best known as 'resurrection plants' due to their extreme poikilohydrous capacity. These shrubs with their small, fan-like leaves can dry out and revive again. Important works for shaping our picture about the systematic relationships of the family are on reproductive structures and embryology (Jäger-Zürn 1966), vegetative anatomy (Grundell 1933; Carlquist 1976; Puff 1978*a*), and on various biological aspects (Puff 1978*b*).

The family has been variously assigned to groups of the Rosidae, Hamamelididae, and Magnoliidae. At present, the tendency is to associate it in some way with Hamamelidales, often next to the Hamamelidaceae. This is done by Cronquist (1981) (together with Cercidiphyllaceae, Eupteleaceae, Platanaceae, and Hamamelidaceae)

Evolution, Systematics, and Fossil History of the Hamamelidae, Volume 1: Introduction and 'lower' Hamamelidae (eds P. R. Crane and S. Blackmore), Systematics Association Special Volume No. 40A, pp. 193–200. Clarendon Press, Oxford, 1989.

and by Dahlgren (1983) (together with Hamamelidaceae and Platanaceae). Takhtajan (1987) puts his Myrothamnales into Hamamelidanae (together with Hamamelidales, Daphniphyllales, Balanopales, Didymelales, Buxales, Simmondsiales, Casuarinales, Fagales, and Betulales). In contrast, Thorne (1983) places the Myrothamnaceae in his Rosiflorae, as opposed to the Hamamelidiflorae.

Studies on the reproductive structures have provided new aspects that support an inclusion in the 'lower' Hamamelididae but show a possible closer association with the Trochodendrales than the Hamamelidales. Jäger-Zürn (1966), Zavada and Dilcher (1986), Barabé *et al.* (1987), and Hufford and Endress (1989) point to similarities especially with the Cercidiphyllaceae.

Discussion

This discussion contains some aspects that have not previously been considered (or only recently mentioned), which indicate a closer relationship of the Myrothamnaceae with the Trochodendrales rather than the Hamamelidales, and especially with the Cercidiphyllaceae. Features of the Myrothamnaceae that are present in some but not in all families of the 'lower' Hamamelididae are included in Table 10.1.

Special features in *Myrothamnus* of general occurrence in the 'lower' Hamamelididae (Trochodendrales and Hamamelidales) include: i) tissues strongly tanniferous, ii) perforation plates of vessel members with numerous bars (in *Myrothamnus* the bars often interconnected, Carlquist 1976), iii) number of floral organs highly variable (Jäger-Zürn 1966), iv) anthers basifixed (Fig. 10.1F), v) connective protrusion present (Fig. 10.1F), vi) endothecium continuous around the anther (Hufford and Endress 1989), vii) carpels fused, but only at the base (Fig. 10.1A,E), viii) stigmas widely decurrent (Fig. 10.1A,E).

Special features in *Myrothamnus* that occur in the Trochodendrales (*Trochodendron*, *Tetracentron*, *Euptelea*, *Cercidiphyllum*) but not (or less frequently) in the Hamamelidales (Platanaceae, Hamamelidaceae) include: i) plants essentially glabrous, ii) secretory idioblasts present, as in *Tetracentron* and *Trochodendron* (Fig. 10.1D) (cf. Endress 1986), iii) leaf teeth chloranthoid (Fig. 10.2) (Grundell 1933, cf. Hickey and Wolfe 1976), iv) perianth absent or only loosely integrated in the flower (Jäger-Zürn 1966, cf. Endress 1986), v) stigmatic papillae unicellular (Fig. 10.1C) (cf. Endress 1986; but also in Platanaceae), vi) chromosome base number $x = 10$ (Puff 1978*b*, as in *Trochodendron*, *cf.* Ratter and Milne 1973), vii) chalcones/dihydrochalcones present (Giannasi 1986), viii) proanthocyanidins absent (Giannasi (1986). In addition, the peculiar leaves of *Myrothamnus* (Fig. 10.2) should also be compared with

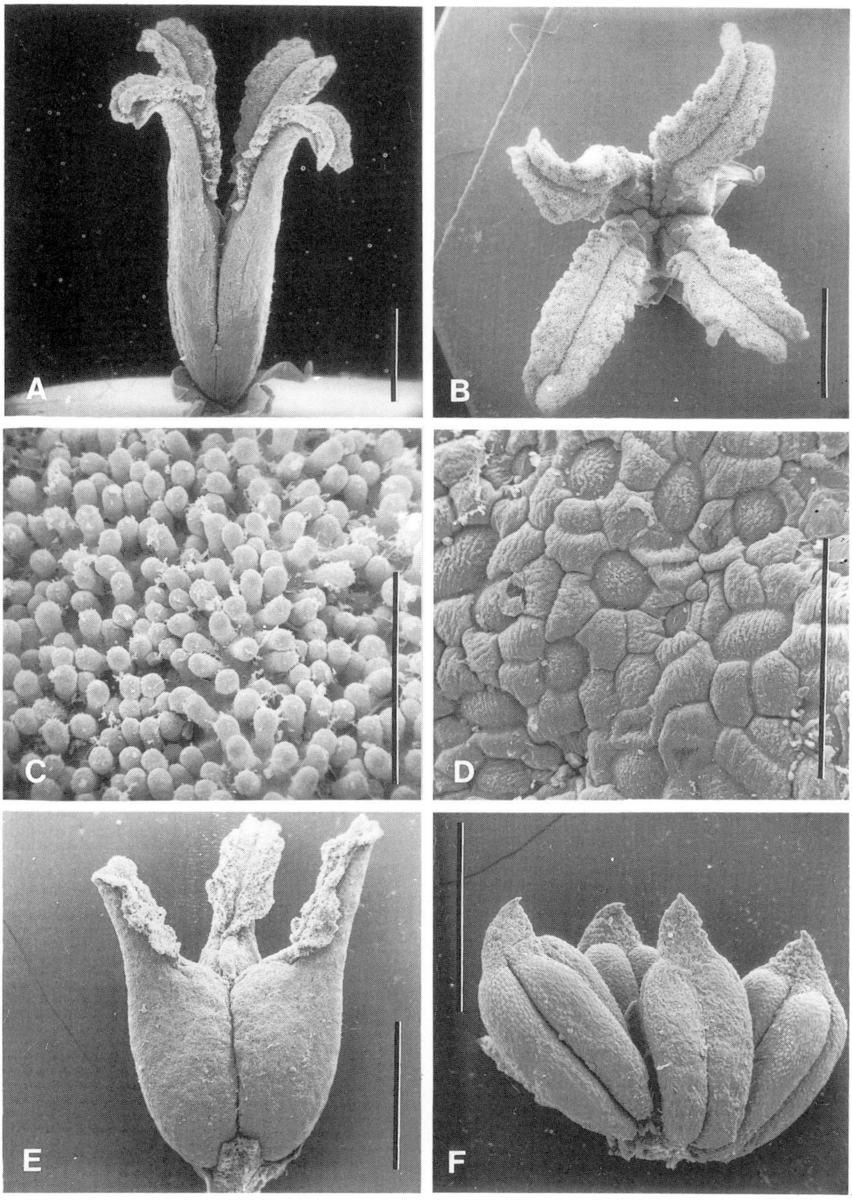

Fig. 10.1 A–C, *Myrothamnus moschata* Baill. (Rauh 7473). D–F, *M. flabellifolia* Welw. (Eller 1987). A, Female flower, from the side. B, Female flower, from above. C, Stigmatic surface. D, Surface of carpel flank with secretory idioblasts. E, Female flower. F, Male flower. (Scale line 1 mm in A, B, C, E, F, 0.1 mm in C, D.)

Table 10.1 Comparison of Myrothamnaceae with 'lower' Hamamelididae.

Tetracentron	*Trochodendron*	*Euptelea*	*Cercidiphyllum*	*Myrothamnus*	*Platanus*	*Hamamelidaceae*
+	+	+	+	Plants essentially glabrous	–	–
–	–	+	+	Wood with vessels	+	+
–	–	–	+	Leaf position decussate	–	(–)
–	+	–	–	Leaves persistent	(–)	(+)
+	–	–	+	Leaf venation palmate	+	(–)
+	–	–	+	Stipules present	+	+
–	–	–	+	Stipules intrapetiolar	+	–
+	+	–	+	Leaf teeth chloranthoid	–	–
–	–	–	+	Flowers unisexual	+	(–)
–	–	–	+	Dioecious	–	–
–	–	–	+	Flower position decussate	–	(–)
–	–	+	+	Anemophilous	+	(–)
–	–	+	+	Anthers long, connective protrusion long	+	(+)
–	–	+	+	Anther with only proximal stomial bifurcation	–	(–)
–	–	+	+	Stigma enlarged, wing-like, decurrent	+	(+)

+	−	−	−	Flowers often tetramerous (but labile in *Myrothamnus*)	+	(−)
+	+	−	−	Phyllomes between the two prophylls and stamens (sometimes) present	+	+
−	+	+	+	Stable, sizeable perianth absent	−	(−)
+	+	−	+	Pollen triaperturate	+	(+)
−	−	−	+	Apertures not well delimited	−	−
+	+	−	−	Gynoecium syncarpous at base	−	+
−	−	+	+	Ovary superior	+	(−)
+	+	−	+	Fruits dehiscent, ± follicular	−	+
−	−	+	−	Seeds without appendage	+	(+)
−	+	−	−	Chromosome base number $x=10$	−	−
+	−	+	+	Chalcones/dihydrochalcones present	−	−
+	−	+	+	Proanthocyanidins present	−	−
−	−	−	+	Ellagic acid present	−	+

+, present; (+), mostly present; (−), rarely present; −, absent.

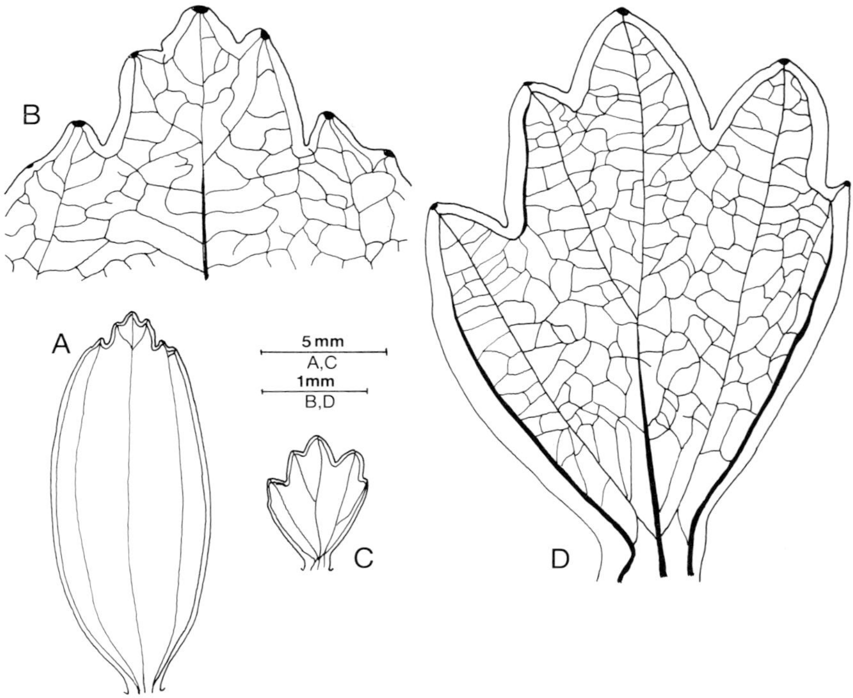

Fig. 10.2 A–B, *Myrothamnus moschata* Baill. (Rauh 7473). C–D, *M. flabellifolia* Welw. (Eller 1987). A and C, Leaves (without stipules). B and D, Venation pattern of the same leaves.

fossil leaves assigned to the Trochodendrales, as surveyed by Iljinskaja (1974), Crane (1984, 1989), and Crane and Stockey (1985). It appears that *Trochodendroides* is especially similar in the acrodromous venation and the chloranthoid teeth with the strong peripheral veins.

Special features of *Myrothamnus* in common with *Cercidiphyllum* but not or less so with other Trochodendrales include: i) leaf position decussate, ii) leaf venation palmate (also in *Tetracentron)*, iii) stipules intrapetiolar, iv) flower position decussate, v) dioecious, vi) anemophilous (also in *Euptelea*) including the related features of anthers long (also in *Euptelea*), connective protrusion long (also in *Euptelea*), anther dehiscence almost slit-like but with a stomial bifurcation at the proximal end (also in *Euptelea*) (Hufford and Endress 1989), stigma enlarged, wing-like (Fig. 10.1B) (also in *Euptelea*) (cf. Endress 1986), vii) Pollen apertures three but not well delimited (Zavada and Dilcher 1986), viii) Ellagic acid present (Giannasi 1986).

Conclusions

These common features strongly point to a closer phylogenetic relationship of the Myrothamnaceae with the Trochodendrales than with the Hamamelidales, and especially with the Cercidiphyllaceae, although there are some superficial resemblances with the genus *Distylium* of the Hamamelidaceae (cf. Endress 1977). However, since all these families of the 'lower' Hamamelididae are very isolated and *Myrothamnus* has undergone unique anagenetic change and is not known in the fossil record, it is difficult to establish the cladogenetic relationships among the families.

Acknowledgements

I thank B. Eller and P. Leins for providing valuable fixed material of both species of *Myrothamnus*. I am grateful to R. Siegrist for microtome sections and to U. Jauch for the use of the SEM.

References

Barabé, D., Bergeron, Y. and Vincent, G. (1987). La répartition des caractères dans la classification des Hamamelididae (Angiospermae). *Canadian Journal of Botany* **65,** 1756–67.

Carlquist, S. (1976). Wood anatomy of *Myrothamnus flabellifolia* (Myrothamnaceae) and the problem of multiperforate perforation plates. *Journal of the Arnold Arboretum* **57,** 119–26.

Crane, P. R. (1984). A re-evaluation of *Cercidiphyllum*-like plant fossils from the British early Tertiary. *Botanical Journal of the Linnean Society* **89,** 199–230.

Crane, P. R. (1989). Paleobotanical evidence on the early radiation of nonmagnoliid dicotyledons. *Plant Systematics and Evolution* **162,** 165–91.

Crane, P. R. and Stockey, R. A. (1985). Growth and reproductive biology of *Joffrea speirsii* gen. et sp. nov., a *Cercidiphyllum*-like plant from the late Paleocene of Alberta, Canada. *Canadian Journal of Botany* **63,** 340–64.

Cronquist, A. (1981). *An integrated system of classification of flowering plants.* Columbia University Press, New York.

Dahlgren, R. (1983). General aspects of angiosperm evolution and macrosystematics. *Nordic Journal of Botany* **3,** 119–49.

Endress, P. K. (1977). Evolutionary trends in the Hamamelidales–Fagales-group. *Plant Systematics and Evolution, Supplement* **1,** 321–47.

Endress, P. K. (1986). Floral structure, systematics, and phylogeny in Trochodendrales. *Annals of the Missouri Botanical Garden* **73,** 297–324.

Giannasi, D. E. (1986). Phytochemical aspects of phylogeny in Hamamelidae. *Annals of the Missouri Botanical Garden* **73,** 417–37.

Grundell, R. (1933). Zur Anatomie von *Myrothamnus flabellifolia* Welw. *Symbolae Botanicae Upsalienses* **2,** 1–17.

Hickey, L. J. and Wolfe, J. A. (1976). The bases of angiosperm phylogeny: vegetative morphology. *Annals of the Missouri Botanical Garden* **62,** 538–89.

Hufford, L. D. and Endress, P. K. (1989). The diversity of anther structures and dehiscence patterns among Hamamelididae. *Botanical Journal of the Linnean Society* **99,** 301–46.

Iljinskaja, I. A. (1974). Trochodendrales, Cercidiphyllales. In *Magnoliophyta fossilia URSS* **I,** (ed. A. Takhtajan), pp. 112–28. Nauka, Leningrad (In Russian).

Jäger-Zürn, I. (1966). Infloreszenz- und blütenmorphologische, sowie embryologische Untersuchungen an *Myrothamnus* Welw. *Beiträge zur Biologie der Pflanzen* **42,** 241–71.

Puff, C. (1978a). The nodal anatomy of *Myrothamnus flabellifolius* (Myrothamnaceae): another example of a 'split-lateral' condition. *Journal of the Arnold Arboretum* **59,** 192–6.

Puff, C. (1978b). Zur Biologie von *Myrothamnus flabellifolius* Welw. (Myrothamnaceae). *Dinteria* **14,** 1–20.

Ratter, J. A. and Milne, C. (1973). Chromosome numbers of some primitive angiosperms. *Notes of the Royal Botanical Gardens Edinburgh* **32,** 423–8.

Takhtajan, A. (1987). *Systema Magnoliophytorum.* Nauka, Leningrad (In Russian).

Thorne, R. F. (1983). Proposed new alignments in the angiosperms. *Nordic Journal of Botany* **3,** 85–117.

Zavada, M. S. and Dilcher, D. L. (1986). Comparative pollen morphology and its relationship to phylogeny of pollen in the Hamamelidae. *Annals of the Missouri Botanical Garden* **73,** 348–81.

11. The floral morphology, vascular anatomy, and ontogeny of the Rhodoleioideae (Hamamelidaceae) and their significance in relation to the 'lower' hamamelids

A. LINN BOGLE

Botany and Plant Pathology Department, University of New Hampshire, Durham, NH 03824, USA

Abstract

The inflorescence and floral morphology, vascular anatomy, and ontogeny of *Rhodoleia championii* Hook. f. (Hamamelidaceae, Rhodoleioideae) are described. The capitate inflorescence is modified as a pseudanthial bird-flower, with an involucre of bracts surrounding the head of flowers. Floral morphology is modified accordingly, with irregular symmetry; reduced calyx; one to four showy petals (or none) only on the exposed outer side of each flower; a cycle of six to ten elongate stamens surrounding a cycle of sterile, glandular phyllomes; and a bicarpellate, multiovulate ovary with filamentous styles about equalling the stamens in length, and bearing minute terminal stigmas. Floral vascular anatomy shows similarity to that of Liquidambaroideae. Hypanthial trunk bundles provide irregular traces to the base of the calyx, a single trace to each petal and stamen, and ramifying systems of traces to the sterile phyllomes. Each carpel receives a dorsal and two ventral bundles; the numerous ovules in each carpel are supplied from the ventrals. The ventrals of adjacent carpel margins may be fused basally for a short distance.

In ontogeny a calyx ring primordium appears early, but its growth is retarded. Petals and stamens appear on the outer side first, then more stamens on the inner side of the floral primordium. Petals lag behind stamens in early

Evolution, Systematics, and Fossil History of the Hamamelidae Volume 1: Introduction and 'Lower' Hamamelidae. (eds P. R. Crane and S. Blackmore), Systematics Association Special Volume No. 40A, pp. 201–20. Clarendon Press, Oxford, 1989.

development. Two carpel primordia appear within the cycle of stamen primordia. The sterile phyllome primordia appear well after the carpels, but at maturity are borne on the hypanthium. They appear comparable in ontogeny and structure to similar organs in Liquidambaroideae; their nature and significance in relation to similar organs in the other subfamilies is discussed.

Introduction

The genus *Rhodoleia* constitutes one of the so-called 'lower' Hamamelidaceae, a group of seven or eight genera in four subfamilies (Table 11.1) which contain the more primitive genera of the family. The subfamilies are easily distinguishable from each other, and from the more advanced genera in the subfamily Hamamelidoideae, on vegetative and floral characteristics, but their evolutionary relationships are not well understood.

Table 11.1 The subfamilies and genera of lower Hamamelidaceae (approximate number of species in parentheses).

1. **Liquidambaroideae**
   Altingia Nor. (7–14)
   Semiliquidambar H. T. Chang (3–4)
   Liquidambar L. (2–5)
2. **Rhodoleioideae**
   Rhodoleia Champ. ex Hook. (1–9)
3. **Exbucklandeoideae**
   Exbucklandia R. W. Brown (2–4)
   Chunia H. T. Chang (1)
   Mytilaria H. Lecomte (1)
4. **Disanthoideae**
   Disanthus Maxim. (1)

All of the genera are small, containing only one to relatively few species. Some of the genera (e.g. *Chunia*, *Mytilaria*), although taxonomically distinct, are poorly known as a function of their restricted distributions in remote, inaccessible areas, and the resulting scarcity of herbarium specimens and research material. Specimens of *Semiliquidambar* are scarce and somewhat intermediate between *Altingia* and *Liquidambar*. It has been suggested that they may represent intergeneric hybrids (Bogle 1986).

Subfamily Rhodoleioideae

The monogeneric subfamily Rhodoleioideae contains only *Rhodoleia* Champion ex Hooker (Hooker 1850), a genus of large, evergreen shrubs or trees with alternate, vestigially stipulate (Endress 1978*a*), leathery, entire margined, pinnately veined leaves. The genus is distributed in southeastern Asia from Yunnan to Hong Kong, and southward through Indo-China and the Malay Peninsula to Sumatra.

Exell (1933) recognized seven species of *Rhodoleia*, but Vink (1957) provided a detailed account of variation in the genus and concluded that only one polymorphic species, *R. championii* Hook. f., is represented. Tardieu-Blot (1965) concurred with Vink, but suggested that some varieties could be recognized. At least two new species have been described subsequent to Vink's work (Chang 1973, 1979). A new investigation of the taxonomy of the genus is needed.

The subfamily is set apart from the other subfamilies of the Hamamelidaceae by its combination of reduced stipules; pseudanthial, few-flowered, capitate inflorescence surrounded by a conspicuous involucre of large bracts; a pseudo-corolla made up of the abaxial petals of the several flowers; and smooth, scrobiculate, or finely reticulate, tricolpate pollen. The combination of the involucre and pseudo-corolla has caused the inflorescence to be mistaken for a single large flower.

Prior to its segregation as a subfamily, *Rhodoleia* was consistently grouped with *Bucklandia* (*Exbucklandia*) and some combination of the genera *Altingia*, *Liquidambar*, and/or *Disanthus*, in most cases as a member of a tribe Bucklandieae or subfamily Bucklandeoideae (Lindley 1853; Bentham and Hooker 1865; Reinsch 1889; Niedenzu 1891; Tong 1930). One of the morphological characters uniting these genera is the presence of several to many ovules per locule, borne on both margins of each carpel. However, the differences of inflorescence and floral morphology among them are so great that the group is now split into four subfamilies. Nakai (1943) elevated all four subfamilies to family status. Shaw (1966) accepted family status for *Rhodoleioideae* but this position has not been adopted by others.

Both McLaughlin (1933) and Tippo (1938) included *Rhodoleia* in their extensive systematic wood-anatomy studies, but did not provide specific data for the genus. Tang (1943) found the wood of *Rhodoleia* to be distinct from that of both *Exbucklandia* (as *Bucklandia*) and *Disanthus*, and agreed with the segregation of *Rhodoleia* as a separate subfamily.

Harjal *et al.* (1984) included *Rhodoleia* in their study of leaf vasculature, but did not provide details for the genus. Wolfe (1973) described, illustrated, and discussed the occurrence of pinnate (*Altingia*, *Rhodoleia*) and palmate (*Liquidambar*, *Exbucklandia*, *Mytilaria*, *Disanthus*) venation patterns in Hamamelidaceae. He found *Rhodoleia* to exhibit several primitive characteristics that made it distinctive from the rest of the Hamamelidaceae, and concluded that the genus should probably be excluded from the Hamamelidaceae and possibly from Hamamelididae. Skvortsova (1960*a*,*b*,*c*), in contrast, found *Rhodoleia* to be comparable to a number of Hamamelidoideae in petiole vasculature, leaf venation, and stomatal type, but distinct from *Exbucklandia*, *Mytilaria*, *Disanthus*, *Altingia*, and *Liquidambar*. Skvortsova (1960*c*) described paracytic stomatal structure in *Rhodoleia* and *Altingia*, but encyclocytic stomata in *Exbucklandia* and *Mytilaria*. Baranova (1983, 1985) found laterocytic, encyclocytic, and intermediate forms in *Exbucklandia*.

Melikian (1971, 1972, 1973*a*,*b*) found the seed-coat anatomy of *Rhodoleia* to be similar to that of *Exbucklandia* and *Chunia*, and concluded that the three genera could be included in a single subfamily Rhodoleioideae, perhaps as tribes Exbucklandieae and Rhodoleieae. *Altingia* and *Liquidambar* were considered as a separate family, Altingiaceae, by Melikian.

Pollen morphology at the levels of light microscopy and electron microscopy has been described and illustrated by Chang (1964), Endress (1977), Bogle and Philbrick (1980), and Zavada and Dilcher (1986). Chang (1964) described the pollen as tricolporate, while the other authors cited here report it as tricolpate. The pollen of *Rhodoleia* is distinct from that of all other hamamelid genera in having a smooth, scrobiculate, or finely reticulate exine, perhaps adapted for bird pollination (Porsch 1923, Leeuwen 1938).

Goldblatt and Endress (1977) summarized the published information on chromosome numbers in the family, and reported the first chromosome count for *Rhodoleia* as $n=12$, similar to the Hamamelidoideae, and differing from Liquidambaroideae ($n=15$, 16), Exbucklandioideae ($n=32$) and Disanthoideae ($n=8$).

The paleobotanical record of hamamelidaceous fruits, seeds, and pollen indicates the first appearance of '*Rhodoleia*'-like seeds in the Maastrichtian (as *Klikovispermum*), and *Rhodoleia* seeds in the late Eocene (Tiffney 1986). For other lower Hamamelidaceae Tiffney (1986) cites reports of *Liquidambar* for the late Paleocene, *Disanthus* from the early Oligocene, and *Exbucklandia* (as *Bucklandia*) from the Pliocene (perhaps extending back to mid-Miocene).

Floral vasculature and ontogeny in other lower Hamamelidaceae

Few of the lower Hamamelidaceae have been studied in detail with regard to floral vascular anatomy or organogeny. Bogle (1967, 1968) described the floral morphology and vascular anatomy of five genera of lower Hamamelidaceae, including *Altingia*, *Liquidambar*, *Rhodoleia*, *Exbucklandia*, and *Disanthus*, but the Liquidambaroideae, especially *Liquidambar*, have received the most attention. Schmitt (1966) studied the pistillate partial inflorescence of *Liquidambar styraciflua* and described interfascicular cambial activity within the head; absence of a perianth; carpels with three major bundles originating from three gaps in the marginal meristem; ramification of the ventral bundles; and formation of 'papillae' (presumably the phyllomes, staminodia, or disk lobes of various authors) on the surface of the capsules, each containing a vascular trace which differentiates basipetally.

Wisniewski and Bogle (1982) described inflorescence and flower development in *Liquidambar styraciflua*, establishing the compound structure of the functionally pistillate or staminate heads (partial inflorescences); the bisexual potential of both pistillate and staminate flowers in early ontogeny; the absence of perianth even in early ontogeny; and the late, centrifugal initiation of the cycle of sterile phyllomes (disc lobes, staminodes, papillae) in the pistillate flowers, and their absence in staminate flowers. Subsequently, Bogle (1986) described some aspects of the inflorescence and floral morphology and vascular anatomy of *Liquidambar* and *Altingia*, including observations on the little-known segregate genus *Semiliquidambar*, and discussed their similarities and relationships with the other subfamilies.

Among the Exbucklandioideae only the genus *Exbucklandia* (formerly *Bucklandia*, *Symingtonia*) has been studied in detail. Griffith (1836) provided numerous illustrations of floral organs and dissections. He described and illustrated pentamerous hermaphrodite, and tetramerous female flowers; petaloid stamens; the cycle of conspicuous, broad, fleshy phyllomes (interpreted as calyx); dissections of ovules; and the mature fruiting inflorescence.

Kaul and Kapil (1974) studied the embryology of *Exbucklandia populnea*. They also described both hermaphrodite and pistillate flowers, the former lacking sepals, the latter lacking both sepals and petals. A ring of five to nine 'lobules' (disk lobes, phyllomes) was reported to surround the gynoecium in all flowers. The transformation of stamens into petals was illustrated in detail.

There are no published accounts of the floral vascular anatomy of *Disanthus*, apart from a reference by Endress (1978*b*, p. 297; Abb. 48) to a peripheral bundle system in the ovary; nor is there any information on floral ontogeny. The petals of *Disanthus* have nectariferous surfaces at their base (cf. Endress 1977), but the flowers generally lack an inner cycle of sterile organs. However, Bean (1917) illustrates a cycle of small 'disk glands', and Harms (1930) and Mizushima (1968) refer to a cycle of five tiny 'staminodia', alternating with the stamens. These organs may correspond to a cycle of staminodial nectaries such as those in *Hamamelis*, or to the cycles of sterile organs, which I refer to as phyllomes, present in the flowers of *Altingia*, *Liquidambar*, *Rhodoleia*, and *Exbucklandia*. The ovary of *Disanthus* is often described as being superior, in contrast to the other lower Hamamelidaceae, which have semi-inferior ovaries. However, my observations (unpublished data) suggest that this is a secondarily derived condition, probably related to the effects of inflorescence reduction on floral development.

For Rhodoleioideae a brief description of inflorescence and floral ontogeny in *Rhodoleia* was published (Bogle 1987), but I know of no published data on the floral vascular anatomy of the genus. This paper provides further information on the morphology, vascular anatomy, and ontogeny of the inflorescence and flowers of *Rhodoleia championii.*

Materials and methods

Materials of *Rhodoleia championii* used for this study consisted of alcohol-preserved inflorescence buds, flowers, and fruits, and pressed specimens collected by the author at Klang Gates, near Kuala Lumpur, Selangor, Malaysia (Bogle, Bogle and Kaul 276); in the vicinity of Tanah Rata Village, Cameron Highlands, Pahang, Malaysia (Bogle, Bogle and Kaul 316), and in the Hong Kong Botanic Garden, Hong Kong (Bogle and Bogle 582). Vouchers for these collections are deposited at MIN and A, and in the author's collection. Collections at A and GH were also examined extensively.

The methods of preservation, preparation and staining of sections, and clearing of floral parts have been described previously (Bogle 1970). Inflorescences and flowers at all stages of development were studied. For scanning electron microscopy, specimens were dehydrated through absolute ethanol, critical-point dried for 15 minutes, coated with 200 Å of 60/40 palladium–gold, and photographed in an AMR Model 1000 SEM at the University of New Hampshire. Drawings were made with the aid of the drawing attachments designed for the Wild M-20 and M-5 microscopes.

Results

Morphology

Inflorescence: The inflorescence of *Rhodoleia* is a capitulum, derived from a condensed raceme or spike, consisting of five to eight (up to ten; Vink 1957) sessile, complete, asymmetric flowers fused in a nodding, hemispherical head (Fig. 11.1B–D). The head is surrounded by an involucre of numerous, broadly rounded, spirally arranged, coloured bracts, which are caducous after anthesis. The individual flowers are apparently not subtended by obvious primary bracts. Condensation of

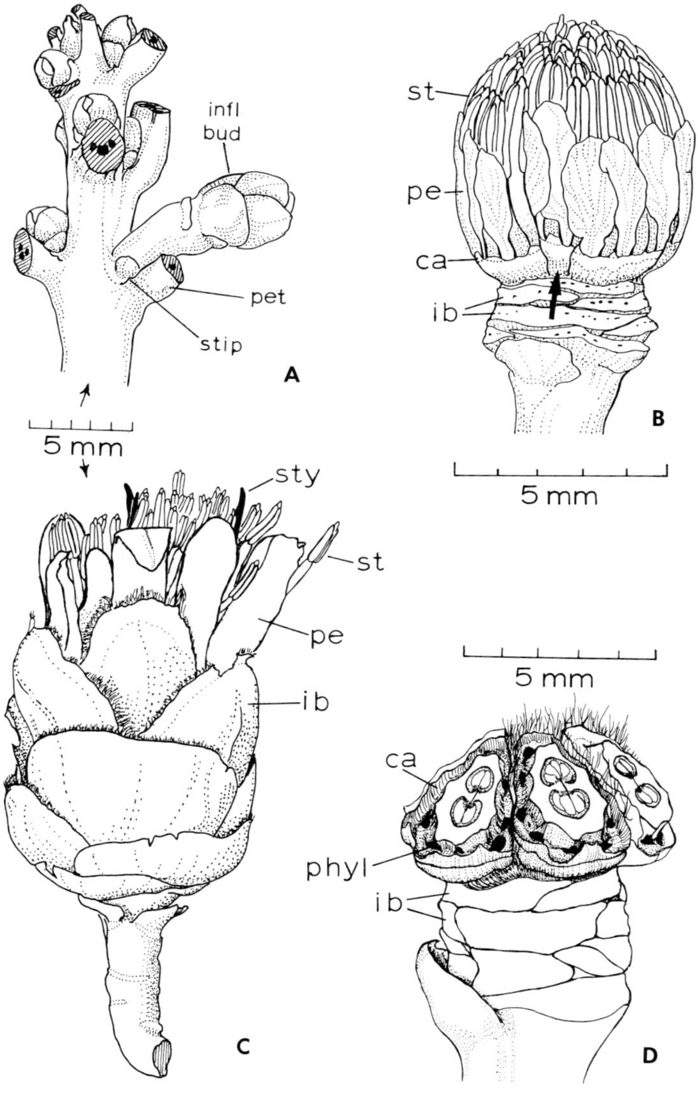

Fig. 11.1 *Rhodoleia championii.* A, Tip of shoot bearing young inflorescence buds. B, Partially developed inflorescence, involucral bracts removed; petals lagging behind stamens in elongation development. C, Mature inflorescence, showing involucre, petals, stamens, styles. D, Mature inflorescence with involucral bracts, petals, stamens, and upper part of ovaries removed to show orientation of flowers, calyx, phyllomes, and placentation of ovary.

(ca, calyx; ib, involucral bract; infl bud, inflorescence bud; pet, petiole; pe, petal; phyl, phyllome; st, stamen; stip, stipule; sty, style.)

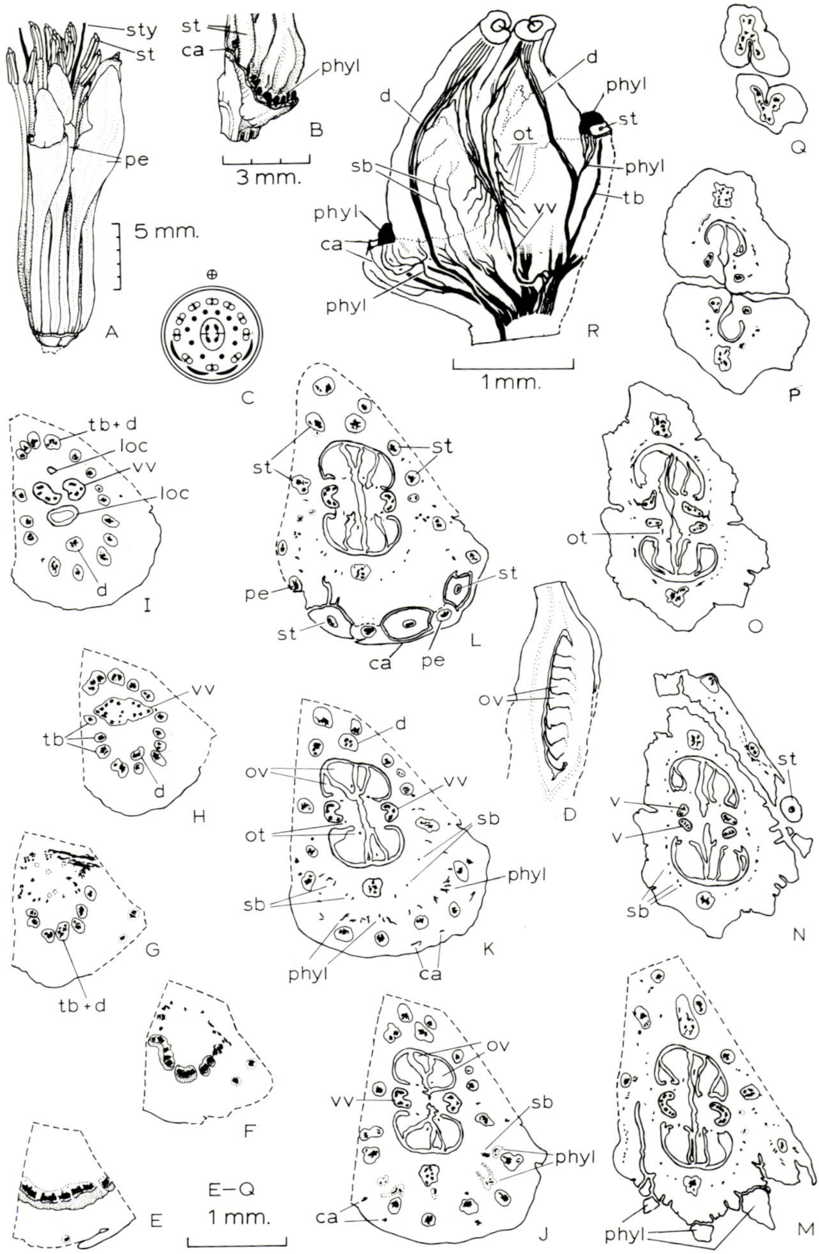

Fig. 11.2 *Rhodoleia championii.* A, Flower, abaxial view, showing petals, stamens, styles, exposed abaxial side of ovary. B, Flower, dissected from head, abaxial petals and stamens removed, showing position of sterile phyllomes on inner rim of hypanthium

the inflorescence into a head has resulted in the flowers being tightly packed and laterally compressed to varying degrees. The outline of a flower on the periphery of (and presumably 'lower' in) the head may vary from broadly to narrowly triangular in cross-section, with the median plane of the flower and gynoecium lying along a radius of the head (Fig. 11.1D). In fewer-flowered inflorescences all of the flowers may be radially orientated. However, in larger inflorescences one or more peripheral flowers may approach a tangential orientation of the gynoecium (Fig. 11.4B, C), while flowers crowded toward the centre (presumably 'higher' in the inflorescence), with minimal exposure to the periphery (Fig. 11.1B, middle flower indicated by arrow), may be radial in orientation but narrowly oblong to polygonal in outline. The common radial orientation, coupled with the presence of petals only on the outer side of a flower, produces a floral asymmetry or zygomorphy that is apparent even in the early stages of flower development (Fig. 11.3B–F). The inflorescence is nodding at maturity.

Flower: The perianth and androecium are incorporated into a hypanthium, which is fused for most or all of its length with the ovary. A rudimentary calyx limb is present in each flower at anthesis in the form of a short, inconspicuous, undulating, continuous or discontinuous, membranous collar on the outer rim of the hypanthium (ca in Figs 11.1B, D; 11.2R). Between the flowers the calyx is often elaborated into relatively long, scale-like lobes. Definite numbers of sepals or calyx lobes

---

around base of superior portion of carpels. C, Floral diagram (radial symmetry and number of organs arbitrary). D, Arrangement of ovules on one carpel margin. E–Q, Serial sections of peduncle, ovary and hypanthium at anthesis. E, Peduncular stele. F, Lobing of peduncular stele as floral traces emerge. G, Lower receptacle: hypanthial trunk bundles formed on abaxial side. H, Receptacle: compound ventral bundles forming; one carpel dorsal and numerous trunk bundles formed. I, Base of locules: compound ventrals and most trunk bundles distinct. J, Branching of trunk bundles to form traces to calyx, sterile phyllomes and supplementary carpel bundles; ovule insertion on partial septum in lower part of ovarian cavity. K, Elaboration of dendritic trace system to base of sterile phyllomes; supplementary carpel bundles in more centripetal position. L, Insertion of calyx, petals and stamens on abaxial side of flowers. M, Sterile phyllomes distinct on abaxial side. N, Part of the hypanthium shown separate from ovary on adaxial side of flower; four ventral bundles now distinct. O, Lateral sutures beginning to appear between carpels. P, Free style bases, carpel margins not sealed; attachment of highest ovules. Q, Styles. R, Cleared flower, lateral view: forward part removed, showing vasculature in posterior half. Note: ovule traces departing from both compound and free portions of ventral bundles; sterile phyllome supply from hypanthial trunk bundle on abaxial (lower) side, from carpel dorsal bundle on adaxial (upper) side.

(ca, calyx; d, dorsal bundle; loc, locule; ot, ovule trace; ov, ovule; sb, supplementary bundle; pe, petal; phyl, phyllome; st, stamen; sty, style; tb, trunk bundle; v, ventral bundle; vv, compound ventral bundle.)

could not be distinguished in my material. One to four unequal, linear, narrowly spathulate to obovate, clawed, pink to rose-red petals are conspicuous on the abaxial side of each flower (Figs 11.1B, C; 11.2A, C), but petals are usually lacking on the abutting sides of adjacent flowers, and some flowers may lack petals entirely. In each flower, six to ten elongated stamens with narrowly oblong, basifixed anthers are inserted around the rim of the hypanthium, alternating with the petals on the abaxial side (Fig. 11.2L). The anthers open laterally by means of two narrow valves on each side (Fig. 11.5C, D), and the connective is scarcely produced. Alternate with the bases of the stamen filaments and next to the ovary is inserted a cycle of small, amorphous sterile bodies, here termed phyllomes (phyl in Figs 11.2B, M, R; 11.4D–F). These usually occur at the base of the strong vertical ridges in the carpel walls, and are similar to but smaller than the sterile organs seen in the pistillate flowers of the Liquidambaroideae (Bogle 1986). They may appear as distinct, entire or lobed structures, or be interconnected by a low ridge of smaller lobes, thus forming a more or less continuous ring of lobes on the inner rim of the hypanthium (Fig. 11.4D). They are reported to function as nectaries (van Leeuwen 1938), and are presumably the glands mentioned in the original description of the genus (Hooker 1850).

The ovary is half-inferior, being fused below the middle with the hypanthium. The gynoecium consists of two carpels whose involute margins and opposing surfaces are fused below to form a thick partial septum in the base of the locule (Fig. 11.2I, J). Both Baillon (1871) and Harms (1930) have noted variation in the degree of development of the septum. In my material the septum divides at about the level of insertion of the lowest ovules to form two compound parietal placentae, each made up of two fused adjacent carpel margins (Fig. 11.2K–P). Each margin bears a row of about nine imbricate, anatropous, winged ovules (Fig. 11.2D).

The two carpels begin to separate from each other just above the level of separation of the ovary from the hypanthium. A cleft appears in the outer wall on each side of the ovary and becomes progressively deeper at higher levels, finally reaching the stylar canal and separating the style bases (Fig. 11.2O, P). Above this level the carpels are more or less conduplicate and unsealed. Ovules are inserted on the carpel margins above the level at which the style base becomes separate (Fig. 11.2P). Each carpel bears an elongate, filiform style with no obvious stigmatic surfaces along the ventral margins, but with a slightly flared to shallowly bilobed stigmatic tip, on which pollen grains are deposited (Fig. 11.5A, B). The styles are caducous in fruit.

At maturity the ovary forms a two-celled, four-valved woody capsule that dehisces along dorsal and ventral lines to expose the ranks of flattened to winged seeds (Fig. 11.4H), of which only one or two toward the base of each margin are fertile (Vink 1957).

Vascular anatomy: Well-defined pedicellar vascular cylinders were not apparent in the inflorescences of *Rhodoleia*. Vascular bundles entering the base of a flower radiate from the margins of a large gap in the peduncular stele (Fig. 11.2E–G). Trunk bundles (tb), more or less equal in number to the number of stamens and petals, are formed in the base of the flower (Fig. 11.2G–I). At higher levels they supply traces to the perianth and androecium, as well as to the fleshy, glandular phyllomes (disk lobes).

Within the hypanthium each trunk bundle divides to form several branches (Fig. 11.2J, K). A small lateral or abaxial branch (ca) is usually produced which ascends toward the periphery of the hypanthium to supply the base of the calyx (Fig. 11.2J, K). These calyx bundles may end blindly as a single strand; may dichotomize to form two bundles which diverge horizontally; or may trichotomize to form a short, central, terminal strand and two laterally divergent branches. No regularity is apparent in the number or origin of the calyx bundles or their branches, all of which end blindly at the base of the calyx.

One or more branches from each trunk bundle may be produced laterally or adaxially to supply the phyllomes (phyl) and to provide supplementary bundles (sb) to the carpel walls. The phyllomes are irregular in size and may be somewhat subdivided (Fig. 11.2M). The major lobes are opposite the petals and alternate with the stamens (Fig. 11.2I–M). The vascular bundles supplying the phyllomes ramify to form dentritic systems whose branch tips end blindly in the bases of the phyllomes (phyl in Fig. 11.2K, R). On the crowded, adaxial side of the flower the supply to the phyllomes may be derived from one of the carpel dorsal bundles, separating from it near the top of the hypanthium (Fig. 11.2R).

Supplementary bundles (sb) to the carpel walls (Fig. 11.2K, R) may be derived directly from the trunk bundles, from the minor bundles of the hypanthium, or from the dendritic bundle systems which supply the phyllomes. The supplementary bundles may pass upward into the carpel wall unbranched, or may dichotomize several times. At anthesis the tips of the supplementary bundles may be free, or may form anastomoses with each other and with lateral branches from the dorsal and ventral carpel bundles (Fig. 11.2R).

After providing the various traces to other organs the remainder of each trunk bundle ascends to the rim of the hypanthium to form the single trace to the base of a petal (pe) or stamen filament (st) (Fig. 11.2L, N). Small trunk bundles which do not supply a petal or stamen usually provide a branch to the base of the calyx and a ramifying dendritic bundle system to the base of an adjacent phyllome (Fig. 11.2R). Vestigial petal traces were not observed in the hypanthium on the abutting flanks of the flowers, where petals were absent.

The vascular supply to the gynoecium consists of the carpel dorsal bundles, ventral bundles to each carpel margin, and the supplementary bundles to the carpel walls discussed above. Each carpel dorsal bundle (d) may originate as a distinct bundle in the receptacle, particularly on the less crowded abaxial side, or may be fused basally with a trunk bundle of the hypanthium (Fig. 11.2G–I, R). As the dorsal bundle passes upward in the ovary wall it produces two to five pairs of sub-opposite lateral branches which pass into the carpel walls where, through branching and anastomoses, they become incorporated into the vascular network of the carpels. The dorsals finally pass into the style bases (Fig. 11.2P).

The ventral bundles of the adjacent fused carpel margins are also fused basally (Fig. 11.2I, R) to form two compound ventral bundles (vv) in the base of the ovary, below the level of the locule bases. The way in which these compound ventrals are formed is somewhat variable. Each may arise as a single bundle, or be formed by the fusion of two to three bundles which move centripetally from the periphery toward the center of the receptacle (Fig. 11.2R). Two opposing compound ventral bundles become apparent just above the level of the locule bases, each containing several strands of protoxylem and appearing arcuate in transverse section (Fig. 11.2J). The compound ventrals ascend in the septum and placentae to a level near the middle of the ovary, where each divides to form individual ventral bundles (v) to each of the still fused carpel margins (Fig. 11.2N). At various levels the compound ventrals may appear to be collateral, bicollateral, or amphicribral. Each ventral bundle provides one to three minor lateral branches to the upper portion of the carpel wall, or it may divide at about the level of the highest ovule, forming two bundles which then provide minor traces to the carpel walls before passing into the base of the style (Fig. 11.2N–Q).

Ovule bundles are produced from both the compound and free portions of the ventral bundles (Fig. 11.2J–O, R). The ovular traces arch sharply before descending into the pendent ovules. In both cleared material and serial sections the ventral bundles may appear to be double, with the ovular traces being derived from the more marginal of the two strands. This phenomenon is not encountered consistently in all margins, however, and the additional marginal strand is interpreted here as the minor placental bundle derived from the ventral bundle.

Pronounced lateral or compound lateral bundles, such as occur in the gynoecia of the tribe Hamamelidoideae (Bogle 1970) are not present in the ovaries of *Rhodoleia*.

Ontogeny

The inflorescence of *Rhodoleia championii* begins development as a lateral axillary bud (Fig. 11.1A). Its axis elongates to form the short peduncle of the inflorescence. Two prophylls flank the bud, and a few short, broad bracts are spirally inserted along the peduncle, intergrading distally with the bracts of the involucre which, in the absence of well-developed sepals, enclose and protect the developing flowers.

Within the young bud the terminal meristem differentiates into a broad, shallow dome (Fig. 11.3A), on the flanks of which a cycle of five to eight floral primordia initiate almost simultaneously (Fig. 11.3B). No obvious primary bracts subtend the flowers, and it was not possible in the available material to determine whether certain primordia around the base of the apical dome, such as those seen in Fig. 11.3A, represent abortive primordia of bracts transitional between the involucre and the inflorescence, or floral bracts that fail to develop. Trichome primordia develop from protodermal cells between the flower primordia (Fig. 11.3B, C). Variation in floral outline (ranging from rounded, to broadly triangular, to elliptic) as a function of congestion and limited growing space, becomes apparent early in floral development (Fig. 11.3B, C, E, F).

Figure 11.2C is a diagrammatic representation of a flower having a shallow, collar-like calyx, four petals on the abaxial (outer) side, and cycles of ten stamens, ten sterile phyllomes, and two carpels. The radial symmetry and number of organs shown are arbitrary, however, since floral shape is variable, as are the numbers of petals, stamens, and sterile phyllomes in a given flower.

Each floral primordium forms a broad, shallow dome, on the periphery of which a concentric calyx ring primordium initiates around a central primordial dome, the center of which becomes depressed (Fig. 11.3B–D), forming an inner ring primordium. The development of the calyx lags behind that of the inner ring, eventually becoming little more than an irregular, membranous collar on the hypanthium (Fig. 11.3G; 11.4E, F). Initiation of petal and stamen primordia begins first, and more or less simultaneously, on the abaxial side of the inner ring primordium (Fig. 11.3D), and progresses toward the flanks and adaxial side of the primordium. Petal primordia lag behind stamen primordia in development (Fig. 11.1B; 11.3E–G; 11.4A). The young petals are at first rounded to ovate in outline, but gradually take on a clawed shape as the blade of the petal expands (Fig. 11.1B; 11.3G; 11.4A). The stamens develop rapidly, in early stages of development consisting mostly of the large anther, which quickly takes on a quadrangular configuration as the four sporangia develop.

Within the ring of stamen primordia two dome-shaped carpel primordia initiate (Fig. 11.3F); the abaxial primordium differentiates first and for a while is considerably larger than the adaxial one. The primordia gradually become crescentic, and finally folded in configuration (Fig. 11.3H). The apocarpous portions of the carpels are tightly

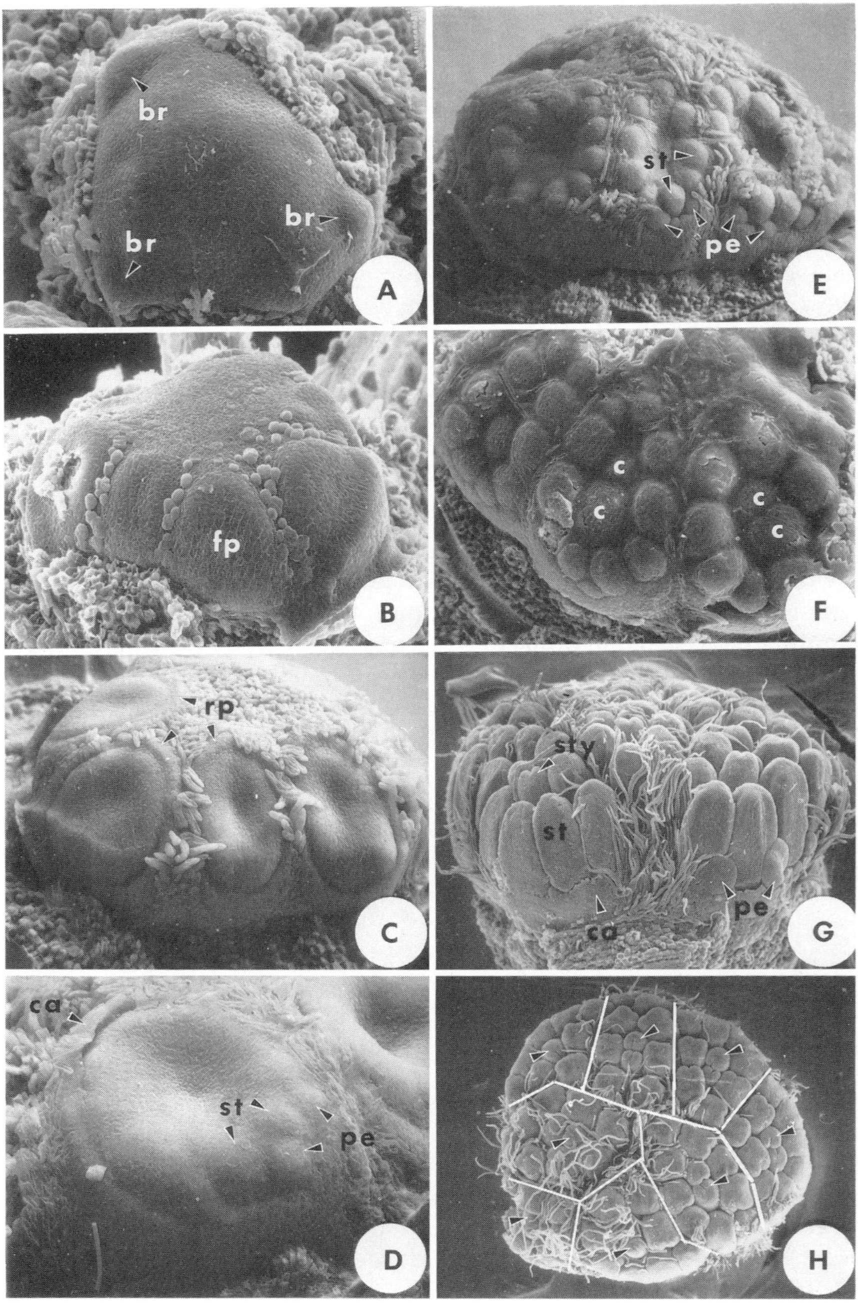
br
br
br
A
fp
B
rp
C
ca
st
pe
D
st
pe
E
c
c
c
c
F
sty
st
pe
ca
G
H

packed among, and appear to elongate in concert with, the stamens, their tips all remaining at about the same level (Fig. 11.3G, H; 11.4A, B).

The cycle of glandular phyllomes that occupies the inner rim of the hypanthium initiates very late in flower development. The carpels are well developed before the phyllomes become apparent as small protrusions at the base of ridges in the carpel walls (Fig. 11.4B–F). They may appear quite distinct (Fig. 11.4E), or appear to be connected by a line of smaller lobes (Fig. 11.4D). In fruit they become incorporated into the indurated ridge formed by the rim of the hypanthium around the base of the capsule (Fig. 11.4G).

Expansion in length of the petal bases, stamen filaments, and filamentous styles to their final exserted condition at anthesis takes place in the latest stages of development. The involucral bracts appear to provide the only protection for the developing flowers of the inflorescence.

Discussion

Inflorescence: Inflorescence condensation to form spikes or heads is common in all five subfamilies of Hamamelidaceae, but is perhaps most pronounced in the lower Hamamelidaceae, among which flowers have become embedded in fleshy, spicate axes (*Chunia*, *Mytilaria*), connate in simple heads (*Rhodoleia*, *Disanthus*), or incorporated into compound heads (*Altingia*, *Liquidambar*). *Rhodoleia* is unique among these in its adaptation as a pseudanthial 'bird-flower' (Porsch 1923; van Leeuwen 1938). *Altingia* and *Liquidambar* are wind pollinated; *Exbucklandia* and *Chunia* may be wind pollinated (observations are lacking in the

---

Fig. 11.3 *Rhodoleia championii.* Developmental stages in inflorescence and flower development. A, Broad, shallow inflorescence dome prior to floral initiation, × 40. B, Individual flower primordia form a ring on flanks of inflorescence dome; trichome primordia developing between flowers, × 113. C, Ring primordium of calyx forming on margin of each floral primordium, × 109. D, Petal and stamen primordia initiating on abaxial side of central dome of floral primordium; petal primordia closer to periphery, × 157. E, A ring of distinct petal and stamen primordia apparent in each flower; petals only abaxial and lagging in development, × 64. F, Carpel primordia forming within ring of stamens; the upper, adaxial carpel lagging in initiation and development, × 80. G, Lateral oblique view of whole inflorescence, showing irregular calyx, small petals, stamens, styles of two outer flowers, × 36. H, Polar view of whole inflorescence of eight flowers at similar stage of development as G; pair of folded carpels defines center of each flower, × 24.

(br, bract; c, carpel; ca, calyx; fp, floral primordium; pe, petal; rp, ring primordium; st, stamen; sty, style.)

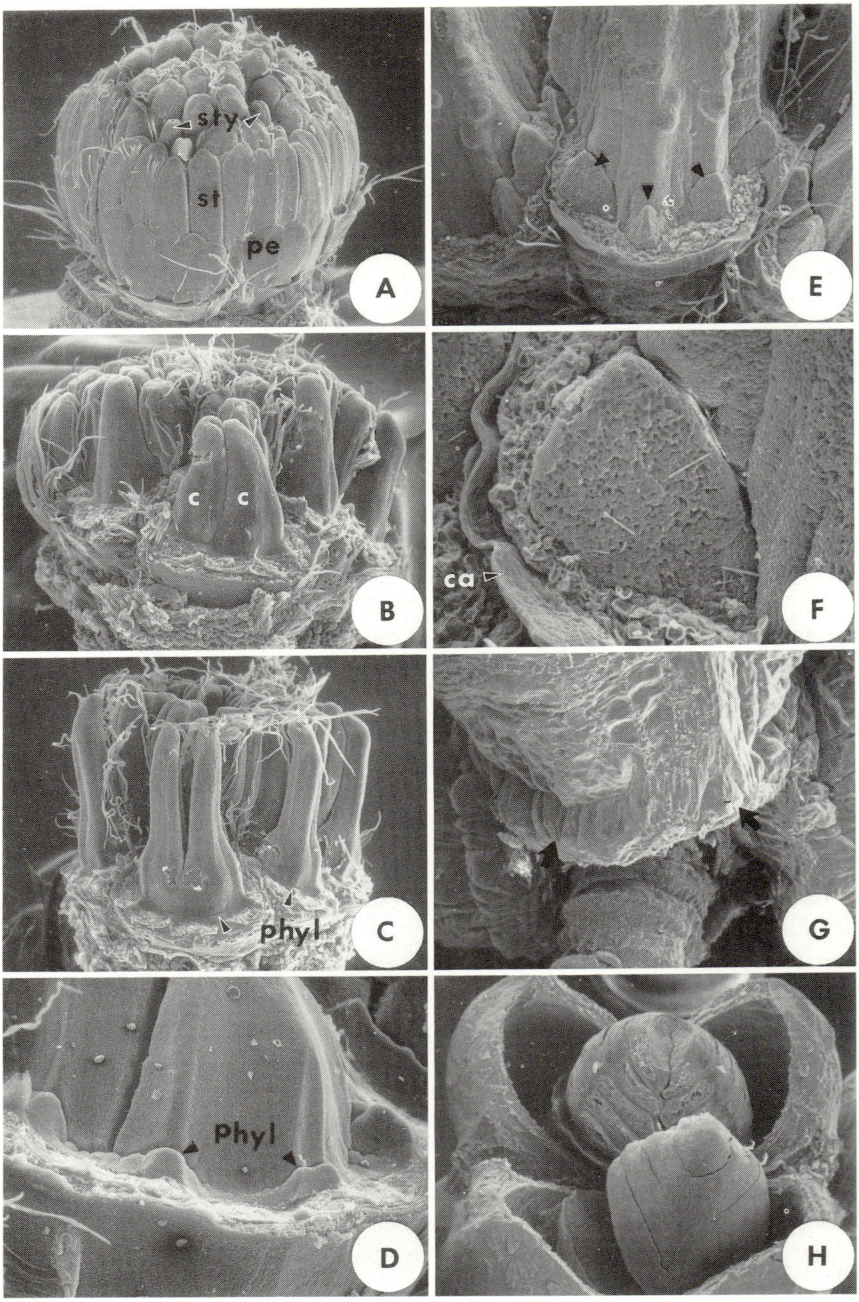
sty
st
pe
A
E
c c
B
ca
F
phyl
C
G
Phyl
D
H

literature); *Disanthus* is insect pollinated, and *Mytilaria* probably is also (observations lacking).

The morphological adaptations of the *Rhodoleia* inflorescence to bird pollination include its pendent habit, the calyx-like involucre and reduced calyx, the pseudo-corolla of reddish petals, the stamens and styles long and more or less exserted, and the nectar-secreting activity of the sterile phyllomes. Exell (1933) and Vink (1957) provide data for size ranges found in the inflorescences, petals, stamens, anthers, styles, and for ovary pubescence in the various species or geographic races of *Rhodoleia*. The involucre serves both to protect the developing flowers, and possibly to play some role in maintaining the presentation of the petals, stamens, and styles as a single unit ('flower') at pollination. The stamens are reported by Exell and Vink to be slightly shorter to slightly longer than the petals, depending on the species, and the styles are said to slightly exceed the stamens in length. Porsch (1923) and van Leeuwen (1938) name specific bird, insect, and animal visitors to the flowers. The classic illustration of the inflorescence (Champion ex Hooker 1850) shows the pseudanthium as open and cup-shaped, with the petals spreading (it is often referred to as being camellia-like). More extensive field observations on the pollination ecology of *Rhodoleia* in different parts of its range are desirable to determine what constitutes anthesis, when pollination takes place, and whether different pollinators may function at different stages of anthesis.

Flower

Calyx: Although the calyx appears as a distinct ring primordium early in ontogeny, at maturity it is so reduced and irregular that it appears to be little more than vestigial. Its vascular supply is irregular and the traces appear to terminate in the hypanthium before entering the calyx limb. Its development becomes retarded at an early stage, and at no

---

Fig. 11.4 *Rhodoleia championii.* Developmental stages in inflorescence and flower development. A, Young whole inflorescence at later stage of development; small petals becoming clawed, ×33. B, Development stage similar to A; some petals and stamens removed to show young, folded carpels; phyllome primordia not yet apparent, ×65; C, Late developmental stage; style bases elongating; phyllome primordia beginning to develop, ×40. D, Later stage of development; phyllomes now apparent as discrete structures, or with intervening small lobes between them, ×93. E, Mature flowers with petals and stamens removed to show discrete or bilobed, angular phyllomes around carpels, ×24. F, Single phyllome enlarged to show rough epidermis, ×190. G, Base of a mature fruit to show indurated ridge of capsule incorporating phyllomes, ×16. H, Mature dehisced capsule to show packing of mature seeds, ×14.
(c, carpel; ca, calyx; pe, petal; phyl, phyllome; st, stamen; sty, style.)

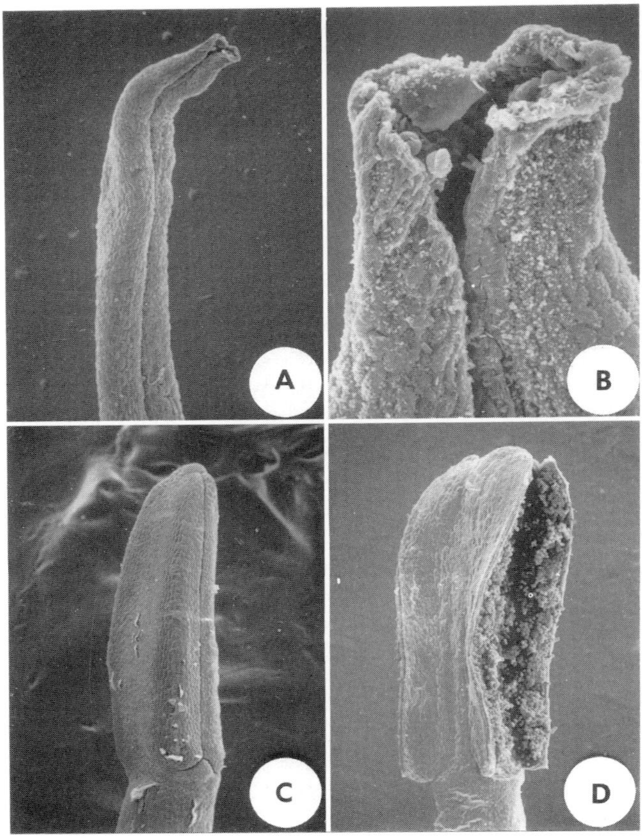

Fig. 11.5 *Rhodoleia championii*. Stigma and anther. A, Terminal portion of style; margins involute, without marginal stigmatic surfaces, × 30. B, Tip of style enlarged to show slightly flared, stigmatic area with trapped pollen grains, × 250. C, Anther prior to dehiscence, × 24. D, Anther dehiscence via two narrow valves on each side; numerous pollen grains, × 28.

time does it appear to have a distinct morphology in terms of regular lobing, vascularization, or any kind of protective function. Its reduction may represent a transition stage toward complete loss of the calyx as a function of inflorescence condensation and fusion of flowers within heads, as in the Liquidambaroideae. A similar tendency is seen in the Exbucklandioideae, where a calyx is well developed in *Mytilaria*, present as a ring primordium early in floral ontogeny but absent in mature flowers of *Exbucklandia*, and absent in mature flowers of *Chunia* (personal observations).

Corolla: The petals of *Rhodoleia* are probably staminodial in origin, although I have not observed any transitional petaloid stamens bearing partial anthers. Petaloid stamens are common in *Exbucklandia* (Kaul and Kapil 1974), and have been observed in *Disanthus* (Bogle 1967). Various accounts indicate none to four petals on the abaxial margin of individual flowers. Petals may be absent even in flowers with the maximum abaxial margin. The loss of petals on the abutting flanks and adaxial margins of the flowers may be the result of the same forces of inflorescence condensation, and floral congestion and fusion, as is the reduction of the calyx.

Each petal receives a single vascular trace which then branches in the petal blade. Vestigial petal traces were not observed on the flanks of the flowers where petals were absent.

The petals appear to initiate simultaneously with the abaxial stamens, but to be smaller and slower in development than the stamen primordia, which soon exceed them in length. They develop as straight organs, as do the petals of *Mytilaria* and *Dicoryphe*, rather than being circinnately coiled, like the petals of *Disanthus* and a number of hamamelidoids (e.g., *Maingaya*, *Loropetalum*, *Hamamelis*).

Androecium: The seven to ten stamens in an individual flower form a single cycle, and do not appear to be affected in size or form by the crowding and fusion of flowers. Each stamen receives a single vascular bundle. The stout, almost fleshy filaments may play a role in holding nectar within the inflorescence via capillary action. According to van Leeuwen (1938), who made the most detailed original observations on the flowers and pollinators on trees of *Rhodoleia teijsmannii* Miq. at Tjibodas, western Java: 'When the open heads are shaken thick drops of a sweet and very mucous liquid fall out. This honey is kept back by the capillaire force of the filaments which stand very close together, and although the openings of the heads face downwards the honey does not run out. I failed to find any special capillaire constructions or hairs to hold or conduct the honey such as have been described for other bird-flowers. When greatly enlarged the cuticula of the epidermis of the filaments did indeed show extremely small stripes, and a transverse section showed that the cuticula is finely grooved. I do not venture to decide whether or not these extremely small grooves can assist in retaining the honey'.

The length of the stamens varies regionally such that the anthers may or may not be exserted beyond the corolla. The basifixed anthers produce abundant sticky pollen (van Leeuwen 1938) that could be carried by either the birds or bumble bees that visit the flowers.

Sterile phyllomes: The sterile structures that form a cycle on the inner rim of the hypanthium, surrounding the ovary, are apparently the secretory organs that produce the floral nectar described by van Leeuwen (1938), as follows: 'Round the base of the stamens grow small, flattened globular papillae; the cuticula is but slightly thickened, there being no cavity of liquid between the cuticula and the cellulose wall. The honey when secreted sticks as sparkly, thick drops. This honey is very sweet with a slightly bitter, after-taste . . . but it does not give forth any scent. The flower-heads do indeed give forth a very delicate flavour, something like that of pounded rose petals, but at some distance this cannot be noticed.' I have not observed papillae actually 'surrounding' the bases of the stamens, as indicated by van Leeuwen, in addition to the cycle of organs described and illustrated above.

These organs appear to be comparable to similar cycles of sterile organs in Liquidambaroideae (Bogle 1986) and *Exbucklandia* (the 'lobules' of Kaul and Kapil 1974). However, in these genera they have not been demonstrated to be glandular. They may also be comparable to cycles of 'disk lobes', 'disk glands', 'nectary scales', 'staminodes', etc., occurring in some members of the Hamamelidoideae, e.g., *Maingaya*, *Dicoryphe*, *Loropetalum*, *Tetrathyrium*, *Corylopsis* (Bogle 1984, 1988; Endress 1967; Mione and Bogle 1987; Tong 1930). In the last three genera they often appear to occur as pairs more or less opposite the petals. The members of each pair may show varying degrees of fusion into single, deeply divided or bilobed to entire, truncate organs. The five 'disk glands' or 'staminodes' which occasionally appear in *Disanthus*, the nectaries of *Hamamelis*, and the five staminodia of *Ostrearia* may represent a different cycle of staminodial organs. At present there is ontogenetic evidence for this supposition only for *Hamamelis* and *Corylopsis* (see below).

Tong (1930) discussed the presence and distribution of sterile floral organs of various forms in the family and considered all of them to represent staminodia (*Altingia*, *Liquidambar*, *Maingaya*, *Dicoryphe*, *Loropetalum*, *Hamamelis*, *Corylopsis*, *Eustigma*). He also cites transitional stages from stamens to scale-like 'staminodia' in several of these genera.

Endress (1967) presented a detailed review of the literature concerning such floral organs in relation to his description of a cycle of previously overlooked staminodia alternate with the stamens in six species of *Corylopsis*, and distinct from the inner cycle of 'nectar scales'. He noted that this pattern also occurs in *Maingaya*, and that in most Hamamelideae one cycle or the other is absent. He discussed instances of *Corylopsis* flowers having ten stamens as a result of petals or staminodes being replaced by stamens, and of nectar scales producing reduced anthers. In ontogeny he found the petals and staminodia of *Corylopsis* to initiate simultaneously, while the nectar scales initiated after the gynoecium had already grown to a considerable height. In contrast he cited evidence (Baillon 1871; Shoemaker 1905) that the staminodial nectaries of *Hamamelis* arise ontogenetically after the stamens and before the carpels, thus being comparable to the staminodia of *Corylopsis* and different from the nectar scales. He concluded that structures comparable to nectar scales are not found in *Hamamelis*.

Mione and Bogle (1987, and in mss.), in a comparative ontogenetic study of *Loropetalum chinense* (flowers tetra- or pentamerous) and *Hamamelis virginiana* (flowers strictly tetramerous), confirmed the centripetal origin of the staminodial nectaries of *Hamamelis*, but found that the eight to ten sterile phyllomes in *Loropetalum* are comparable to the nectar scales of *Corylopsis* in that they initiate centrifugally, after the gynoecium is well developed. Depending on whether the flowers of *Loropetalum* are tetra- or pentamerous, the sterile phyllomes may appear as eight or ten distinct organs, or show various degrees of fusion to form four or five bilobed organs. In *Loropetalum* Mione (1987) reported petals transformed into partial stamens and, rarely, sterile phyllomes developed as sterile stamens. At maturity the sterile phyllomes of *Loropetalum* are inserted on the hypanthium, and their vascular supply is derived (Bogle 1967) in a manner similar to that described by Endress (1967) for *Corylopsis*.

These observations on the time of initiation, form, and vascular supply of sterile floral organs support the possibility that two different cycles of sterile organs are basic not only to the Hamamelideae, as indicated by Endress (1967), but to all five subfamilies; i.e., a cycle of staminodia equal in number and alternating with the functional stamens (e.g., *Maingaya*, *Dicoryphe*, *Hamamelis*, *Corylopsis*, possibly *Disanthus*), and/or a cycle of sterile phyllomes, double (or equal through fusion of pairs) the number of stamens, on the inner rim of the hypanthium (e.g., *Altingia*, *Liquidambar*, *Rhodoleia*, possibly *Exbucklandia*, *Maingaya*, *Dicoryphe*, *Loropetalum*, *Corylopsis*). The time of initiation of the cycle(s) of sterile organs, either before or after the gynoecium, may be the determining factor in their interpretation.

The two distinct cycles of organs may both have originated through modification of stamens, and thus be basically staminodial. However, I have not yet observed sterile phyllomes transitional to stamens in any of the lower Hamamelidaceae such as those reported in *Corylopsis* (Tong 1930, Endress 1967) and now in *Loropetalum* (Mione 1987).

Further ontogenetic work is needed in such genera as *Disanthus*, *Exbucklandia*, *Maingaya*, *Dicoryphe*, and *Tetrathyrium* to determine the sequence of initiation and function (if possible) of their sterile phyllomes, how they relate to those of the other genera, and perhaps the significance of their number in terms of cycles; i.e., does an inner cycle of ten members represent a single cycle or two pentamerous cycles, morphologically?

Gynoecium: The gynoecium of *Rhodoleia* is typically bicarpellate, and is closest to that of the Liquidambaroideae in terms of ovule number and vasculature; it is therefore among the least specialized in the family. *Rhodoleia* thus exhibits a highly specialized inflorescence and floral morphology and pollen morphology, as a function of a specialized mode of pollination, in combination with relatively primitive gynoecial morphology and vascular anatomy in comparison with other Hamamelidaceae.

Conclusions

It is generally accepted, especially through the observations of van Leeuwen (1938), that the inflorescences and flowers of *Rhodoleia* are strongly adapted for bird pollination and in this respect are unique in the family. However, the details of floral morphology, vascular anatomy, and ontogeny for *Rhodoleia* fall within the range of conditions seen in the family as a whole; this would indicate a relatively close relationship to the Liquidambaroideae, and support subfamilial status for the genus rather than isolation as a separate family, as has been suggested by a few authors.

Although largely ignored or barely mentioned (and frequently omitted from floral diagrams of *Rhodoleia*), the presence, form, and belated ontogenetic appearance of the cycle of sterile phyllomes, in particular, prove a strong link with the Liquidambaroideae and Exbucklandioideae on the one hand, and with the more primitive Hamamelidoideae on the other. These organs appear to represent one of two potential cycles of sterile appendages that may appear in flowers of the lower Hamamelidaceae and constitute one of the more significant unifying features of the family. One of these is a cycle of staminodia appearing before the gynoecium in ontogeny and equaling the stamens in number (e.g., the nectaries of *Hamamelis*, the staminodia of *Corylopsis*, possibly the reported staminodia of *Disanthus*). The other is a cycle of sterile phyllomes appearing after the gynoecium in ontogeny and equalling or exceeding (to double) the number of stamens (*Liquidambar*, *Rhodoleia*, *Corylopsis*, *Loropetalum*). In *Corylopsis* and *Loropetalum* the phyllomes may rarely appear as partially developed stamens, suggesting a possible staminodial origin for this cycle as well. Flowers of *Maingaya* and some species of *Dicoryphe* contain both cycles, but their ontogenetic sequences are not yet known. The fact that the number of phyllomes is consistently (e.g. *Maingaya*) or often (e.g., *Loropetalum*, *Tetrathyrium*) double the number of stamens raises the question of whether one or two cycles of floral appendages is represented in the cycle. Further ontogenetic studies in genera such as *Exbucklandia*, *Mytilaria*, *Disanthus*,

Maingaya, *Dicoryphe*, and *Tetrathyrium* are needed to determine how their ontogenetic patterns compare with those already described in *Liquidambar*, *Rhodoleia*, *Corylopsis*, *Hamamelis*, and *Loropetalum*. The data derived would provide a better basis for interpreting the number of cycles of floral organs basic to the hamamelidaceous flower, their morphological relationships, and their significance to the phylogeny of the family.

Acknowledgements

This paper represents Scientific Contribution Number 1573 from the New Hampshire Agricultural Experiment Station. Financial aid from the University of New Hampshire Central University Research Fund, Grant no. 980, is gratefully acknowledged. Translations of the Russian literature by A. Baranov, and the Chinese literature by Shiu-Ying Hu, were invaluable in completing this work. I wish to thank the Director of the Harvard University Herbaria (A and GH) for access to their herbarium and library collections, Hollis Bedell for help in the library (A and GH), and the Director of the Hong Kong Botanic Garden for access to its living collections. I also wish to thank Margaret Bogle, David Boufford, Arthur Cronquist, Garrett Crow, Thomas Lee, Thomas Mione, Thomas Philbrick, and Shirley Tucker for reading and criticizing the manuscript.

References

Baillon, H. (1871). Nouvelles notes sur les Hamamelidees. *Adansonia* **10,** 120–37.

Baranova, M. (1983). On the laterocytic stomatotype in angiosperms. *Brittonia* **35,** 93–102.

Baranova, M. A. (1985). Classifications of the morphological types of stomata. *Botaničeskij Žurnal (Moscow & Leningrad)* **70,** 1585–95, pls. 1–4.

Bean, W. J. (1917). *Disanthus cercidifolia. Curtis's Botanical Magazine* **143,** pl. 8716.

Bentham, G. and Hooker, J. D. (1865). Hamamelidae. In *Genera Plantarum*, Vol. **1,** pp. 664–9. L. Reeve, London.

Bogle, A. L. (1967). Floral vascular anatomy and the nature of the hamamelidaceous flower. Unpublished D.Phil. thesis, University of Minnesota, St. Paul.

Bogle, A. L. (1968). Major features of morphology and vascular anatomy of the hamamelidaceous gynoecium. *American Journal of Botany* **55,** 715. (Abstr.)

Bogle, A. L. (1970). Floral morphology and vascular anatomy of the Hamamelidaceae: the apetalous genera of the Hamamelidoideae. *Journal of the Arnold Arboretum* **51,** 310–66.

Bogle, A. L. (1984). Floral morphology and vascular anatomy of *Maingaya* Oliv. (Hamamelidaceae, Hamamelidoideae, Hamamelideae). *American Journal of Botany* **71,** 19. (Abstr.)

Bogle, A. L. (1986). The floral morphology and vascular anatomy of the Hamamelidaceae: subfamily Liquidambaroideae. *Annals of the Missouri Botanical Garden* **73,** 325–47.

Bogle, A. L. (1987). Inflorescence and flower ontogeny in the pseudanthium of *Rhodoleia* (Hamamelidaceae). *American Journal of Botany* **74,** 607–8. (Abstr.)

Bogle, A. L. (1988). Floral morphology and vascular anatomy of *Dicoryphe* Thouars. (Hamamelidaceae, Hamamelidoideae, Hamamelideae). *American Journal of Botany* **75,** 18 (Abstr.).

Bogle, A. L. and Philbrick, C. T. (1980). A generic atlas of hamamelidaceous pollens. *Contributions from the Gray Herbarium*, No. **210,** 29–103.

Chang, C. T. (1964). The pollen morphology of the families Hamamelidaceae and Altingiaceae. *Trudy Botaničeskogo instituta Akademii nauk SSSR. Ser. 1, Flora i sistematika vysših rastenij. Acta instituti botanici academiae scientiarum URPSS. Flora et systematica plantae vasculares* **13,** 173–232. pls. 1–21 (In Russian).

Chang, H. T. (1973). A revision of the hamamelidaceous flora of China. *Sun Yatsen University Bulletin* **1,** 54–71 (In Chinese).

Chang, H. T. (1979). Hamamelidaceae. In *Flora Reipublicae Popularis Sinicae* **35**(2). (ed. H. T. Chang), pp. 36–116. Science Press, Beijing.

Endress, P. K. (1967). Systematische Studie über die verwandtschaftlichen Beziehungen zwischen den Hamamelidaceen und Betulaceen. *Botanische Jahrbücher für Systematik, Pflanzengeschichte und Pflanzengeographie* **87,** 431–525.

Endress, P. K. (1977). Evolutionary trends in the Hamamelidales-Fagales group. *Plant Systematics and Evolution, Supplement* **1,** 321–47.

Endress, P. K. (1978*a*). Stipules in *Rhodoleia* (Hamamelidaceae). *Plant Systematics and Evolution* **130,** 157–60.

Endress, P. K. (1978*b*). Blütenontogenese, Blütenabgrenzung und systematische Stellung der perianthlosen Hamamelidoideae. *Botanische Jahrbücher für Systematik, Pflanzengeschichte und Pflanzengeographie* **100,** 249–317.

Exell, A. W. (1933). A revision of the genus *Rhodoleia*. *Sunyatsenia* **1,** 95–101.

Goldblatt, P. and Endress, P. K. (1977). Cytology and evolution in Hamamelidaceae. *Journal of the Arnold Arboretum* **58,** 67–71.

Griffiths, W. (1836). Description of two genera of the family Hamamelideae, two species of *Podostemon* and one species of *Kaulfussia*. In *Asiatic Researches (Calcutta)* **19,** 94–114, pls. XIII–XVIII.

Harjal, N., Bisht, P. S. and Paliwal, G. S. (1984). Foliar vasculature as an aid to the classification of Hamamelididae. In *The vegetational wealth of the Himalayas* (ed. G. S. Paliwal), pp. 282–300. University of Garhwal, Srinagar.

Harms, H. (1930). Hamamelidaceae. In A. Engler and K. Prantl, *Die Natürlichen Pflanzenfamilien* (2nd edn), **18a,** (ed. H. Harms), pp. 330–43. W. Englemann, Leipzig.

Hooker, W. J. 1850. *Rhodoleia championi. Curtis's Botanical Magazine* **76,** t. 4509.

Kaul, U. and Kapil, R. N. (1974). *Exbucklandia populnea*—from flower to fruit. *Phytomorphology* **24,** 217–28.

Leeuwen, W. M. D. van. (1938). Observations about the biology of tropical flowers. *Annales du jardin botanique de Buitenzorg* **48,** 27–68. pl. X. [*Rhodoleia*, pp. 48–52.]

Lindley, J. (1853). *The vegetable kingdom* (3rd edn). Bradbury and Evans, London.

McLaughlin, R. P. (1933). Systematic anatomy of the woods of the Magnoliales. *Tropical Woods* **34,** 3–39.

Melikian, A. P. (1971). Anatomical structure of the spermodermis in the members of the genera *Liquidambar* L. and *Altingia* Nor. in relation to their taxonomy. *Izvestija: Biologičeskie nauki. Tełekagir: Biologiakan gitouthjounner. Erivan* **24,** 50–5 (In Russian).

Melikian, A. P. (1972). On the place of the genera *Rhodoleia*, *Exbucklandia*, and *Chunia* in the system of the family Hamamelidaceae. *Izvestija: Biologičeskie nauki. Tełekagir: Biologiakan gitouthjounner. Erivan* **25,** 39–45 (In Russian).

Melikian, A. P. (1973*a*). Seed coat types of Hamamelidaceae and allied families in relation to their systematics. *Botaničeskij Žurnal (Moscow & Leningrad)* **58,** 350–9 (In Russian).

Melikian, A. P. (1973*b*). The anatomy of seed coat and systematics of the family Hamamelidaceae. *Izvestija: Biologičeskie nauki. Tełekagir: Biologiakan gitouthjounner. Erivan* **26,** 104–5 (In Russian).

Mione, T. (1987). Comparative ontogeny of the inflorescences and flowers of *Hamamelis virginiana* and *Loropetalum chinense* (Hamamelidaceae). Unpublished M.S. thesis, University of New Hampshire, Durham.

Mione, T. and Bogle, A. L. (1987). Comparative ontogeny of the flowers of *Hamamelis virginiana* and *Loropetalum chinense* (Hamamelidaceae). *American Journal of Botany* **74,** 620–21. (Abstr.)

Mizushima, M. (1968). On the flower of *Disanthus cercidifolius* Maxim. *Journal of Japanese Botany* **43,** 522–44.

Nakai, T. (1943). *Ordines, familiae, tribi, genera, sectiones, species, varietates et combinationes novae.* Imperial University, Tokyo.

Niedenzu, F. (1891). Hamamelidaceae. In *Die Natürlichen Pflanzenfamilien*, Teil **3,** Abteilung 2a, (eds. A. Engler and K. Prantl), pp. 115–30. W. Engelmann, Leipzig.

Porsch, O. (1923). Blütenstände als Vogelblumen. *Oesterreichisches Botanisches Zeitschrift. Gemeinnütziges Organ für Botanik* **72,** 125–49. T. I–III. [Hamamelidaceae, pp. 147–8.]

Reinsch, A. (1889). Über die anatomische Verhältnisse der Hamamelidaceae mit Rücksicht auf ihrer systematisches Gruppierung. *Botanische Jahrbücher für Systematik, Pflanzengeschichte und Pflanzengeographie* **2,** 347–95.

Schmitt, D. (1966). The pistillate inflorescence of sweetgum (*Liquidambar styraciflua* L.). *Silvae Genetica* **15**(2), 33–5.

Shaw, H. K. A. (ed.) (1966). In Willis, J. C., *A dictionary of the flowering plants and ferns* (7th edn). Cambridge University Press.

Shoemaker, D. N. (1905). On the development of *Hamamelis virginiana. Botanical Gazette (Crawfordsville)* **39,** 248–66.

Skvortsova, N. T. (1960*a*). The anatomical structure of the conducting system of leaf petioles of representatives of families Hamamelidaceae and Altingiaceae. *Doklady Akademii nauk SSSR* **133,** 1231–3. (In Russian).

Skvortsova, N. T. (1960*b*). On the types of leaf venation in the members of the family Hamamelidaceae. *Trudy Leningradskogo Khimiko-farmatzevticheskogo Instituta Ministerstva zdravookhraneniia RSFSR* **12,** 75–83 (In Russian).

Skvortsova, N. T. (1960*c*). The structure of epidermis in the members of the family Hamamelidaceae. *Botaničeskij Žurnal (Moscow & Leningrad)* **45,** 712–7 (In Russian).

Tang, Y. (1943). Systematic anatomy of the woods of the Hamamelidaceae. *Bulletin of the Fan Memorial Institute of Biology* II **1,** 8–63, tables 1–3.

Tardieu-Blot, M. L. (1965). Hamamelidaceae. In *Flore du Cambodge de Laos et du Vietnam* No. **4,** (eds. A. Aubreville and M. L. Tardieu-Blot), pp. 75–116. Muséum National d'Histoire Naturelle, Paris.

Tiffney, B. H. (1986). Fruit and seed dispersal and the evolution of the Hamamelidae. *Annals of the Missouri Botanical Garden* **73,** 394–416.

Tippo, O. (1938). Comparative anatomy of the Moraceae and their presumed allies. *Botanical Gazette* **100,** 1–99.

Tong, K. (1930). Studien Über die Familie der Hamamelidaceae mit besonderer Berücksichtigung der Systematik und Entwicklungsgeschichte von *Corylopsis*. *Bulletin of the Department of Biology* No. **2,** Sun Yatsen University, Canton.

Vink, W. (1957). Hamamelidaceae. In *Flora Malesiana* No. **5**(1), (ed. C. G. G. J. van Steenis), pp. 363–79. Noordhoff-Kolff N.V., Djakarta.

Wisniewski, M. and Bogle, A. L. (1982). The ontogeny of the inflorescence and flower of *Liquidambar styraciflua* L. (Hamamelidaceae). *American Journal of Botany* **69,** 1612–24.

Wolfe, J. A. (1973). Fossil forms of Amentiferae. *Brittonia* **25,** 334–5.

Zavada, M. S. and Dilcher, D. L. (1986). Comparative pollen morphology and its relationship to phylogeny of pollen in the Hamamelidae. *Annals of the Missouri Botanical Garden* **73,** 348–81.

12. Phylogenetic relationships in the Hamamelidoideae

PETER K. ENDRESS

Botanischer Garten und Institut für Systematische Botanik der Universität Zürich, CH-8008 Zürich, Switzerland

Abstract

Different amounts of plasticity and stability of features at various levels of the systematic hierarchy, from individuals to tribes, are shown for the subfamily Hamamelidoideae. Such differences are in some instances correlated with different biological situations. On the basis of unique anther dehiscence patterns and exclusive geographic distributions, together with other features, the tribe Hamamelideae can be subdivided into three subtribes. The separation of the apetalous genera into two tribes is artificial and cannot be maintained. *Fortunearia* and *Sinowilsonia* should be associated with *Eustigma* rather than with *Corylopsis*, which forms a tribe of its own. A phylogenetic tree is provided for all 22 genera of the Hamamelidoideae. The reconstruction is most probably correct for the subtribe Dicoryphinae and for several pairs of genera. The remaining ramifications are less certain due to a large amount of parallel evolution because of plasticity of characters and the lack of key features.

Introduction

The Hamamelidoideae are the largest and most diverse subfamily of the Hamamelidaceae, with 22 genera and about a hundred species. They have a main centre of diversity in eastern Asia but have a very scattered global distribution due to the antiquity of the family. Most genera are monotypic or have only two or three species. The few genera with a somewhat higher number of species are difficult to treat taxonomically,

Evolution, Systematics, and Fossil History of the Hamamelidae Volume 1: Introduction and 'Lower' Hamamelidae. (eds P. R. Crane and S. Blackmore), Systematics Association Special Volume No. 40A, pp. 227–48. Clarendon Press, Oxford, 1989.

possibly a reflection of more recent diversification at the generic level. They include *Corylopsis* (Morley and Chao 1977; Yamanaka 1986), *Dicoryphe* (Rakotobe 1987), *Distyliopsis* (Endress 1970; Chang 1973, 1979; Anonymous 1977), *Distylium* (Walker 1944), *Hamamelis* (Bradford and Marsh 1978), and *Trichocladus* (Verdcourt 1971). The last infrafamilial classifications were those by Harms (1930) and Schulze-Menz (1964).

If one compares features of all genera of the Hamamelidoideae it becomes evident that there is immense plasticity in the group. This is correlated, at the evolutionary level, with a large amount of parallel evolution. In fact, such plasticity and resulting parallel evolution is not a kind of evolutionary background noise, as it is sometimes treated, but seems one of the most central aspects of evolution, not only in Hamamelidoideae.

Although I have not done this, it would be highly informative to plot the entire range of plasticity (variability) of critical features for every level of the systematic hierarchy within our group: individuals, populations, species, genera, and tribes, and not just simple, average character states. However, I have recorded some examples of plastic features, especially those where the amount of plasticity varies from group to group. In doing so, it is much easier to understand that zig-zag evolution must have occurred again and again.

So, when we assess polarity of a character set for a group such as a family, this does not mean that polarity is the same at every point of the phylogenetic tree of the family members: polarity may be reversed many times. If one were to look from the other side and follow the pedigree of a single individual back for a character such as the number of stamens per flower or the length of the anther connective protrusion, one would certainly find some fluctuation back and forth as one retraced the pedigree. This must be all the more true for the phylogenetic tree of entire species, genera, or tribes.

This prevailing undirected, erratic component in the process of evolution becomes evident in many attempts of evolutionary reconstruction by the resulting large number of equally parsimonious phylogenetic trees. A possible way to proceed in this case is to try to trace those rare circumstances in the particular history of a given group, which provide key features for major evolutionary pathways. Such circumstances may be created by extremely rare and strange features, or by the detection of certain key fossils, or by extreme geographic isolation, which has been conserved by lack of long-distance dispersal. Therefore, in trying to find traces of the evolutionary history of the Hamamelidoideae I pursued two very different questions: i) what is the amount of plasticity in significant features in the Hamamelidoideae and in their systematic environment, and how does this plasticity vary from group to group? ii) what are outstanding circumstances that could help in delimiting natural groups within the Hamamelidoideae?

Plasticity and parallel evolution

Number of floral organs

It is obvious that the flowers of the petaliferous groups of the Hamamelidoideae are relatively constantly pentamerous (or tetramerous in *Hamamelis* and *Dicoryphe*), while in apetalous groups with additional reduction of the calyx the number of stamens is mostly larger and very variable (cf. Endress 1970, 1989). But also in the petaliferous genera the degree of (comparatively smaller) variability is different from group to group. In *Hamamelis* one finds almost no variability, while in many other genera stamen number may be increased in many flowers of certain species. In *Dicoryphe* decrease to trimery may be observed occasionally (Rakotobe 1987). Even within species there can be different behaviour. In *Trichocladus grandiflorus* I found constantly pentamerous flowers in one specimen, while in another stamen number was increased in most flowers, up to ten in some, where the stamens occurred partly as staminodes. The same situation was found in *Corylopsis willmottiae* (Endress 1967). Thus, one and the same species may exhibit what is, at other places in the Hamamelidoideae, an important generic difference.

Flower colour

Petaliferous Hamamelidoideae have white, yellow, or red petals. Plasticity is high within genera and even within species. In *Dicoryphe stipulacea* they are white, then become cream, in *D. viticoides* they are dark red. In *Trichocladus crinitus* and *T. ellipticus* they are yellow (cf. Marloth 1925), In *T. grandiflorus* white with a red base (Hooker 1895). *Hamamelis japonica*, *H. vernalis*, and *H. virginiana* are yellow but all have red morphs, which are common in horticulture (Harms 1930; Anderson 1933). *Loropetalum chinense* is mostly white but has red morphs (Yu 1983). *Maingaya* is yellow (Ng 1984). In *Embolanthera glabrescens* petals are yellow (Tardieu-Blot 1965), but white in *E. spicata* (Vink 1957). *Corylopsis*, on the other hand, is uniformly yellow. With this in mind, it is not surprising that in the closely related Australian genera next to *Ostrearia* and *Neostrearia* with yellow petals comes *Noahdendron* with white petals that have a reddish base (like *Trichocladus grandiflorus*).

Stamens are often red, also in taxa with non-red (or lacking) petals, but often not consistently within genera (e.g. *Distylium*, *Parrotia*,

Molinadendron p.p., Corylopsis p.p., Hamamelis p.p., Noahdendron, Trichocladus p.p.).

Indumentum on floral organs

Most genera have stellate hairs on various floral organs, mainly on those that are exposed for longer phases during development, such as the outer surface of sepals and the ovary (including the part that is exposed only after anthesis). In *Trichocladus crinitus* and *T. ellipticus* this is also true for the outside of the petals (also in some *Dicoryphe* species, where they are, however, not exposed, cf. Rakotobe 1987), in *Matudaea*, species of *Distylium* and *Molinadendron* also for the anthers. However, other organs that are exposed longer remain glabrous, such as the massive coiled petal tips in *Noahdendron* and the anther tips in *Parrotia* and *Sycopsis*.

Ovaries that are covered by a floral cup tend to develop stellate hairs with long arms that are directed upwards, while unenclosed ovaries are covered by a dense mat of short-armed stellate hairs. This contrast can be seen in closely related pairs of genera, such as *Sinowilsonia vs. Fortunearia, Sycopsis vs. Parrotia*, and *Distyliopsis vs. Distylium* and *Matudaea*. Glandular fimbriate bracts (stipules) occur in *Embolanthera* and *Tetrathyrium* (both in Loropetalinae), but also in *Matudaea*. In *Matudaea hirsuta* even the anther connective terminates in a gland.

Petals (Fig. 12.1).

There is considerable variation in petal size, also among closely related groups. The largest petals of the Hamamelidoideae (and Hamamelidaceae) seem to occur in *Trichocladus goetzei* (more than 3 cm long), while

---

Fig. 12.1 Petals of Hamamelidoideae. Stippled: fused parts. In 3, vertical and lateral view. Collection number in brackets, E: Endress. 1a *Trichocladus goetzei* Engl. (Stolz 1067), 1b *T. grandiflorus* Oliv. (Rauh and Schlieben 9737), 1c *T. ellipticus* Eckl. et Zeyh. (E 4.10.1983), 1d *T. crinitus* Pers. (E 4908), 2a *Dicoryphe stipulacea* Jaume St. Hil. (E 7801), 2b *D. viticoides* Bak. (E 7807), 3 *Noahdendron nicholasii* Endress et al. (Tracey 14945), 4 *Neostrearia fleckeri* Smith (E 4220), 5 *Ostrearia australiana* Baill. (Irvine 2081), 6a *Embolanthera spicata* Merr. (after Vink 1957), 6b *E. glabrescens* Merr. (after Tardieu-Blot 1965), 7 *Maingaya malayana* Oliv. (Ng 29.3.1985), 8 *Loropetalum chinense* Oliv. (E 3246), 9 *Tetrathyrium subcordatum* Benth. (Chun 7458), 10a *Hamamelis mollis* Oliv. (E 7296), 10b *H. japonica* Sieb. et Zucc. (E 26.2.1968), 10c *H. vernalis* Sarg. (E 7291), 10d *H. virginiana* L. (E 3170), 11a *Corylopsis spicata* Sieb. et Zucc. (E 1966), 11b *C. platypetala* Rehd. et Wils. (E 958), 12 *Eustigma oblongifolium* Gardn. et Champ. (Hongkong, Urban Services 1969), 13 *Fortunearia sinensis* Rehd. et Wils. (E 4472), 14 *Sinowilsonia henryi* Hemsl. (E 4451), 15 *Molinadendron sinaloense* Endress (E 1423), 16 *Parrotiopsis jacquemontiana* Rehd. (E 718), 17 *Fothergilla major* Lodd. (E 8.5.1978), 18 *Parrotia persica* C. A. Mey. (E 3567), 19 *Sycopsis sinensis* Oliv. (E 3199), 20a *Distyliopsis dunnii* Endress (Steward and Cheo 170), 20b *D. laurifolia* Endress (Henry 11365A), 21 *Distylium racemosum* Sieb. et Zucc. (E 3244), 22a *Matudaea trinervia* Lund. (E 1236), 22b *M. hirsuta* Lund. (E 1420).

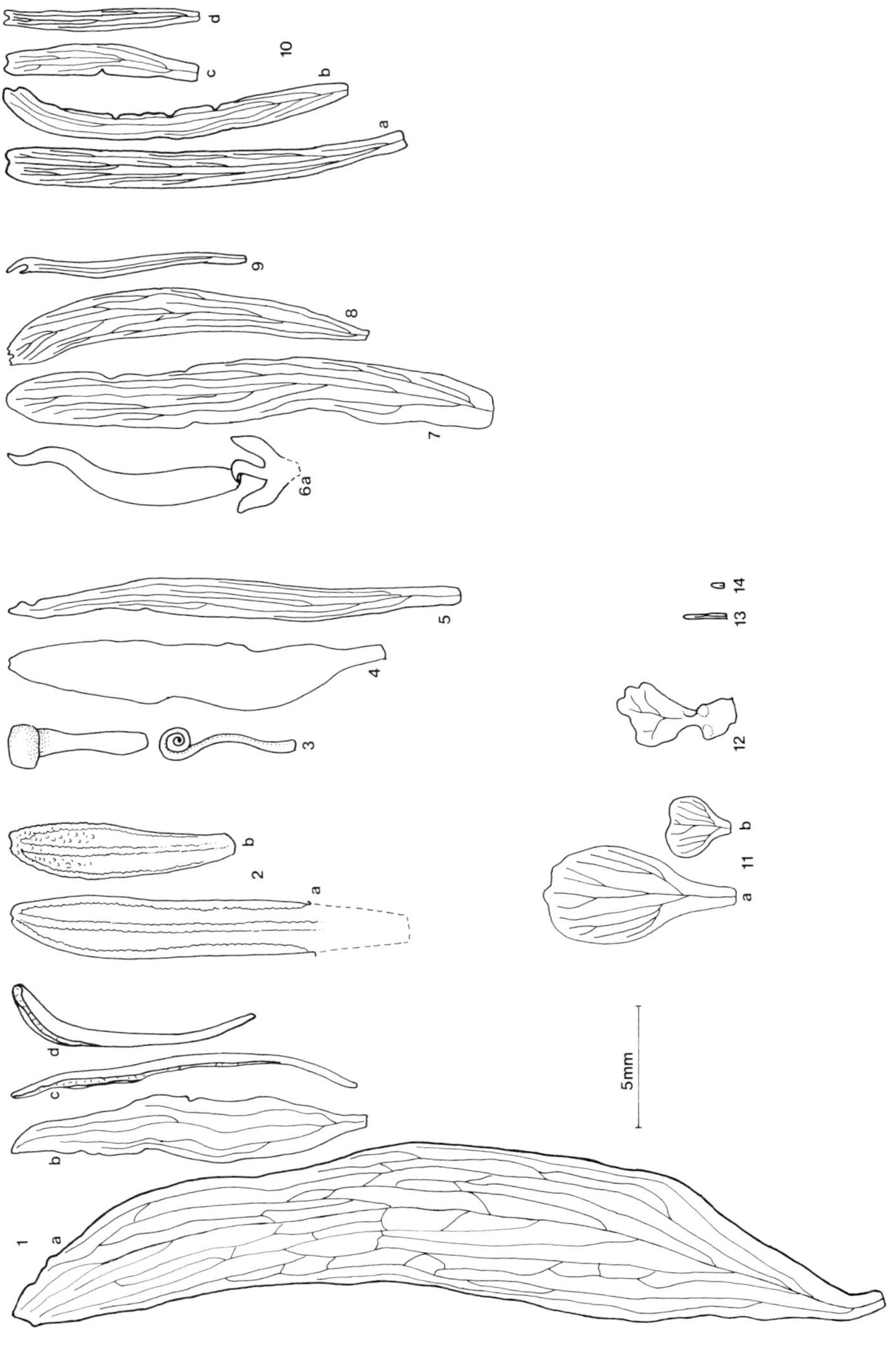
1 a
b
c
d
2 a
b
3
4
5
6a
7
8
9
10 a
b
c
d
11 a
b
12
13
14
5mm

T. crinitus has relatively small ones (Fig. 12.1:1a,d). In *Sinowilsonia* the petals are tiny rudiments and are often absent (Fig. 12.1:14, and Fig. 12.4B, cf. also Endress 1967).

Petals are mostly linear. In Hamamelideae they are usually long and circinately coiled in bud. Geniculate petals with paired lateral (dorsal) effigurations are shared by the phylogenetically remote *Embolanthera* (Fig. 12.1:6a) and *Eustigma* (Fig. 12.1:12 and Fig. 12.4C). The dorsal effigurations are (nectariferous?) swellings in *Eustigma* and apparently thin auricles in *Embolanthera*. Only *Corylopsis* has simple more or less orbicular petals (Fig. 12.1: 11a,b).

Stamens (Fig. 12.2)

Stamens vary much in size, even within genera, such as in *Corylopsis* (Fig. 12.2:11a,b). Anther size tends to be large in wind-pollinated genera (e.g. Fig. 12.2:18–22) and relatively small in entomophilous groups with more intimate synorganization of floral parts (Fig. 12.2:1–10).

The anther connective protrusion, a constitutive feature in the family, varies considerably in length. Plasticity occurs in some instances within a genus, for example *Trichocladus* (*T. goetzei* with considerable, *T. crinitus* with minute protrusion, Fig. 12.2:1a,d) or even within a single flower, for example in *Parrotia persica*, where the connective apex may be blunt or have a longer protrusion (cf. Endress 1968). On the other hand, connective protrusion length may be relatively constant in certain genera or even in groups of genera. It is invariably very long in the Loropetalinae (Fig. 12.2:6–9).

There are also functional correlations: long connective protrusions occur only in those anemophilous genera with long anthers (part of Fothergilleae, Fig. 12.2:18,19,22) and in those entomophilous genera with short styles (part of Hamamelideae, Fig. 12.2:3,6–9), but not in all of them. In these Hamamelideae the connivent anther protrusions cover the stigmas at anthesis and so probably play some role in pollination biology. Therefore, it is necessary to study the form of the connective protrusion in the context of the entire floral architecture. In these entomophilous Hamamelideae protrusion length is relatively constant because of the synorganization of parts. In contrast, the greater variability within individuals in the mentioned anemophilous groups may be due to the lack of synorganization; the relative length of the connective protrusion may be a mere by-product of general anther elongation. In contrast to these plastic features, the pattern of anther dehiscence is highly stable throughout genera and often even higher categories (cf. Fig. 12.2 and next section).

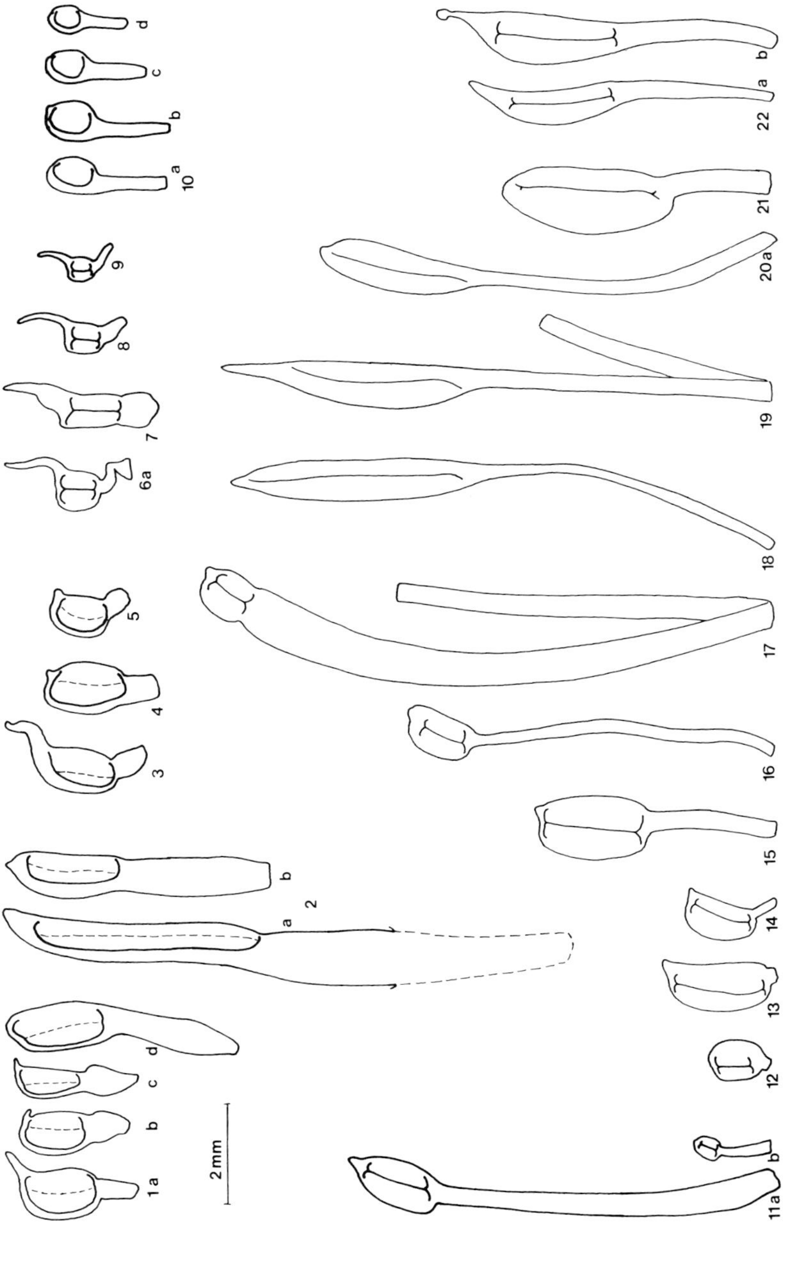

Fig. 12.2 Stamens of Hamamelidoideae. Stippled: septum between the two pollen sacs of a theca in 1–5, and fused base of filament in 2a. All with dorsal side on the left. In 17 and 19 lower part of filament artificially bent upwards. See legend to Fig. 12.1 for key to genera and species.

Staminodes

Inner staminodes that alternate with the stamens occur in *Hamamelis*, *Maingaya*, *Ostrearia*, and *Dicoryphe*. In *Corylopsis*, the situation is again very variable in that some species rarely have staminodia (e.g. *C. spicata*), while in others they are quite frequent (e.g. *C. willmottiae*) (cf. Endress 1967). Supernumerary inner androecium members differentiated as staminodia are not uncommon in various genera (Fig. 12.4B,D). It seems that the ability to produce supernumerary stamens or staminodes towards the gynoecium is inherent in many Hamamelideae but not always expressed. The degree of expression is sometimes labile.

Nectary disc

A nectary disc in the form of about ten (in pentamerous flowers) fleshy knobs occurs in *Corylopsis*, in the Loropetalinae (except *Embolanthera*), and in *Dicoryphe* (Fig. 12.4E). In *Dicoryphe* they are covered by hairs but they show the same rich vasculature as it typical for secretory organs (Rakotobe 1987; see also Bogle 1984 for *Maingaya*). These organs arise late in ontogeny (Endress 1967 for *Corylopsis*; Mione and Bogle 1987 for *Loropetalum*). This may be the reason for the great plasticity in shape and number within individuals, highly influenced by the formation of the androecium in the individual flowers (Fig. 12.4E).

Gynoecium (Fig. 12.3)

Style length and stigma size of the dimerous gynoecium is plastic. In long-styled carpels the stigma is often widely decurrent. In some entomophilous groups it is more restricted to the apical region (*Hamamelis*, *Corylopsis*, Fig.12.3:10b,11a). In contrast, in *Eustigma* (Fig. 12.3:12), it is exceedingly large and optically attractive.

The position of the ovary is unusually labile in the subfamily. The entire range between superior and inferior occurs. Even within a genus this lability is considerable: *Corylopsis henryi* has an almost superior and *C. spicata* an almost inferior ovary (Harms 1930; Morley and Chao 1977), *Fothergilla* has a superior to one-third inferior ovary (Weaver 1969; Bogle 1970). On the other hand, in some groups of genera there is a tendency of a more stabilized position; more or less inferior in the Loropetalinae (except *Embolanthera*), superior in *Sycopsis*, *Distyliopsis*, *Distylium*, and *Matudaea* (Endress 1970).

There is a peculiar trend of ovary retardation resulting in a conspicuous delay of fertilization after anthesis, as shown by several temperate genera, such as *Sycopsis*, *Parrotia* of the Fothergilleae,

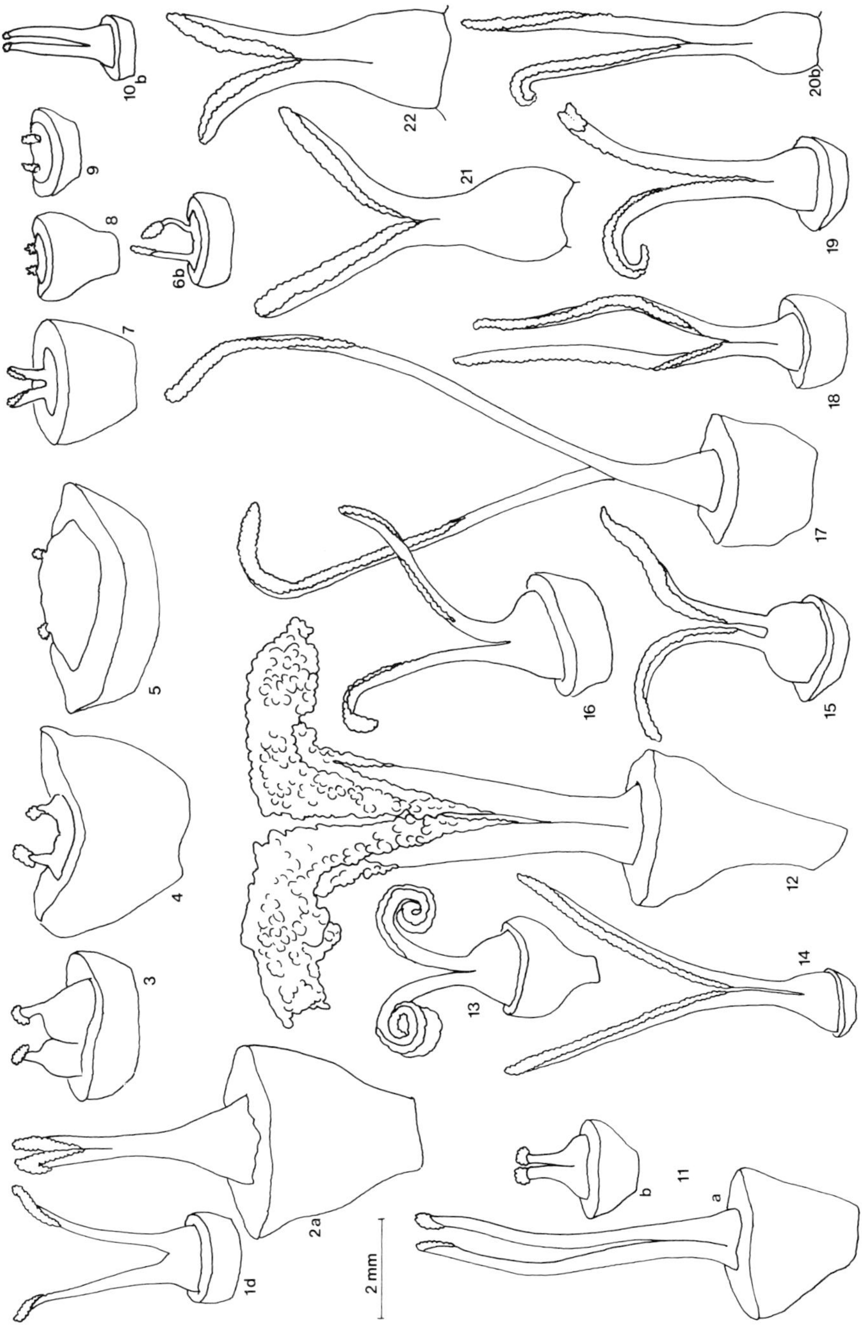

Fig. 12.3 Gynoecia of Hamamelidoideae. See legend to Fig. 12.1 for key to genera and species.

Sinowilsonia of the Eustigmateae, and *Hamamelis* of the Hamamelideae (cf. small ovaries in Fig. 12.3:10b,14,18,19, see also Endress 1977). This is always correlated with a protraction of anthesis to early spring or winter. This feature is well established in the Betulaceae (and other 'higher' Hamamelididae) and may well be an important preadaptation among the 'lower' Hamamelididae for the successful evolution of these families (Endress 1967, 1977).

Fusions in perianth and androecium

There is a trend to fusion of sepals into a tubular structure that ruptures at anthesis or falls off by a circumscissile detachment. This occurs among the Loropetalinae in *Maingaya* and *Embolanthera*, among the Dicoryphinae in *Neostrearia*, *Dicoryphe*, and in *Trichocladus grandiflorus* and *T. goetzei*, but apparently not in the other species of *Trichocladus*. Sepals are partly fused also in Fothergilleae. In *Parrotia persica* differences between individuals in the amount of fusion are considerable.

A very notable anagenetic trend is the congenital fusion of petals and stamens (and staminodes) to form a tube and the tendency to produce salverform flowers as a further elaboration of mere sepal fusion. This occurs in *Embolanthera*, especially in *E. glabrescens* (Tardieu-Blot 1965), and still more in *Dicoryphe*, especially *D. stipulacea*, while *D. viticoides* for example show almost no fusion (Fig. 12.1 and 12.2:2a,b, see also Rakotobe 1987).

Fruits

Fruits are uniform through the Hamamelidoideae in that they apparently all show the same explosive seed discharge mechanism of the one-seeded carpels. Perhaps also the high constancy of bicarpellate gynoecia is due to a constraint imposed by this highly specialized seed discharge mechanism. A certain plasticity is shown in the size, indumentum, and direction and length of the beaks formed by the stylar remains. In most cases these features do not seem to be constant for groups above the generic level.

Leaves

Tropical and subtropical representatives have persistent, brochidodromous leaves that are sometimes semicraspedodromous toward the apex. Temperate genera have deciduous, craspedodromous leaves with platanoid teeth (terminology after Hickey and Wolfe 1976). Both types occur in all three tribes Fothergilleae, Eustigmateae, and Hamamelideae but they are more or less consistent within a genus. The easy

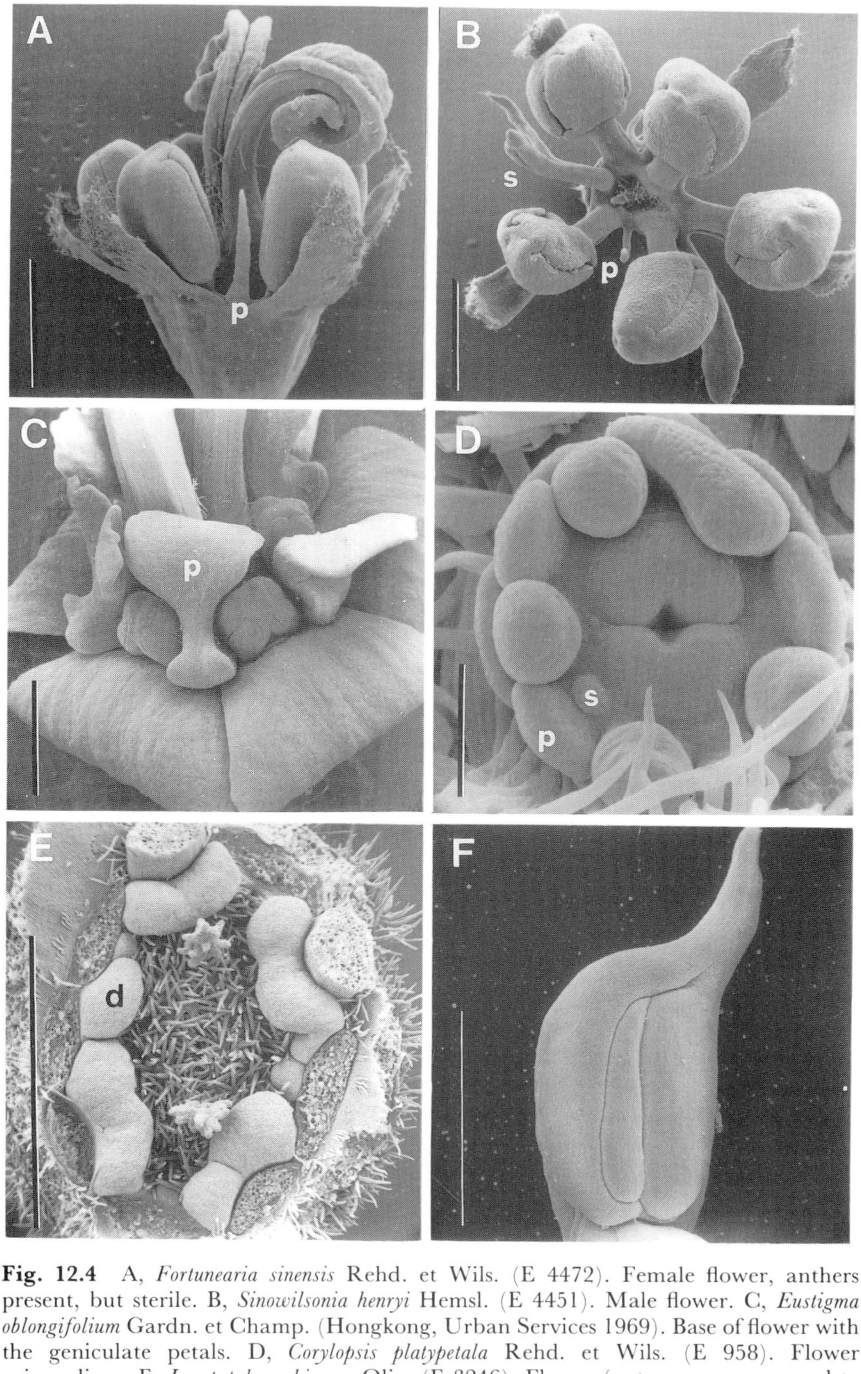

Fig. 12.4 A, *Fortunearia sinensis* Rehd. et Wils. (E 4472). Female flower, anthers present, but sterile. B, *Sinowilsonia henryi* Hemsl. (E 4451). Male flower. C, *Eustigma oblongifolium* Gardn. et Champ. (Hongkong, Urban Services 1969). Base of flower with the geniculate petals. D, *Corylopsis platypetala* Rehd. et Wils. (E 958). Flower primordium. E, *Loropetalum chinense* Oliv. (E 3246). Flower (outer organs removed to show nectary disc). F, *Noahdendron nicholasii* Endress et al. (Tracey 14945). Anther before dehiscence.

(d, nectary disc; p, petal; s, staminode; E. Endress; scale bars = 1 mm, in D 0.1 mm.)

transition between the two types is nicely shown by the intermediate pattern in the hybrid × *Sycoparrotia semidecidua* whose parents *Sycopsis sinensis* and *Parrotia persica* are representatives of the two types (Endress 1968).

Loropetalum chinense has the smallest leaves of the Hamamelidaceae, which often attain only 1–2 cm in length. In contrast, the closely related *Maingaya* has large leaves that may reach 25 cm in length (cf. Vink 1957). Similarly large leaves are found in *Noahdendron* and *Sinowilsonia*. A tendency to form peltate leaves, mainly in rapidly growing long shoots, occurs in *Molinadendron*, *Trichocladus*, and *Dicoryphe*.

Wood anatomy

Wood anatomy is relatively uniform in the subfamily and at the same time shows plasticity to such a degree that it does not provide substantial new aspects for phylogenetic considerations among the subfamily (cf. Reinsch 1890; Tang 1943; Huang 1986).

Reconstructing phylogenetic relationships (Fig. 12.5)

The most extensive taxonomic treatment covering the entire subfamily (and family) is that by Harms (1930). It was slightly amended, although less comprehensively, by Schulze-Menz (1964). I recognize four subfamilies in the Hamamelidaceae: Hamamelidoideae, Exbucklandioideae (incl. Disanthoideae), Altingioideae, and Rhodoleioideae (Endress 1989b).

It is somewhat arbitrary whether one takes Exbucklandioideae or Rhodoleioideae as an outgroup for the Hamamelidoideae within the Hamamelidaceae. It does not matter much for the phylogenetic interpretation of the Hamamelidoideae. I chose here Exbucklandioideae.

The Hamamelidoideae are characterized by one ovule per carpel (very rarely two additional reduced ones) and ballistic dispersal (*vs.* five to eight ovules per carpel and non-ballistic dispersal in Exbucklandioideae), by pinnately veined, simple leaves (*vs.* palmately veined, with a tendency toward a tricuspidate shape), by moderately small, deciduous stipules (*vs.* large, persistent stipules), and by a chromosome base number of $x=12$ (*vs.* $x=8$) (cf. Goldblatt and Endress 1977). In the other two subfamilies, Liquidambaroideae and Rhodoleioideae, the ovule number is still higher, and the flowers are arranged densely in globular or semiglobular heads (cf. Bogle 1986). There is possibly a functional correlation between these two features in that the explosive mechanism, as present in Hamamelidoideae, can only work with one-seeded carpels and with fruits that are sufficiently spaced and not crowded together and partially immersed in compact heads.

Hamamelideae

The tribe Hamamelideae is characterized (among the subfamily) by its long, linear petals that are circinately coiled in bud. For its phylogenetic interpretation we are fortunate in having some key features.

A plesiomorphic feature for the family Hamamelidaceae (and the entire subclass Hamamelididae) is that each anther theca opens with two valves like two window shutters (cf. Endress 1986, 1989a; Hufford and Endress 1989). This feature is fairly widespread throughout the family, but there is a unique exception in five genera of the tribe Hamamelideae (*Ostrearia*, *Neostrearia*, *Noahdendron*, *Dicoryphe*, and *Trichocladus*). Here, the thecae open by a single valve that serves both pollen sacs (Fig. 12.1:1–5 and Fig. 12.4F; cf. Endress *et al.* 1985). This construction is outstanding because it occurs nowhere else in the angiosperms, except in some way in *Podophyllum* (Berberidaceae) (Endress and Hufford 1989). A second peculiarity is that all these five genera of the Hamamelideae have an exclusive Gondwanan distribution (Australia, Madagascar, Comores, Africa), while all other genera of the family are Laurasian.

The coincidence of these two unique features (in addition to other similarities) makes it highly probable that the five genera form a phylogenetic unit among the family (subtribe Dicoryphinae, Endress 1989*b*). Further critical tendencies in this unit are i) especially thick, fleshy petals with an uneven surface that may, in some cases, be eaten by pollinating insects (*Dicoryphe*, *Trichocladus p.p.*, *Noahdendron*, slightly in *Neostrearia*), and ii) persistent and often large stipules that cover subsequent buds (*Dicoryphe p.p.*, *Trichocladus p.p.*, *Noahdendron*).

The Dicoryphinae consist of two natural subunits: a west Gondwanan group (*Dicoryphe*: Madagascar, Comores: *Trichocladus*: Africa) and an east Gondwanan group (*Ostrearia*, *Neostrearia*, *Noahdendron*: Northern Queensland). Again some critical features or tendencies distinguish these two subgroups: the western genera (*Dicoryphe*, *Trichocladus*) show both a strong trend to opposite leaf arrangement (unique in the family!) and to peltate leaves. They further tend to have the flowers in heads (up to about 20, but usually less) (*vs.* spikes with more than about 24 flowers in the eastern genera). The petals tend to be only incurved in bud (except some *Trichocladus* species) (*vs.* strongly circinate in the eastern genera). Some species of both genera have hairs on the petals (*vs.* glabrous petals in the eastern genera). The ovary is semi-inferior to inferior (*vs.* superior to semi-inferior in the eastern genera). The styles are moderately long (*vs.* very short in the eastern genera).

Of the three genera of the eastern subgroup *Neostrearia* and *Noahdendron* are closer to each other than to *Ostrearia*. Both have three ovules (two of them sterile) and this is almost unique in the subfamily (but also in *Corylopsis*, cf. Endress 1989*a*). Both genera also lack staminodes (*vs.* *Ostrearia*) (Smith 1958; Endress *et al.* 1985). All this

---

Fig. 12.5 Phylogenetic tree of the Hamamelidoideae. Thick branches: well-established relationships. Thin branches: less certain relationships. Thick bars: key features or remarkable anagenetic trends. Thin bars: other features. Features: 1 Ovules 5–8 per carpel, 2 Leaves palmately veined, 3 Stipules large, ±persistent, 4 Chromosome base number $x=8$, 5 Ovule 1 per carpel, 6 Leaves pinnately veined, 7 Stipules moderately sized, deciduous, 8 Flowers with sepals and petals, ±pentamerous, 9 Chromosome base number $x=12$, 10 Petals absent, 11 Flower organs variable in number, 12 Stigma large, 13 Anemophilous, 14 Inflorescence capitate, 15 Petals small, 16 Stigma large, 17 Anthers almost sessile, 18 Fruit stalked, large, lenticellate, with robust vertical beaks, 19 Inflorescence spicate, 20 3 ovules per carpel (2 sterile), 21 Sepals reduced, subtending bracts protecting flower buds, 22 Petals simple, orbicular, 23 Staminodes often present, 24 Nectary disc present, 25 Leaves deciduous, 26 Leaf venation intermediate between pinnate and palmate, 27 Inflorescence spicate, 28 Petals linear, circinately coiled in bud, 29 Sepals reduced, subtending bracts protecting flower buds, 30 Vegetative prophylls 1, 31 Leaves often peltate, 32 Andromonoecious, 33 Pollen finely reticulate and verrucate, 34 Anther dehiscence by longitudinal slits, 35 Centrifugal polyandry present, 36 Entomophilous, 37 Leaves deciduous, 38 Sepals absent, 39 Ovary superior, 40 Inflorescence paniculate (or botryoid), 41 Anthers long, 42 Floral cup absent, 43 Pollen pluriaperturate, 44 Anther dehiscence ±valvate, 45 Stamens with hairs, 46 Centripetal polyandry present, 47 Floral bract pair below flower fused, 48 Flowers bisexual, 49 Vegetative prophylls 2, 50 Ovary superior, 51 Leaves deciduous, 52 Filaments thick, conspicuous, 53 Inflorescence spicate, 54 Bracts below inflorescence large, conspicuous, 55 Stigma exceedingly large, conspicuous, 56 Petals obcordate, geniculate, with 2 dorsal swellings, 57 Petals linear, reduced, 58 Flowers unisexual, 59 Anemophilous, 60 Leaves deciduous, 61 Leaves with large number of teeth, 62 Petals minute, often lacking, 63 Fruit not lenticellate, 64 Male and female inflorescences dimorphic, 65 Thecae 1-locular, 1-valvate, 66 Flowers tetramerous, 67 Staminodes present, nectariferous, 68 Inflorescence capitate, 69 Leaves deciduous, 70 Temperate North America and eastern Asia, 71 Connective protrusions long, 72 Styles short (much shorter than stamens), 73 Tropical and subtropical Asia, 74 Thecae 2-locular, 1-valvate, 75 Gondwana, 76 Ovary semi-inferior to inferior, 77 Nectary disc present, 78 Inflorescence capitate, 79 Petals and stamens fused, 80 Sepals fused, 81 Petals geniculate, with two lateral auricles, 82 Inflorescence spicate, 83 Bracts (stipules) fimbriate, glandular, 84 Inflorescence few-flowered, 85 Bracts (stipules) fimbriate, glandular, 86 Sepals fused, 87 Staminodes present, 88 Australia, 89 Styles short, 90 Inflorescence spicate, 91 West Gondwana, 92 Leaves often decussate, 93 Stipules persistent, 94 Sepals often fused, 95 Styles moderately long, 96 Inflorescence capitate, 97 Staminodes present, 98 3 ovules per carpel (2 sterile), 99 Sepals fused, 100 Petal apex tightly coiled at anthesis, 101 Connective protrusions long, 102 Stipules large, persistent, 103 Petals and stamens fused, 104 Petals not incurved in bud, valvate, 105 Flowers tetramerous, 106 Staminodes present, 107 Nectary disc present, 108 Madagascar, Comores, 109 Africa.

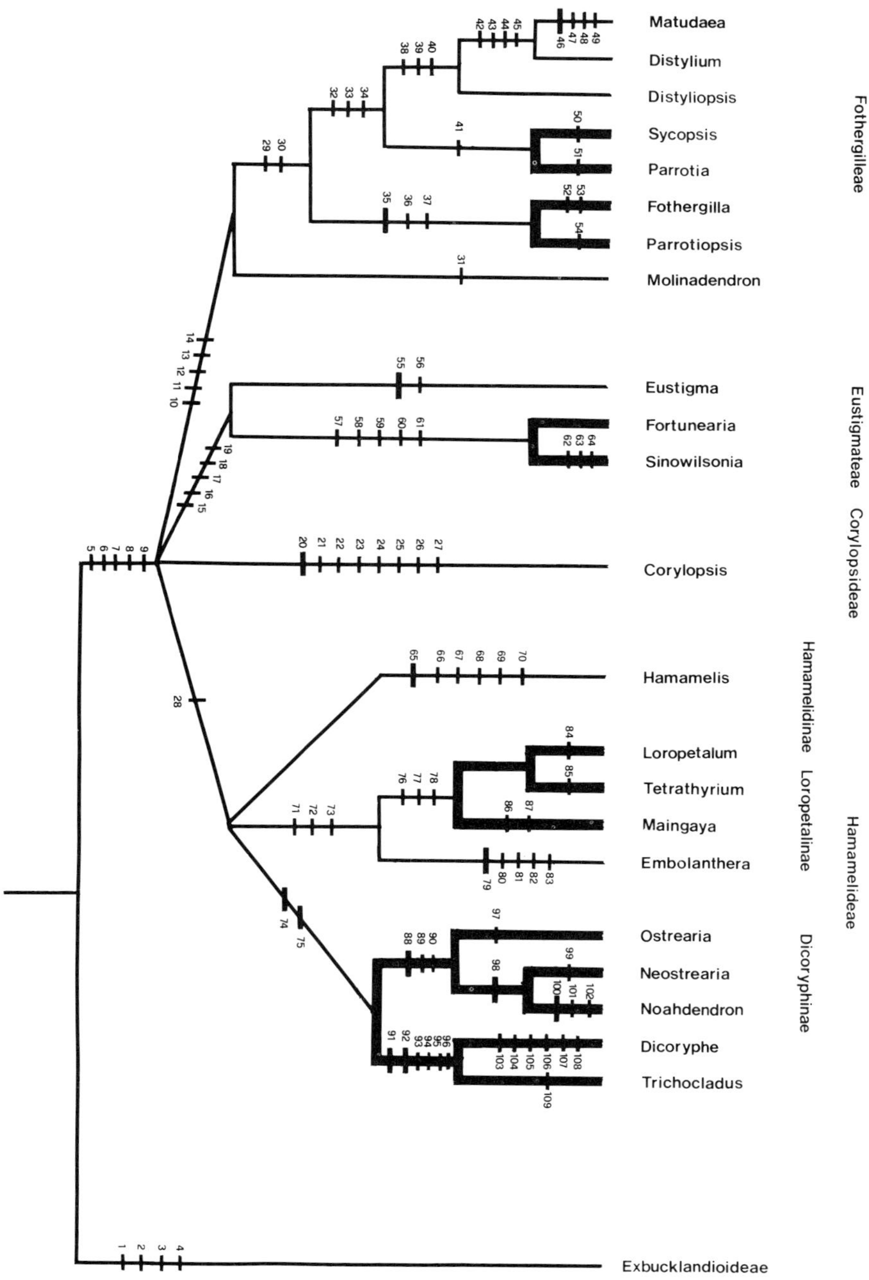
Matudaea
Distylium
Distyliopsis
Sycopsis
Parrotia
Fothergilla
Parrotiopsis
Molinadendron
Fothergilleae
Eustigma
Fortunearia
Sinowilsonia
Eustigmateae
Corylopsis
Corylopsideae
Hamamelis
Hamamelidinae
Loropetalum
Tetrathyrium
Maingaya
Embolanthera
Loropetalinae
Hamamelideae
Ostrearia
Neostrearia
Noahdendron
Dicoryphe
Trichocladus
Dicoryphinae
Exbucklandioideae

evidence allows the construction of a fairly well-founded phylogenetic tree for the subtribe Dicoryphinae.

At this point it may be noted that some parallel features between Dicoryphinae and Exbucklandioideae occur: i) Presence of more than one ovule per carpel (*Neostrearia*, *Noahdendron*, as in all Exbucklandioideae); ii) presence of large, persistent stipules (*Noahdendron*, *Dicoryphe p.p.*, as in all Exbucklandioideae; cf. Endress *et al.* 1985, Rakotobe 1987); iii) very coarsely reticulate pollen (*Dicoryphe*, as in *Exbucklandia*; cf. Chang 1964, Bogle and Philbrick 1980); iv) densely packed gynoecia immersed in short spikes and with short stigmas (*Ostrearia*, as in *Mytilaria*; cf. Chang 1948); v) nectar on the petal base (*Noahdendron*, as in *Disanthus*). These features are not consistent and are somewhat erratic in the two groups. It is, therefore, uncertain whether they point to a closer phylogenetic relationship of the Dicoryphinae and the Exbucklandioideae. If so, the relationship would be based on homoplasies and therefore not be on the cladogenetic side but rather on the anagenetic side of phylogeny.

The remaining five genera of the tribe Hamamelideae can be allocated to two further subtribes. *Hamamelis* is very isolated and forms a subtribe of its own (Hamamelidinae, Endress 1989*b*). The thecae have only one pollen sac and open by a single valve. This is unique for the subfamily but is shared with *Exbucklandia* (Exbucklandioideae). Further, the connective tip is minute (Fig. 12.2:10), and the dorsiventrally flat, nectariferous staminodes are unique. *Hamamelis* is the only temperate, deciduous genus of its tribe.

The third subtribe, Loropetalinae (Endress 1989*b*) is made up by the (sub)tropical Asiatic genera *Loropetalum*, *Tetrathyrium*, *Maingaya*, and *Embolanthera*. In contrast to the two other subtribes, they have retained the primitive anther opening pattern with two valves per theca (Fig. 12.2:6–9). All have short styles (Fig. 12.3:6–9) but very long connective protrusions (Fig. 12.2:6–9) that cover the stigmas at anthesis and make them accessible (together with the disc nectaries, Fig. 12.4E) only through narrow slits between the anthers (in *Maingaya* additional laterally flattened staminodes contribute toward concealing of the floral centre). The ovaries are semi-inferior to inferior. The group has highly heterogeneous uniseriate rays and scanty wood parenchyma in the stem (Tang 1943, *Embolanthera* not studied).

The two closest genera are *Loropetalum* and *Tetrathyrium*, both with disc nectaries but lacking staminodes, and both with free sepals. *Maingaya* comes next with additional staminodes and with fused sepals. *Embolanthera* seems to stand somewhat more apart (although lacking staminodes) with fused sepals and fused petals and stamens, the petals being geniculate and auriculate (Fig. 12.1:6a). Apparently, anagenesis has somewhat obliterated the traces of relationship and its exact place within the Loropetalinae is more difficult to establish than for the other three genera.

Fothergilleae

The second of the two larger tribes, the Fothergilleae, comprising eight genera, is less diverse and is more difficult to resolve into systematic subunits (cf. Bogle 1970; Endress 1968, 1970, 1978). The only clear subgroups are two pairs of genera (*Fothergilla/Parrotiopsis*, and *Parriotia/Sycopsis*). The tribe is characterized by its Laurasian, mostly subtropical and temperate distribution, and by its strong tendency towards anemophily and andromonoecious condition. In general, number of floral organs is highly variable, even within individuals. Petals are absent. Styles are long with large, decurrent stigmas (Fig. 12.3:15–22). Anthers are large, often long (Fig. 12.2:15–22). The valvate opening pattern is somewhat obliterated in some long-anthered genera and reversed to almost simple longitudinal slits (Fig. 12.2:18–21). The calyx is small, not covering the inner floral organs in bud (except *Molinadendron*); in three genera (*Distyliopsis*, *Distylium*, *Matudaea*) it has completely disappeared (Endress 1970, 1978). Lateral shoots in the vegetative region have only one prophyll (two in *Molinadendron* and *Matudaea*) (Endress 1970).

In the pair *Fothergilla/Parrotiopsis* secondary entomophily is enhanced by inflation of the filaments in *Fothergilla* and enlargement of white bracts in *Parrotiopsis* (both conspicuous secondary optical attractions), and further by a decrease of anther size (Fig. 12.2:16,17) and increase of stamen number per flower by centrifugal polyandry (Endress 1976). In contrast, polyandry is centripetal in the (probably anemophilous) *Matudaea* (Endress 1976). Therefore, polyandry (around 20 stamens) in Fothergilleae has clearly arisen independently more than once. Also, *Fothergilla* and *Parrotiopsis* are both deciduous and have similar leaf structure.

In *Parrotia* and *Sycopsis* floral structure is very similar. The intergeneric hybrid × *Sycoparrotia* (Endress 1968) documents their intimate relationship and militates against the traditional separation of the apetalous Hamamelidoideae just between these two genera into Fothergilleae (or Parrotieae) and Distylieae (cf. Harms 1930). The anemophilous taxa are further characterized by finely reticulate and slightly verrucate pollen, some with increase in aperture number (Endress 1977; Bogle and Philbrick 1980).

The largest anagenetic change in the tribe has resulted in some way in *Matudaea*. Polyandry, loss of the entire perianth, and, in its place, involucral elaboration and fusion of the uppermost bract pair below each flower; also the extremely fine reticulation of the pluriaperturate pollen grains are characteristics of the genus.

Eustigmateae

The remaining genera of the Hamamelidoideae are more isolated, except the closely related pair *Fortunearia*/*Sinowilsonia*. They are both temperate, deciduous shrubs from China. They have thin leaves with an unusually large number of fine, setose teeth, since not only the side veins of the first order but also of higher orders terminate in teeth. Flowers are borne on long spikes and are unisexual. The occurrence of female flowers is exclusive for the subfamily. (Earlier claims for other genera are erroneous, as shown for *Sycopsis*, cf. Endress 1968, *Trichocladus*, cf. Verdcourt 1971, and *Distylium*, cf. Endress 1978; these misinterpretations are based on postanthetic bisexual flowers with the stamens fallen but the stigmas still turgescent). The flowers are pentamerous with protecting sepals but very small, linear petals, which are often even absent in *Sinowilsonia* (Fig. 12.1:13,14 and Fig. 12.4A,B). *Sinowilsonia* has pendent, catkin-like inflorescences with a moderate dimorphism between larger male and smaller female ones, thus exhibiting a striking parallel to more elaborate catkin-bearing families of the 'higher' Hamamelididae.

Eustigma seems to be distantly related to *Fortunearia*. Although it deviates by its exceedingly large purple stigmas (Fig. 12.3:12) that probably play a role in optical attraction of pollinators (flies?) and the geniculate petals with two dorsal (nectariferous?) swellings (Fig. 12.1:12 and Fig. 12.4C), there are a number of common features: pentamerous, pedicellate flowers, covering sepals, small petals, almost sessile anthers with markedly valvate dehiscence, large lenticellate fruits (cf. Anonymous 1977) with robust vertical beaks, phloem in the axis containing libriform fibre groups (Tong 1930).

Corylopsideae

The last genus, *Corylopsis*, stands quite isolated from all other genera, despite some superficial similarities with *Fortunearia* (Harms 1930 erected Corylopsideae including both genera) (see also Endress 1970). I classify *Corylopsis* in a tribe of its own (Endress 1989*b*). It is the only member of the entire subclass Hamamelididae with petals of a shape that is common in many dicot groups for example in Rosidae: more or less orbicular, and markedly retarded in development before anthesis (Fig. 12.1:11). It has three ovules per carpel (two of them sterile), a nectariferous disc, and very labile staminodes. Fruits are small and have long, horizontally divergent beaks. Petal, stamen, and style length is quite variable, as is the ovary position (Fig. 12.1:11a,b, Fig. 12.2:11a,b, Fig. 12.3:11a,b). The deciduous leaves are more or less cordate and show a tendency toward palmate venation, as is typical for the Exbucklandioideae (see also Skvortsova 1975). The connective protrusion is located in the centre of the anther, much as in Trochodendraceae, and not displaced towards the adaxial side, as in most Hamamelidoideae (Endress 1986; Hufford and Endress 1989). *Corylopsis* has especially numerous traits or tendencies in common with the Betulaceae and seems to occupy a key position for understanding the relationships between the 'lower' and 'higher' Hamamelididae, as has been worked out earlier (Endress 1967, 1977).

Conclusions

The taxonomic treatments of the Hamamelidaceae by Harms (1930) and Schulze-Menz (1964) are a good basis for our understanding of the subdivision of the Hamamelidoideae. However, a number of changes have become necessary: i) the Hamamelideae can be subdivided into three subtribes: Hamamelidinae, Loropetalinae, and Dicoryphinae (cf. Endress 1989b). ii) the separation of the apetalous genera into Fothergilleae (or Parrotieae) and Distylieae is artificial and can not be maintained. iii) *Fortunearia* and *Sinowilsonia* should be associated with *Eustigma* rather than with *Corylopsis*.

Tracing a rough phylogenetic history, it seems that the Hamamelidoideae originated in Laurasia. Fossil flowers with univalvate thecae of a probable Hamamelideae are documented from the Upper Cretaceous of Sweden (Friis 1985). Some ancient Hamamelideae moved into Gondwanaland, where the Dicoryphinae originated somewhere in mid- (or Lower?) Cretaceous.

Among the first Hamamelidoideae probably pentamerous flowers with sepals and petals occurred (including some plasticity of course!). A double perianth probably was not yet present in the first representatives of the subclass Hamamelididae (Ehrendorfer 1977; Endress 1986) but had been acquired at least by the Proto-Hamamelidaceae. This is suggested by the following facts: i) narrow petals occur in all subfamilies, except Altingioideae. ii) the extant apetalous forms of the Hamamelidoideae are limited to Asia and North America and are a much more uniform group than the petaliferous ones. They probably originated there in the temperate zone together with anemophily. Some reverted to entomophily with novel optical attractants (*Fothergilla* and *Parrotiopsis*). Another subgroup spread secondarily into the tropics (especially

Distylium and *Distyliopsis* in Malesia). Possibly Betulaceae (and Fagaceae) originated in Laurasia from some Hamamelidoideae in mid-Cretaceous.

Acknowledgements

For support in the field or for supply of valuable material I am grateful to B. P. M. Hyland, A. Molina, F. S. P. Ng, E. A. Rakotobe, and the Georges-und-Antoine-Claraz-Schenkung. I also thank R. Siegrist for skilful microtomy and U. Jauch for his support with the SEM.

References

Anderson, E. (1933). Variation in flower color in *Hamamelis vernalis*. *Journal of the Arnold Arboretum* **14,** 253–7.

Anonymous (1977). Hamamelidaceae. In *Flora Yunnanica* **1,** pp. 106–40. Academia Sinica, Beijing (In Chinese).

Bogle, A. L. (1970). Floral morphology and vascular anatomy of the Hamamelidaceae: The apetalous genera of Hamamelidoideae. *Journal of the Arnold Arboretum* **51,** 310–66.

Bogle, A. L. (1984). Floral morphology and vascular anatomy of *Maingaya* Oliv. (Hamamelidaceae, Hamamelidoideae, Hamamelideae). *American Journal of Botany* **71,** 19.

Bogle, A. L. (1986). The floral morphology and vascular structure of the Hamamelidaceae: subfamily Liquidambaroideae. *Annals of the Missouri Botanical Garden* **73,** 325–47.

Bogle, A. L. and Philbrick, C. T. (1980). A generic atlas of hamamelidaceous pollens. *Contributions of the Gray Herbarium* **210,** 29–103.

Bradford, J. L. and Marsh, D. L. (1978). Comparative studies of the witch hazels, *Hamamelis virginiana* L. and *H. vernalis* Sarg. *Proceedings of the Arkansas Academy of Science* **31,** 29–31.

Chang, H.-T. (1973). A revision of the hamamelidaceous flora of China. *Sunyatsen University Bulletin* **1,** 54–71 (In Chinese).

Chang, H.-T. (1979). Hamamelidaceae. In *Flora Reipublicae Popularis Sinicae* **35(2),** (ed. H.-T. Chang and S. Z. Yau), pp. 36–116. Academia Sinica, Beijing (In Chinese).

Chang, T.-T. (1964). Pollen morphology of Hamamelidaceae and Altingiaceae. *Trudy Akademija Nauk SSSR* (Moscow and Leningrad), Ser. 1, **13,** 173–232 (In Russian; English summary).

Ehrendorfer, F. (1977). New ideas about the early differentiation of angiosperms. *Plant Systematics and Evolution, Supplement* **1,** 227–34.

Endress, P. K. (1967). Systematische Studie über die verwandtschaftlichen Beziehungen zwischen den Hamamelidaceen und Betulaceen. *Botanische Jahrbücher für Systematik* **87,** 431–525.

Endress, P. K. (1968). Zur systematischen Stellung von *Parrotia* C. A. Mey. und *Sycopsis* Oliv. und zum neuen Bastard × *Sycoparrotia semidecidua* P. Endress et J. Anliker. *Schweizerische Beiträge zur Dendrologie* **16–18,** 11–28.

Endress, P. K. (1970). Die Infloreszenzen der apetalen Hamamelidaceen, ihre grundsätzliche morphologische und systematische Bedeutung. *Botanische Jahrbücher für Systematik* **90,** 1–54.

Endress, P. K. (1976). Die Androeciumanlage bei polyandrischen Hamamelidaceen und ihre systematische Bedeutung. *Botanische Jahrbücher für Systematik* **97,** 436–57.

Endress, P. K. (1977). Evolutionary trends in the Hamamelidales–Fagales-group. *Plant Systematics and Evolution, Supplement* **1,** 321–47.

Endress, P. K. (1978). Blütenontogenese, Blütenabgrenzung und systematische Stellung der perianthlosen Hamamelidoideae. *Botanische Jahrbücher für Systematik* **100,** 249–317.

Endress, P. K. (1986). Floral structure, systematics and phylogeny in Trochodendrales. *Annals of the Missouri Botanical Garden* **73,** 297–324.

Endress, P. K. (1989*a*). Aspects of evolutionary differentiation of the Hamamelidaceae and the Lower Hamamelididae. *Plant Systematics and Evolution* **162,** 193–211.

Endress, P. K. (1989*b*). A suprageneric taxonomic classification of the Hamamelidaceae. *Taxon* **38,** 371–6.

Endress, P. K. and Hufford, L. D. (1989). The diversity of stamen structures and dehiscence patterns among Magnoliidae. *Botanical Journal of the Linnean Society* **100,** 45–85.

Endress, P. K., Hyland, B. P. M. and Tracey, J. G. (1985). *Noahdendron*, a new Australian genus of the Hamamelidaceae. *Botanische Jahrbücher für Systematik* **107,** 369–78.

Friis, E. M. (1985). Structure and function in Late Cretaceous angiosperm flowers. *Det Kongelige Danske Videnskabernes Selskab, Biologiske Skrifter* **25,** 1–37.

Goldblatt, P. and Endress, P. K. (1977). Cytology and evolution in Hamamelidaceae. *Journal of the Arnold Arboretum* **58,** 67–71.

Harms, H. (1930). Hamamelidaceae. In *Die natürlichen Pflanzenfamilien* **18a** (2nd edn) (ed. A. Engler and K. Prantl), pp. 303–45, 487. Engelmann, Leipzig.

Hickey, L. J. and Wolfe, J. A. (1976). The bases of angiosperm phylogeny: vegetative morphology. *Annals of the Missouri Botanical Garden* **62,** 538–89.

Hooker, J. D. (1895). *Trichocladus grandiflorus*. *Curtis's Botanical Magazine* **51,** t. 7418.

Huang, G.-L. (1986). Comparative anatomical studies on the woods of Hamamelidaceae in China. *Acta Scientiarum Naturalium Universitatis Sunyatseni* **1,** 22–8 (In Chinese; English summary).

Hufford, L. D. and Endress, P. K. (1989). The diversity of anther structures and dehiscence patterns among Hamamelididae. *Botanical Journal of the Linnean Society* **99,** 301–46.

Marloth, R. (1925). *The flora of South Africa* **II.** Darter, Capetown.

Mione, T. and Bogle, A. L. (1987). Comparative ontogeny of the flowers of *Hamamelis virginiana* and *Loropetalum chinense* (Hamamelidaceae). *American Journal of Botany* **74,** 620–1.

Morley, B. and Chao, J.-M. (1977). A review of *Corylopsis* (Hamamelidaceae). *Journal of the Arnold Arboretum* **58,** 382–414.

Ng, F. S. P. (1984). Portraits of threatened plants. 7. *Maingaya malayana* Oliv. *Malayan Naturalist* **38**(2), 6.

Rakotobe, E. A. (1987). Approche systématique du genre *Dicoryphe* Dup.-Th. (Hamamélidacées) de Madagascar: Morphologie, anatomie, xylologie, palynologie, phytochimie flavonique, phytogéographie. Unpublished D. Phil. thesis. Université de Madagascar, Antananarivo.

Reinsch, A. (1890). Über die anatomischen Verhältnisse der Hamamelidaceae mit Rücksicht auf ihre systematische Gruppierung. *Botanische Jahrbücher für Systematik* **11,** 347–95.

Schulze-Menz, W. (1964). Rosales. In *A. Engler's Syllabus der Pflanzenfamilien* **II** (ed. H. Melchior), pp. 193–242. Borntraeger, Berlin-Nikolassee.

Skvortsova, N. T. (1975). Comparative morphological studies on representatives of the family Hamamelidaceae and their phylogenetic relationships. In *Problems of comparative morphology of the seed plants* (ed. L. J. Budantsjev), pp. 7–24. Nauka, Leningrad (In Russian).

Smith, L. S. (1958). New species of and notes on Queensland plants 3. *Proceedings of the Royal Society of Queensland* **69,** 43–51.

Tang, Y. (1943). Systematic anatomy of the woods of the Hamamelidaceae. *Bulletin of the Fan Memorial Institute of Biology,* New Series, **1,** 8-63.

Tardieu-Blot, M.-L. (1965). Hamamelidaceae. In *Flore du Cambodge, Laos et Vietnam* **4** (ed. A. Aubréville), pp. 75–116. Museum National d'Histoire Naturelle, Paris.

Tong, K. (1930). Studien über die Familie der Hamamelidaceae, mit besonderer Berücksichtigung der Systematik und Entwicklungsgeschichte von *Corylopsis. Bulletin of the Department of Biology, College of Science, Sun Yatsen University* **2,** 1–72.

Verdcourt, B. (1971). Hamamelidaceae. In *Flora of tropical East Africa,* (ed. E. Milne-Redhead and R. M. Polhill). Crown Agents for Overseas Governments and Administrations, London.

Vink, W. (1957). Hamamelidaceae. In *Flora Malesiana* **I, 5,** (ed. C. G. G. J. van Steenis), pp. 363–379. Nijhoff, Groningen.

Walker, E. H. (1944). A revision of *Distylium* and *Sycopsis* (Hamamelidaceae) *Journal of the Arnold Arboretum* **25,** 319–41.

Weaver, R. E. Jr. (1969). Studies in the North American genus *Fothergilla* (Hamamelidaceae). *Journal of the Arnold Arboretum* **50,** 599–619.

Yamanaka, T. (1986). *Corylopsis* in Japan and Korea. *Acta Phytotaxonomica Geobotanica* **37,** 97–105.

Yu, T.-T. (1983). *The Botanical Gardens of China.* Science Press, Beijing (In Chinese).

13. A survey of the Liquidambaroideae (Hamamelidaceae) with a view to elucidating its fossil record

DAVID K. FERGUSON

Department of Botany, University of Antwerp (RUCA), B-2020 Antwerp, Belgium

Abstract

An account is given of the morphology and anatomy of those plant parts of the Liquidambaroideae encountered in the fossil record, i.e. leaves, wood, staminate inflorescences, pollen, and fruiting heads. This information is then applied in a critical assessment of the fossil remains attributed to the Liquidambaroideae.

Characteristics of the Liquidambaroideae

The Liquidambaroideae occupy an isolated position within the Hamamelidaceae. Compared with the other representatives of the family the stems have relatively large vessels, comparatively broad wood-rays, and perimedullary secretory ducts. These canals extend into the leaves where they occupy a central position in the ring-shaped vascular bundle (U-shaped in the other genera; Metcalfe and Chalk 1950, p. 588). Although a few rudimentary stamens are found among the female florets, the flowers are essentially unisexual with the staminal section of the inflorescence occupying a terminal position. The Liquidambaroideae are anemophilous, with no perianth and a large number of stamens grouped into capitula. The anthers dehisce

Evolution, Systematics, and Fossil History of the Hamamelidae Volume 1: Introduction and 'Lower' Hamamelidae. (eds P. R. Crane and S. Blackmore), Systematics Association Special Volume No. 40A, pp. 249–72. Clarendon Press, Oxford, 1989.

longitudinally, releasing pantoporate pollen (most but not all the Hamamelidaceae have tricolpate pollen). The staminal racemes are then shed, leaving the essentially spheroidal carpellate inflorescences to develop into indurated fruiting heads, composed of anything up to 50 connate fruits. These bilocular fruits contain an indefinite number of seeds, most of which are infertile. The capsules open by septicidal and occasionally loculicidal dehiscence, releasing the mature winged seeds.

While the morphology and anatomy of the Liquidambaroideae display a number of unique features, this subfamily also differs from the rest of the Hamamelidaceae in its cytology. While chromosome numbers of $2n=16$, 24, 36, 48, 64, and 72 are encountered in the other subfamilies, those in the Liquidambaroideae prove to be $2n=30$, 32 (Goldblatt and Endress 1977, Morawetz and Samuel Vol 40A, Ch. 7). Since a basic number of $x=8$ is found not only in the Liquidambaroideae, but in the Exbucklandioideae and Disanthoideae as well, it would appear that the counts of $2n=30$ reported from both *Liquidambar styraciflua* L. and *L. orientalis* Mill. (Ernst 1963; Kesercioğlu 1973) are the result of chromosomal loss.

What little is known of the phytochemistry of the subfamily (only *Liquidambar* has been investigated to date) further underscores its distinctive nature. Thus certain features of the Shikimic Acid Pathway appear to differ from that found in other genera of the Hamamelidaceae. Shikimic acid itself can be derived from or converted into quinic acid by the loss or addition of a water molecule respectively. While the other members of the Hamamelidaceae demonstrate a build-up of free quinic acid, this compound is essentially absent in the Liquidambaroideae. Instead there is a certain amount of free shikimic acid (Hegnauer 1966, p. 243). Via a number of intermediates shikimic acid eventually gives rise to phenylalanine (see Turner 1971, p. 35). The enzyme phenylalanine ammonia lyase is responsible for converting this into cinnamomic acid. Caffeic acid can be derived from cinnamomic acid by the substitution of hydroxy groups in the aromatic ring. In *Liquidambar* caffeic acid is either not synthesized or is rapidly converted into other compounds, for it cannot be detected in this genus. Bate-Smith (1962) found it in all the other Hamamelidaceae he investigated. C6–C3 compounds such as caffeic acid can be converted into flavonoids, lignin, or C6–C1 compounds, including gallic acid. In the Liquidambaroideae the turnover is high, for although no gallic acid itself can be isolated, gallic acid derivatives such as ellagic acid and bergenin are present (Hegnauer 1966, p. 242; Giannasi 1986, p. 425). Jay (1968, p. 141) considers the richness in such products to be a primitive feature. By way of contrast, gallic acid is present to a greater or lesser degree in *Corylopsis*, *Distylium*, *Fothergilla*, *Hamamelis*, *Sinowilsonia*, and *Sycopsis* (Giannasi 1986, p. 425).

Another compound of different derivation which would appear to be characteristic of the Liquidambaroideae is monotroposide which Plouvier (1961, 1964) isolated from the branches and bark of *Liquidambar orientalis* and *L. styraciflua* (see also Hegnauer 1966, p. 244). In addition there are the secretory ducts which yield volatile oils or liquid styrax (for a discussion of the correct name for the exudate, see Thomas 1961). The exact composition of these oils has been investigated by Tattje *et al.* (1980) using liquid chromatography, gas–liquid chromatography, infrared spectroscopy, and nuclear magnetic resonance and mass spectrometry.

Taken as a whole these data constitute a formidable case for placing the Liquidambaroideae in a family of its own. However the arguments for and against the segregate family Altingiaceae Hayne (see Bogle 1986, p. 326 for references) need to be placed in a broader context and as such are outside the scope of this review.

A survey of the genera and species

The Liquidambaroideae consists of three genera, *Altingia* Noronha, *Liquidambar* L., and *Semiliquidambar* H.-T. Chang. While *Liquidambar* has representatives in Central and North America, Greece and Turkey, and the Far East, *Altingia* and *Semiliquidambar* are restricted to the Asian continent. Bogle (1986, pp. 344–345) has suggested that *Semiliquidambar*, which is intermediate in many ways between *Altingia* and *Liquidambar* and occupies the area of overlap between these two genera, may have originated by intergeneric hybridization. Be this as it may, the genus gives the impression of being somewhat heterogeneous and is badly in need of revision.

In his latest account of *Semiliquidambar*, H.-T. Chang (1979) recognizes three species, *S. cathayensis* H.-T. Chang found in the Chinese provinces of Fujian, Guangdong, Guangxi, Guizhou, and Jiangxi, *S. caudata* H.-T. Chang (Fujian and Zhejiang Provinces) and *S. chingii* (Metcalf) H.-T. Chang (Fujian, Guangdong, and Jiangxi Provinces).

The genus *Altingia* is composed of about eight species, which were grouped by H.-T. Chang (1959, see H.-T.Chang 1979) into two sections (Section *Altingia*, Section *Oligocarpa*) on the basis of their fruiting heads. The type species, *A. excelsa* Noronha (i.e. Section *Altingia*) has the most extensive distribution from the Himalayas (Bhutan, southern Tibet) in the north, extending down through Burma, Yunnan, and Malaysia into Sumatra and western Java. The Thai material often referred to this species is probably all *A. siamensis* Craib. Closely related to *A. excelsa*, but apparently distinct from it, is *A. multinervis* Cheng, still only known from the type collection (Tsoong 256) made in Guizhou (China). Chang compared his material with *A. yunnanensis* Rehder et Wilson (Yunnan, northern Vietnam) which, along with *A. poilanei* Tardieu-Blot (a single collection from northern Vietnam), is probably better placed in synonymy under *A. chinensis* (Champion ex Bentham) Oliver ex Hance (Fujian, Guangdong, Guangxi, Guizhou, Hong Kong, Hunan, Jiangxi, Yunnan, and Zhejiang). Closely allied to *A. chinensis* is *A. obovata* Merrill et Chun, which is endemic to Hainan Island (Guangdong). The remaining two species in Section *Altingia* are only known from their type material, *A. cambodiana* Lecomte (Poilane 263, southern Cambodia) and an as yet undescribed species from southern Vietnam (Poilane 3779). H.-T. Chang (1979) also mentions *A. angustifolia* H.-T. Chang under Section *Altingia*. Examination of the type shows this to be an immature fruiting specimen of *A. gracilipes* Hemsley. As such it should be placed in Section *Oligocarpa* H.-T. Chang, typified by its hemispheroidal to turbinate fruiting heads, bearing no more than nine mature fruits. Section *Oligocarpa* is represented by two vicarious species: *A. gracilipes* (Fujian, Guangdong, Hong Kong, Jiangxi, Zhejiang) and *A. siamensis* Craib (including *A. takhtajanii* Thai van Trung, *A. tenuifolia* Chun ex H.-T. Chang) from Thailand, Laos, Vietnam, and the Chinese provinces of Guizhou and Yunnan.

Liquidambar can be divided into two sections (Section *Liquidambar*, Section *Cathayambar*) on the ornamentation of the fruiting heads. Section *Cathayambar* Harms (1930) is represented by *L. formosana* Hance (Vietnam, southern China, and Taiwan) and the enigmatic *L. edentata* Merrill (1927) from Fujian. While Samorodova-Bianki (1957) treats this species in her account of the genus, no mention is made of it in H.-T. Chang's revision (H.-T. Chang 1979). Section *Liquidambar* consists of four allopatric species: *L. styraciflua* Linnaeus (southeastern United States), *L. macrophylla* Oersted (Mexico, Guatemala, Belize, El Salvador, Honduras, and Nicaragua), *L. orientalis* Miller (Turkey, Rhodes) and *L. acalycina* H.-T. Chang (southern China). *L. macrophylla* is frequently placed in synonomy with *L. styraciflua* and even *L. orientalis* is sometimes referred to this species (e.g. Meikle 1977). Another taxon, *L. rosthornii* Diels (1900), which was based on sterile material (Bock and Rosthorn 763), is apparently an *Acer* to judge from the opposite leaf arrangement.

Organography of the recent Liquidambaroideae

Before attempting a critical assessment of the fossil record, it is necessary to consider the diagnostic value of the various isolated plant parts liable to be encountered.

Leaf morphology and anatomy

Leaf shape: The Liquidambaroideae comprises species with simple leaves and those with palmate leaves. Once Noronha's genus *Altingia* (1790) gained acceptance in systematic circles in the course of the nineteenth century, it proved a simple matter to distinguish this genus from the type genus *Liquidambar* Linnaeus. The species with simple leaves were referred to *Altingia*, while those with typically palmate leaves were ascribed to *Liquidambar*. However in 1931 Metcalf described *Altingia chingii*, which combined a number of *Altingia* characteristics with *Liquidambar*-like leaves. These plants were later assigned to *Semiliquidambar* by H.-T. Chang (1962).

The lamina of *Semiliquidambar cathayensis* bears the closest resemblance to that of *Liquidambar* and more particularly *L. acalycina*. Like *L. acalycina* it is either simple or two- to three-lobed. However, in *Semiliquidambar cathayensis* the lamina is usually longer than broad. Thus the length/breadth ratio combined with data on the relative length of the petiole is sufficient to distinguish three-lobed specimens of *Semiliquidambar* from those of *Liquidambar* (Fig. 13.1). However, distinguishing the simple

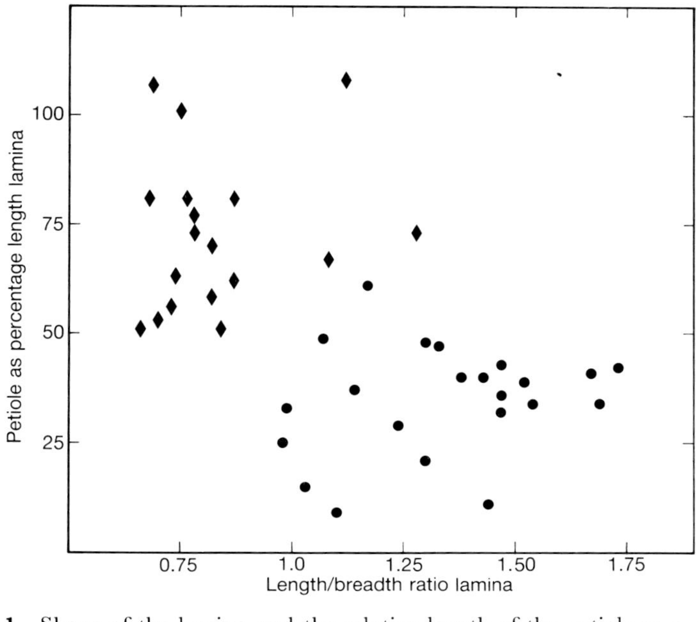

Fig. 13.1 Shape of the lamina and the relative length of the petiole as a means of differentiating three-lobed leaves of *Liquidambar acalycina* (diamonds) from those of *Semiliquidambar cathayensis* (dots).

leaves of *Semiliquidambar* from those of the chartaceous species of *Altingia* is much more problematic because not all *Semiliquidambar* leaves have actinodromous venation. Sometimes an extra large intercostal field at the base of the leaf betrays a *Semiliquidambar* leaf, but this is by no means always the case. H.-T. Chang (1962, 1979) makes extensive use of leaf shape and size in distinguishing the species of *Semiliquidambar* (five in 1962; three in 1979). The reliability of these characters is, however, open to question and requires further study.

The establishment of *Semiliquidambar* as a distinct taxon means that *Altingia* is once again characterized by simple leaves. Within this genus the shape of the leaves varies from ovate over elliptical to obovate. As its name suggests, *A. obovata* has an obovate lamina, while its closest ally, *A. chinensis*, is generally considered to have similar leaves. While leaf shape is employed to distinguish *A. yunnanensis* from *A. chinensis* (e.g. Anonymous 1977, p. 117; H.-T. Chang 1979, p. 62) a biometric analysis by the present author of *A. chinensis*, *A. yunnanensis*, and *A poilanei* proves that there is so much overlap among them in this and other characters that they are better referred to a single species. This means that the leaves in *A. chinensis* must now be considered to display more variation than was hitherto accepted (ovate to obovate). The remaining *Altingia* species all have an ovate leaf-blade and can at best only be distinguished by the shape of their leaf apices and bases.

In *Liquidambar* the species are notoriously heterophyllous (Holm 1930; Duncan 1959; Smith 1967), leading to considerable overlap among the species. However, when the full range of leaf shapes is known, it is possible to discern a morphocline leading from *L. acalycina* to *L. orientalis* (Fig. 13.2). In *L. acalycina* the leaves are mostly trilobate with a fair representation of simple and two-lobed forms. While the leaves of *L. formosana* are generally trilobate, simple and two-lobed forms are rare. Moreover, in young plants and suckers there is a tendency to produce pentalobate leaves, which may or may not display secondary lobing. In *L. macrophylla* both three-lobed and five-lobed forms are common along with the occasional four-lobed leaf. Based on herbarium material the ratio between the five- and three-lobed leaves proves to be of the order of 1.5:1, although the individual exsiccata vary quite considerably. Excluding transitional forms, the mean number of lobes (three, four, and five-lobed forms included) is 4.19; in *L. styraciflua sensu stricto* the mean is 5.05, due to the presence of approximately 5 per cent septentate leaves. Such seven-lobed leaves are also common in *L. orientalis*, which tends to display secondary lobing to a greater degree.

Leaf margin: With few exceptions the leaf margin in the Liquidambaroideae is glandular crenulate–serrulate (Fig. 13.3). In some species these teeth are only poorly developed (e.g. *Altingia siamensis*) or even absent

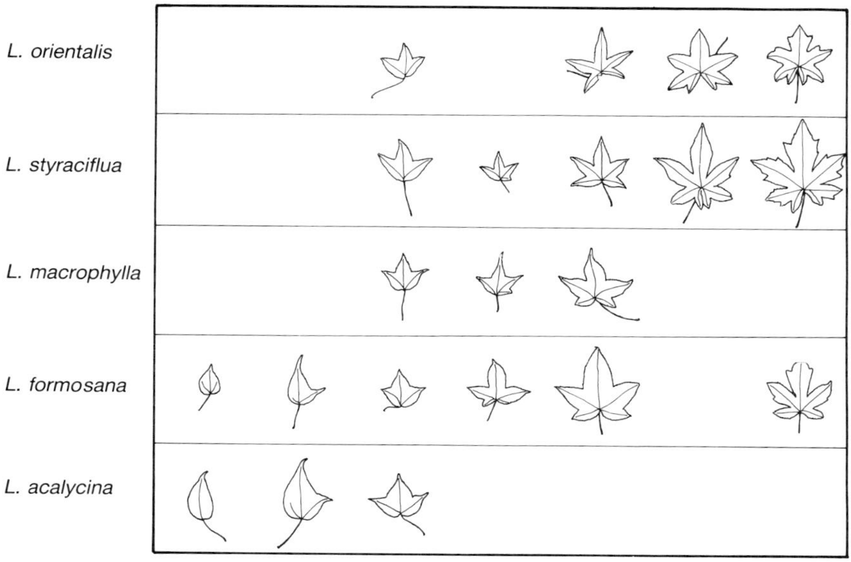

Fig. 13.2 Leaf shape in *Liquidambar* taxa.

(*Altingia gracilipes* var. *gracilipes*, *Liquidambar edentata*). The latter is known only from the type material and may be no more than an unusual specimen of *L. formosana*.

Venation: Two major types of venation are encountered in the subfamily: actinodromous and camptodromous. Actinodromous venation (sometimes but not always palinactinodromous) is characteristic of *Liquidambar* and certain specimens of *Semiliquidambar*, while *Altingia* and other *Semiliquidambar* leaves display a camptodromous (brochidodromous) pattern. While the secondary veins locally extend to the leaf margin in secondarily lobed forms of *Liquidambar*, this is exceptional. In general the teeth at the leaf margin are fed by veinlets arising from the loops which connect the secondary veins (Fig. 13.3). Within the genera this venation pattern is remarkably uniform.

The size of the fine vein reticulum and the number of free vein-endings is often considered to be of considerable taxonomic significance (Carlquist 1961). It is therefore somewhat surprising that so little attention has been paid to this feature. Tanai (1976) noted that the *Liquidambar formosana* which he examined has a much finer vein network than *L. styraciflua*. My own observations show that this character can be used to differentiate herbarium material of *L. macrophylla*, *L. styraciflua*, and *L. orientalis* on the one hand from the eastern Asiatic taxa, *L. acalycina* and *L. formosana*. However, leaves of *L. formosana*, *L. orientalis*,

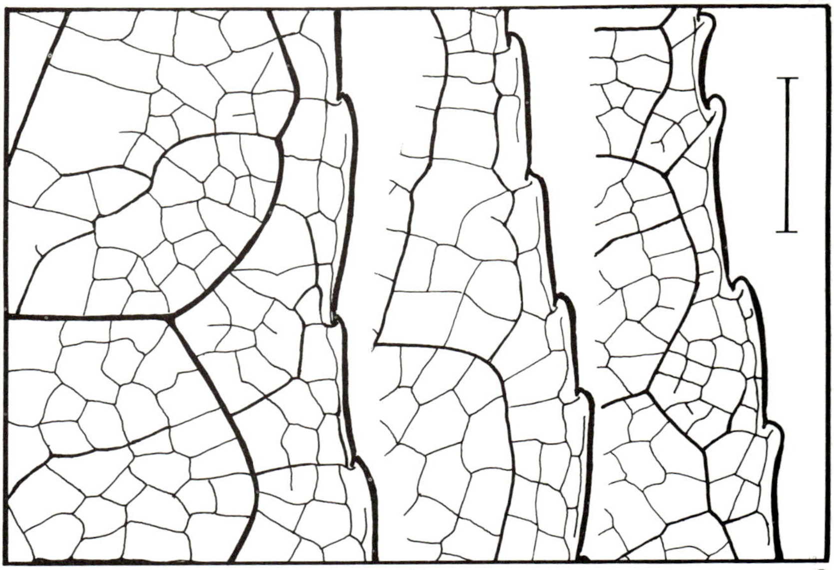

Fig. 13.3 A comparison of the leaf margin in *Altingia chinensis* (A), *Semiliquidambar cathayensis* (B) and *Liquidambar formosana* (C). Bar = 3 mm.

and *L. styraciflua* growing in Kalmthout Arboretum (northern Belgium) demonstrate that the above differences are simply under microclimatic control. The fine venation in the Kalmthout material was not only remarkably similar, but displayed a large number of free vein-endings, a feature not usually seen in herbarium material. Likewise, the size of the areoles and the number of free vein-endings in *Altingia* was found to vary considerably, even within the same species (Fig. 13.4). As such, little taxonomic importance can be attached to this character.

Leaf indumentum: The fact that many species are glabrous, even at the juvenile stage, makes the sporadic occurrence of trichomes all the more interesting. When present these trichomes consist of long soft unicellular or bicellular hairs (Fig. 13.5).

In *Liquidambar* such hairs can form a dense indumentum on the leaves of suckers and young trees, being retained the longest at the leaf margin and on the abaxial leaf surface. In contrast, even the young leaves of mature trees tend to be largely glabrous, with the possible exception of a tuft of hairs at the leaf base. This tuft of hairs or its absence has been used to distinguish vicarious species such as *L. styraciflua* and *L. orientalis* (see Harms 1930, 1932). However, the degree to which these tufts are developed varies within the same species. In *L. orientalis*, which is often

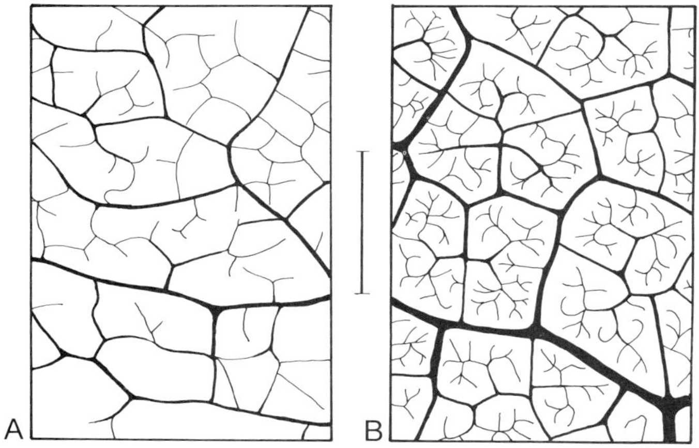

Fig. 13.4 Variation in the fine venation of *Altingia excelsa*. A, J. F. Rock 7174; B, G. Forrest 8763. Bar = 2 mm.

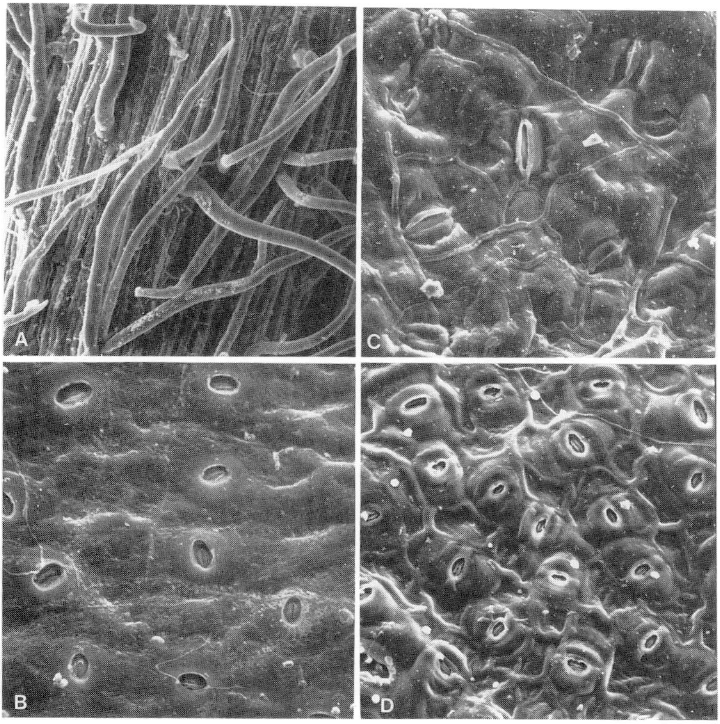

Fig. 13.5 Epidermal features in *Liquidambar* and *Semiliquidambar*. A, *S. cathayensis*: unicellular hairs over the midrib (W. T. Tsang 21257, abaxial leaf surface, × 375) B, *S. cathayensis*: stomata on the abaxial leaf surface (W. T. Tsang 21257, × 750). C, *L. macrophylla*: abaxial leaf surface with stomata and epiphyllous hyphae (J. D. Smith 1855, × 375). D, *L. formosana*: abaxial leaf surface with stomata (T. Sampson s.n., × 375). Note the similarity in stomatal shape in *S. cathayensis* and *L. formosana*.

considered to be entirely glabrous, small tufts of hairs may be present (Peşmen 1972).

At best only a small tuft of hairs is present at the base of *Semiliquidambar* leaves. While sucker leaves of *S. cathayensis* are often distinctly hairy, even these are rapidly lost, being retained the longest over the primary, secondary, and tertiary veins on the lower surface of the leaf. In general the mature leaves are entirely glabrous.

In *Altingia* the leaves are generally glabrous. However, when hairs occur in unusual numbers these have been given due consideration by taxonomists. Thus Koorders and Valeton (1895, p. 212) described a new variety of *A. excelsa* (*A. excelsa* var. *velutina*) on the basis of the dense indumentum on the lower surface of their leaves. In general the leaves of *A. excelsa* are glabrous or no more than sporadically hairy below (Koorders and Valeton 1895; Vink 1957). The fact that the material of *A. excelsa* var. *velutina* is sterile might suggest that the leaves are from young trees.

That the presence of hairs need not be a juvenile character is demonstrated by *A. cambodiana*, which displays a characteristic mat of short woolly hairs restricted to the groove on the adaxial surface of the petiole.

Leaf anatomy: The leaves of the Liquidambaroideae vary from chartaceous to coriaceous. These differences find expression in the degree to which the palisade is developed and the presence and development of a hypodermis. Although Reinsch (1890) and Metcalfe and Chalk (1950, p. 588) suggested that the genera of the Hamamelidaceae are characterized by a specific number of palisade layers, this scheme must not be interpreted too literally. In the Liquidambaroideae the number of superimposed cells varies from one to four. In *Liquidambar* two layers are normal, but this can vary from one to three. In *Semiliquidambar* two or three layers are present, while in *Altingia* still more variation is encountered. Species with chartaceous leaves generally have two layers, although this can vary from one to three, while species with coriaceous leaves tend to have three layers (in a range from two to four). It is not clear how much taxonomic value can be attached to this character, since the number of palisade layers is likely to be affected by ecological factors such as light intensity.

The most valuable anatomical character for distinguishing the taxa would appear to lie in the hypodermis. While this is generally lacking in *Liquidambar*, the eastern Asiatic species (*L. acalycina*, *L. formosana*) do have a few hypodermal cells between the upper epidermis and the palisade. While the biological significance of this discontinuous hypodermis is not clear, it is apparently genetically fixed, being found in *L. formosana* grown in Belgium. In *Semiliquidambar* the hypodermal cells form a continuous layer one cell thick over the palisade. This is also the case in a number of *Altingia* species (*A. gracilipes*, *A. siamensis*, and *A. excelsa p.p.*) with relatively chartaceous leaves, while in the coriaceous leaves of e.g. *A. cambodiana* and *A. obovata* the hypodermis is usually two and sometimes three cell rows thick.

Although characters of the leaf hypodermis are seldom preserved in the fossil state, leaf compressions with a cuticle are more common. While the anticlinal walls of the upper epidermis vary from straight to undulate in each of the *Liquidambar* species, this character may prove to be of some use in distinguishing taxa within *Altingia*. Those species with coarse leathery leaves (*A. cambodiana* and *A. obovata*) invariably have straight anticlinal walls, while species with thinner leaves, such as *A. excelsa*, *A. gracilipes*, and *A. siamensis* display anticlinal walls which are undulate to various degrees. In the material examined so far only *A. chinensis* had either straight or undulate anticlinal walls. In *Semiliquidambar* the anticlinal walls of the upper epidermis are undulate.

Although Reinsch (1890, p. 354) and Sharma and Tyree (1973, p. 180, p. 182) categorically stated that stomata are restricted to the lower leaf surface, a certain number are in fact to be found in the upper epidermis of *Liquidambar* close to the primary veins. These stomata are common in *L. orientalis* and *L. styraciflua*, rare in *L. macrophylla*, and entirely lacking in *L. acalycina* and *L. formosana*.

Further distinctions between the eastern Asiatic and American–western Asiatic representatives of *Liquidambar* become apparent when the stomata on the lower leaf surface are examined. In the eastern Asiatic species the stomatal index (ratio of stomata to the remaining epidermal cells) is considerably higher than in the American-western Asiatic taxa and the shape of the stomata is shorter and broader than the latter (Fig. 13.6). Moreover, the stomatal aperture is comparatively short, being rounded at its extremities (Fig. 13.5). That these characters are not simply the result of environmental factors was demonstrated by an examination of *L. formosana*, *L. orientalis*, and *L. styraciflua* planted in northern Belgium.

While the nature of the stomata can be used to differentiate at least two groups within *Liquidambar*, the lower epidermis in *Altingia* proves to be remarkably homogeneous. The only difference of possible taxonomic significance would appear to lie in a certain variation in the size of the stomata. These stomata are similar to those of eastern Asiatic *Liquidambar*, only larger.

Petiole: As mentioned above, the length of the petiole can be of some help in distinguishing *Semiliquidambar* from *Liquidambar*. The considerable variability in petiole length encountered in the various *Liquidambar* species, however, precludes its use as a diagnostic character within the

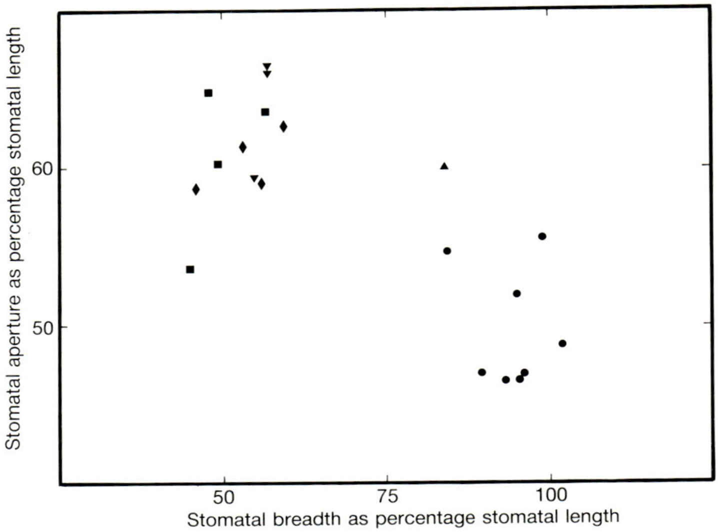

Fig. 13.6 Stomatal shape as a tool for distinguishing eastern Asiatic *Liquidambar* from the American–western Asiatic taxa. The means of 30 measurements are presented. Inverted triangles, *L. orientalis*; squares, *L. macrophylla;* diamonds, *L. styraciflua s.s.*; triangle, *L. acalycina*; dots, *L. formosana*.

genus. Much the same applies to *Semiliquidambar*. While H.-T. Chang (1962, 1979) used petiole length at the varietal level, I am unable to endorse the opinions stated in these accounts. A re-examination of Chang's material shows that he underestimated the full extent of variation. In individual herbarium specimens the longest petiole is usually two to three times as long as the shortest one. It is significant that between 1962 and 1979 Chang himself changed his mind as to the exact limits to be adopted.

While the diagnostic value of the petiole in *Semiliquidambar* has been overstated, its utility in *Altingia* remains unquestioned. Using this character it is possible to distinguish two well-delimited groups within the genus. The length and thickness is related to the nature of the lamina it bears. Those species with large coriaceous leaves have comparatively short (3–25 mm), broad (0.9–2.6 mm) petioles, being at most 15 times as long as broad, e.g. *A. chinensis sensu lato* and *A. obovata*. In species with smaller and/or chartaceous leaves a more delicate petiole (0.5–1.2 mm across) suffices. In these species (*A. excelsa*, *A. gracilipes*, *A. multinervis*, *A. siamensis*) the petiole may be as little as 7.5 times as long as broad, but is usually much longer (anything up to 47 times as long as broad, see Fig. 13.7).

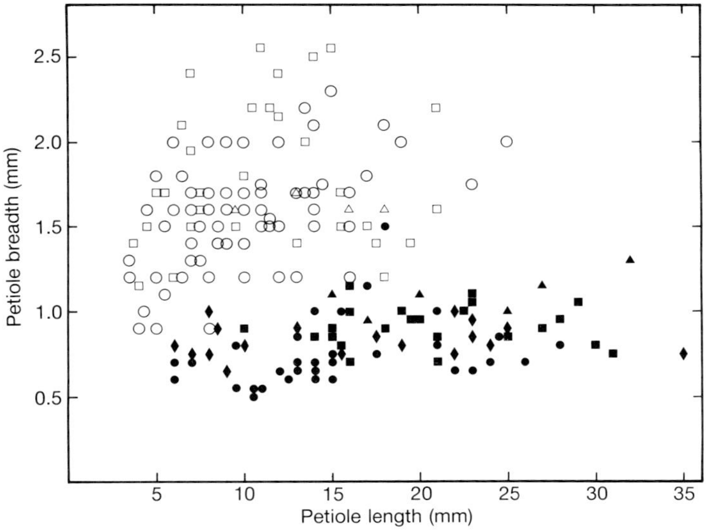

Fig. 13.7 The length/breadth ratio of the petiole as a diagnostic character within *Altingia*. Open triangles, *A. cambodiana*; open squares, *A. obovata*; open circles, *A. chinensis* s.l.; triangles, *A. multinervis*; squares, *A. excelsa*; dots, *A. gracilipes*; diamonds, *A. siamensis*.

Wood anatomy

The wood of only a limited number of species of *Altingia* and *Liquidambar* has been examined to date (e.g. Reinsch 1890; Moll and Janssonius 1914; Tippo 1938; Tang 1943; Metcalfe and Chalk 1950; Greguss 1959; Rao and Purkayastha 1972). The structure is diffuse-porous with vessels occurring singly or in rows of two to three (or as many as six) cells set in ground-tissue consisting of tracheids, fibres, and a few scattered parenchymatous cells. The length of these elements is not only related to the growth conditions (water relations, light intensity, photoperiod, length of growing season) but also the provenance of the material (Hunter and Goggans 1969; Winstead 1972). The vessels have diameters of 30–120 μm and display scalariform perforation plates with approximately 10–40 narrow bars on their oblique end walls. Sometimes spiral thickenings are present at the extremities of the vessels. The wood is transversed by heterogeneous wood-rays anything up to 100 cells high and one to four cells wide. These cells are often filled with a dark gummy substance. Where the ray cells adjoin a vessel, large tiered or alternate pits are developed in their walls. When the pith is preserved it is possible to observe the secretory ducts characteristic for this subfamily in the immediate proximity of the primary xylem (see Kesercioğlu 1973, fig. 10). Moreover, vertical gum canals of traumatic type are occasionally present in the wood (Moll and Janssonius 1914, p. 321; Tippo 1938, p. 45; Rao and Purkayastha 1972, p. 160).

Considering the number of species still to be investigated, it is probably premature to attempt to subdivide the Liquidambaroideae on the basis of their wood anatomy. Moreover, from what is known of their anatomy, the individual genera are remarkably homogeneous. Watari (1952, p. 121) was only able to distinguish *Liquidambar formosana* from *L. styraciflua* on the basis of its slightly more abundant tracheids.

Staminate inflorescence

Staminate inflorescences lack attractive parts such as petals or sepals. Male flowers are little more than bundles of stamens aggregated into (one to) six to fourteen heads, subtended by caducous bracts, and set on a raceme 4–7 cm long. The stamens consist of filaments bearing longitudinally dehiscing anthers, 0.9–1.7 mm long. In *Altingia* and *Semiliquidambar* the filaments are short, almost sessile, rarely more than 0.2–0.4 mm, while those of *Liquidambar* are 0.8–2.3 mm long. Being anemophilous plants, the latter must be considered the more advanced state.

Pollen

The pollen morphology of the Liquidambaroideae has been described by a number of authors (e.g. Erdtman 1943; Ikuse 1956; C.-T. Chang 1958, 1959; Kuprianova 1960; Kapp 1969; T.-C. Huang 1972; Kesercioğlu 1973; Bogle and Philbrick 1980; Anonymous 1982; Zavada and Dilcher 1986), but pollen of *Semiliquidambar* has not yet been examined.

The pollen grains are pantoporate and spheroidal to ellipsoidal in shape. The exine is tectate–columellate with relatively small perforations perhaps related to wind-dispersal (Endress 1977, p. 334). The tectum itself is provided with scabrate processes. The underlying columellae are arranged in a reticulate fashion, thereby producing a characteristic LO-pattern under the light microscope. The columellae rest on a well-developed foot-layer. Over the pores the tectate exine has a partly eroded appearance (Fig. 13.8).

Chinese authors have keyed out *Altingia* and *Liquidambar* on the basis of pollen diameter and pore number (Anonymous 1982, p. 20). However, the differences given are by no means absolute. This is partly the result of the very real similarity of the genera and partly a consequence of a certain amount of variation encountered within and between species. Thus individual grains are difficult to assign to one or other of the genera. By and large the pollen of *Liquidambar* has pores

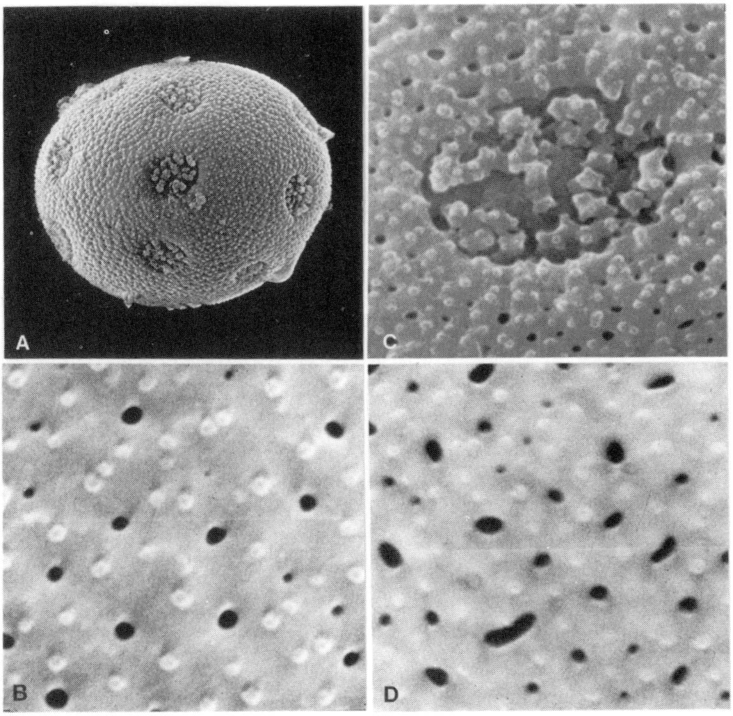

Fig. 13.8 The pollen of the Liquidambaroideae. A, A pantoporate pollen grain of *L. formosana* (B. Balansa 1156, × 1000). B. detail of the tectal surface in *L. styraciflua* (Carey s.n., × 10000) displaying scabrae and regularly shaped perforations; C, A pore in *L. orientalis* (P. H. Davis 40317, × 5000) exhibiting the characteristic disintegration of the tectum. D, irregularly shaped perforations in *A. excelsa* (unknown collector, × 10000).

without an annulus and fairly uniformly shaped tectal perforations. *Altingia*, on the other hand, is characterized by pores surrounded by an annulus and tectal perforations of varying outline (Fig. 13.8). Zavada and Dilcher (1986, p. 351) have shown that pollen of *Liquidambar styraciflua* and *Altingia obovata* can be distinguished on the breadth of their columellae, but transmission electron microscopy on more material is required before this character can be assumed to be diagnostic for the genera.

Attempts to distinguish a number of the species within each of the genera on the basis of characters such as the size and shape of the pores, the pore margin, and the granulation of the pore membrane have been made (see C.-T. Chang 1958, 1959), but based on my own palynological studies this takes too little account of the variation encountered within the individual species.

Fruiting heads

Members of the subfamily bear their fruits in a capitulum at the end of a fairly long peduncle. In some cases the capitula are grouped into racemes and neighbouring capitula may even be fused to some extent. Each carpellate flower normally bears two styles, one for each carpel. In *Liquidambar*, *Semiliquidambar*, and occasionally in *Altingia chinensis* (e.g. A. Henry 10395, an isotype of *A. yunnanensis*) these styles become indurated and are retained in the fruiting heads. In most *Altingia* species, however, these are largely or completely lost. Each fruit normally bears two carpels, although one, three, or four carpels are sometimes encountered (Wisniewski and Bogle 1982, p. 1616). At maturity the carpels undergo dehiscence. In both *Liquidambar* and *Semiliquidambar* the carpels generally split open along their inner surface only (septicidal dehiscence). By way of contrast, dehiscence in *Altingia* is usually taken a stage further with each carpel dividing into two separate valves (loculicidal dehiscence, see Mai 1968, p. 193). The inner surface of the carpels is either completely smooth or exhibits fine transverse ridges. At their outer boundary the lignified fruits display a variety of surface ornamentation. While the origin of some of these phyllomes is still debatable, developmental studies and comparative morphology suggest that the more spinose elements are probably of carpellary origin (Wisniewski and Bogle 1982, pp. 1621–2). Such spines are retained in the fruiting heads of *Liquidambar formosana* and *Semiliquidambar cathayensis*.

The fruiting heads are sufficiently diverse within the genera to yield valuable characters at the specific and supraspecific levels. The spinose *Liquidambar formosana* is referred to Section *Cathayambar*, of Harms (1930), while the remaining species are placed in Section *Liquidambar*. Within Section *Liquidambar* the ornamentation of the peripheral rim (smooth or lobed), the thickness of the receptacular tissue and the length of the indurated styles can be used to differentiate the species (see Gregor 1978, pp. 34–5).

In *Semiliquidambar* the fruiting heads are either spheroidal or truncated spheroidal. However, while *S. cathayensis* displays a spinose peripheral rim, these structures are no more than knobs in *S. caudata*.

Within *Altingia* the peripheral ornamentation is remarkably constant, being tubercular in all species. In this case the overall shape of the fruiting heads can be used to subdivide the genus into two sections and it is possible to infer a trend from spheroidal via truncated spheroidal to hemispheroidal or even turbinate capitula. Since the proximal part of the hemispheroidal and turbinate fruiting heads is sterile, these capitula bear fewer carpels. However, this is no great disadvantage as those that are present on the pendulous fruiting heads are all directed towards the ground. Being less massive, the fruiting heads of Section *Oligocarpa* are borne on slender peduncles, and the capitula swing to and fro, thereby facilitating the distribution of the seeds. This occurs to a lesser extent in the remaining species which belong to Section *Altingia*.

Fossil record

When Unger (1845) compiled his catalogue of fossil plants, representatives of the Liquidambaroideae were only known from three sites in Europe. Since then our knowledge of the fossil record has expanded considerably. The Liquidambaroideae is now known to have been widespread in the northern hemisphere through the Tertiary sub-era.

The oldest fossils referred to this subfamily are some leaves ascribed to *Liquidambar* from the mid-Cretaceous (uppermost Albian to mid-Cenomanian Dakota Formation) of North America (Lesquereux 1874; Newberry 1898). However, these taxa (*L. integrifolia* Lesquereux, *L. obtusilobata* (Heer) Hollick) are almost always entire-margined and the fact that Recent species almost invariably have crenulate–serrulate leaves casts doubt on the identifications. *Liquidambar fontanella* (Brown 1933) from the Upper Cretaceous Aspen Shales of Wyoming has a crenulate–serrulate margin, but the uniformly three-lobed leaves are abnormally deeply divided for a *Liquidambar*. No cuticle is preserved, and fruiting heads resembling those of *Liquidambar* are lacking in the mudstone. It is possible that all these Cretaceous leaves are variants of the platanoid complex which was widespread at that time. Platanoid–*Sapindopsis* reproductive structures are associated with *L. fontanella* at the Wyoming locality (Crane, personal communication).

From the Paleocene Raton Formation of Colorado, Knowlton (1917) described a fragmentary leaf which he provisionally referred to as *Liquidambar*? *cucharas*. However, the leaf margin is 'perfectly entire or at most slightly undulate' (Knowlton 1917, p. 320). In the absence of other morphological arguments, there seems to be no reason for accepting this taxon as an authentic record of the Liquidambaroideae. According to Wolfe (1973, p. 343) the first sound leaf identifications are from the Lower Eocene.

Studies on dispersed pollen suggest that the Liquidambaroideae were in fact already present in the Paleocene (Kuprianova 1960, p. 74; Wolfe 1973, p. 343; Muller 1981, p. 17). While this pollen is fairly characteristic, it is—as pointed out above—difficult to assign individual grains to a particular genus. Thus Kuprianova's (1960) designation of the pollen as *Liquidambar angaridae* Kupr., *L. parviporata* Kupr., *L. thetidae* Kupr., etc. is to say the least premature. Even the use of the organ-genus *Liquidambarpollenites* for such pollen grains is somewhat misleading.

The generic affinity of the Liquidambaroideae pollen found in the

Paleocene is uncertain. Although Chandler (1961) described compressed fruiting heads from the Reading Beds of southern England as *Liquidambar palaeocenica*, re-examination by Crane has shown these to be platanoid (see Manchester 1986, p. 222). It could be that the pollen was produced by *Altingia*-like plants, for by the Eocene capitula resembling those of *Altingia* are encountered. The capitula have been described under a variety of generic names, including *Eoliquidambar* Chandler, *Morinda* Crié, *Protaltingia* Reid et Chandler, *Steinhauera* Presl, and *Syncarpites* Schmalhausen. These remains were revised by Mai (1968), who came to the conclusion that they all belonged to a single species, *Steinhauera subglobosa* Presl. Like *Altingia* the styles were not retained in the fruiting head and the carpels show both septicidal and loculicidal dehiscence. *Steinhauera* is however distinguished from *Altingia* by its large, elongate, ovoid capitula bearing up to 50 fertile fruits (*Altingia* has no more than 30). Nothing is known of other parts of the *Steinhauera* plant. To date *Steinhauera* appears to have been restricted to the Eocene of Europe (see Mai 1968, pp. 195–7 for a discussion of the Paleocene, Oligocene and Miocene records).

While these *Altingia*-like forms would appear to be restricted to the Paleogene, *Liquidambar* persisted on into the Neogene. As a typical element of the Mixed Mesophytic Forests, it was widespread in the northern hemisphere right up until the onset of colder conditions at the beginning of the Pleistocene.

The Neogene record is remarkably complete, consisting of leaf-compressions, wood-remains, staminate inflorescences, pollen, and fruiting heads. The richness of the finds is reflected in the plethora of names which have been erected to cover the various forms. At least 21 species of *Liquidambar* have been described from the Miocene on the basis of isolated leaves, but thorough comparisons between fossil taxa are rare and there seems to have been little consideration of the full range of variation likely to be encountered in single species.

Secondary lobing in *Liquidambar* leaves is first encountered in the Miocene, which could be taken to indicate the presence of taxa having affinities with *Liquidambar orientalis*, *L. styraciflua*, or even *L. formosana*. That the majority of Neogene material from central Europe is related to the American–western Asiatic species rather than the eastern Asiatic taxa is confirmed by cuticle analysis carried out on the fossil leaves (Litke 1967; Walther 1970; Ferguson 1971; Knobloch and Kvaček 1976; Christensen 1976; Mai and Walther 1978). The small number of stomata (*ca.* 12–15 per 0.1 mm^2; stomatal index 3.3–9.0%) would suggest an affinity with *L. orientalis* and more particularly *L. orientalis* var. *integriloba* Fiori from southwestern Anatolia and the islands of Rhodes and Salakos. Nonetheless, Litke (1967) also illustrated a cuticle taken from a trilobate leaf, which displays tightly packed, almost circular stomata, thereby demonstrating that some of the Miocene plants had undoubted eastern Asiatic affinities.

Fruiting heads ascribed to *Liquidambar* have been known from Neogene sediments for over 140 years. However, in many cases these represent no more than compressions and as such give rise to problems of interpretation. Well-preserved fruiting material has been investigated by a number of authors including Czeczott and Skirgiełło (1959) and Gregor (1978). Their studies demonstrate that the capitula can be referred to *Liquidambar* Section *Liquidambar*. The smooth peripheral rim and the thick layer of receptacular tissue present in the fossils would suggest a comparison with *L. orientalis*, while the long indurated styles show a closer similarity to *L. styraciflua*. While emphasizing their intermediate status, Gregor (1978, p. 35) points out that comparisons with *L. acalycina* still require to be undertaken. Fossil wood which has been referred to the Liquidambaroideae has been described by Félix (1884), Houlbert (1910), Watari (1952), Prakash and Barghoorn (1961), Van der Burgh (1964, 1973), Nagy and Petrescu (1969), and Greguss (1969). While it is probably impossible to separate the individual genera on the basis of their wood anatomy (see above), the fact that in the Neogene the staminate inflorescence with their long staminal filaments (see Hantke 1954, tafel 8, fig. 7), leaves, and fruiting heads are all indubitably *Liquidambar* would suggest that the wood also belonged to this genus. Certain authors refer to the wood under its recent generic name, while others prefer to designate it as *Ambaroxylon* or *Liquidambaroxylon*. Either way, there is a consensus that the anatomy closely resembles that of *Liquidambar formosana* and *L. styraciflua* but in the light of the great similarity of the wood anatomy in these two species (Watari 1952, p. 121) it would seem unwise to attempt to recognize different taxa within *Liquidambaroxylon* as Greguss (1969) did. Van der Burgh (1973, p. 180) has since combined several of Greguss' species under *Liquidambaroxylon speciosum* Félix and transferred others to *Hamamelidoxylon*, *Magnolioxylon* and *Dryoxylon*.

Both Makarova (1957) and Kuprianova (1960) have presented phylogenetic trees of *Liquidambar*. In Kuprianova's diagram *Altingia* is split off in the late Cretaceous. From the resemblance of *Steinhauera* to fruiting heads of *Altingia* Section *Altingia* we may assume that this section is more primitive than *Altingia* Section *Oligocarpa*, as might be expected on functional grounds. The trend from *Steinhauera* via *Altingia* Section *Altingia* to *Altingia* Section *Oligocarpa* is one of reduction in the number of carpels per fruiting head.

Makarova postulates that the two sections of *Liquidambar* began to diverge during the Eocene. By the Miocene we have evidence (leaves, fruiting heads) for a form closely resembling *L. orientalis* but with a few *L. styraciflua* traits (e.g. long indurated styles). Considering the close affiliation between these two taxa, it was probably not until late in the Cenozoic that thoroughbred *L. orientalis* evolved. While Kuprianova would have *L. orientalis* as a distinct lineage at least as far back as the Paleogene, Makarova presents a more realistic assessment, placing the point of divergence no further back than the Pliocene.

Although many gaps in the fossil record remain to be filled, it is by now plainly clear that Takhtajan's statement (Takhtajan 1969, p. 153) to the effect that 'both the phylogenetic connections and the geographical distribution of this subfamily indicate that its origin was somewhere in the south-east part of the Asian continent' is lacking in substance. While evolution is no doubt still taking place in this part of the world, the present distribution of the Liquidambaroideae is of a largely relict nature.

Acknowledgements

The core of the information on which this study is based has of necessity been derived from dried material. I am therefore grateful to the directors and curators of A, B, BM, BR, E, IBSC, K, M, P, PE, and U for various loans and to my parents who made a visit to PE possible. Some of the comparative material was supplied by R. and J. De Belder (Kalmthout Arboretum). Ms H. Van Hemelrijk prepared the pollen and Mr F. Neefs the leaf sections, while Dr H.-J. Gregor aided the author with the morphology of the fruiting heads in the initial stages of his research. I am indebted to Dr J. Van der Burgh for examining my account of the wood anatomy, Dr W. Punt and Professor R. Vanhoorne for their discussions on the pollen morphology, and Dr L. Nagels and Dr L. Van Haverbeke for their help with the phytochemistry. My thanks are also due to Mrs L. De Keersmaecker for typing the bibliography, my wife for her help in the preparation of the figures, and Mrs S. Van Belle-Pooters for processing the final draft.

References

Anonymous (1977). *Flora Yunnanica* **1** (*Spermatophyta*). Academiae Scientiarum Sinicae, Peking (In Chinese).

Anonymous (1982). *Angiosperm pollen flora of tropic and subtropic China*. Academiae Scientiarum Sinicae, Peking (In Chinese).

Bate-Smith, E. G. (1962). The phenolic constituents of plants and their taxonomic significance. *Botanical Journal of the Linnean Society* **58,** 95–173.

Bogle, A. L. (1986). The floral morphology and vascular anatomy of the Hamamelidaceae: subfamily Liquidambaroideae. *Annals of Missouri Botanical Garden* **73,** 325–47.

Bogle, A. L. and Philbrick, C. T. (1980). A generic atlas of hamamelidaceous pollens. *Contributions from the Gray Herbarium of Harvard University* **210,** 29–103.

Brown, R. W. (1933). A Cretaceous sweet gum. *Botanical Gazette* **94,** 611–15.

Carlquist, S. (1961). *Comparative plant anatomy.* Holt, Reinhardt and Winston, New York.

Chandler, M. E. J. (1961). *The Lower Tertiary floras of southern England, I. Palaeocene floras, London Clay flora (Supplement).* British Museum, London.

Chang, C.-T. (1958). The morphological characteristics of some recent and fossil pollen of *Liquidambar, Altingia, Sycopsis, Euptelia,* and *Pistacia. Acta Botanica Sinica* **7,** 215–29 (In Chinese; Russian summary).

Chang, C.-T. (1959). Morfologiya Pal'tsa *Liquidambar* L.i *Altingia* Nor. *Botanicheskii Zhurnal SSSR* **44,** 1375–80 (In Russian).

Chang, H.-T. (1962). *Semiliquidambar,* Novum Hamamelidacearum Genus Sinicum. *Bulletin Sunyatsen University, Natural Sciences* **1,** 35–44 (In Chinese and Latin).

Chang, H.-T. (1979). Hamamelidaceae. *Flora Reipublicae Popularis Sinicae* **35,** 36–116 (In Chinese).

Christensen, E. (1976). The Søby Flora: Fossil plants from the Middle Miocene delta deposits of the Søby–Fasterholt area, Central Jutland, Denmark, Part II. *Danmarks Geologiske Undersøgelse* II Raekke **108,** 1–49.

Czeczott, H. and Skirgiełło, A. (1959). Flora Kopalna Turowa Koła Bogatyni II(I), Dicotyledones. *Prace Muzeum Ziemi* **3,** 93–112 (In Polish; English summary).

Diels, L. (1900). Die Flora von Central-China. *Botanische Jahrbücher für Systematik, Pflanzengeschichte und Pflanzengeographie* **29,** 169–659.

Duncan, W. H. (1959). Leaf variation in *Liquidambar styraciflua* L. *Castanea* **24,** 99–111.

Endress, P. K. (1977). Evolutionary trends in the *Hamamelidales–Fagales*-Group. *Plant Systematics and Evolution, Supplement* **1,** 321–47.

Erdtman, G. (1943). *An introduction to pollen analysis.* Chronica Botanica Co. Waltham, Mass.

Ernst, W. R. (1963). The genera of Hamamelidaceae and Platanaceae in the southeastern United States. *Journal of the Arnold Arboretum* **44,** 193–210.

Félix, J. (1884). Magyarország faopáljai palaeophytologiai tekintetben. *Magyar Kiralyi Földtani Intézet Evkönyve* **7,** 1–42. (In Hungarian; German summary).

Ferguson, D. K. (1971). The Miocene Flora of Kreuzau, western Germany I. The leaf-remains. *Verhandelingen der Koninklijke Nederlandse Akademie van Wetenschappen, Afdeling Natuurkunde,* Tweede Reeks **60**(1), 1–297.

Giannasi, D. E. (1986). Phytochemical aspects of phylogeny in Hamamelidae. *Annals of Missouri Botanical Garden* **73,** 417–37.

Goldblatt, P. and Endress, P. K. (1977). Cytology and evolution in Hamamelidaceae. *Journal of the Arnold Arboretum* **58,** 67–71.

Gregor, H.-J. (1978). Die Miozänen Frucht- und Samen-Floren der Oberpfälzer Braunkohle I. Funde aus den sandigen zwischenmitteln. *Palaeontographica* **167**(*B*), 8–103.

Greguss, P. (1959). *Holzanatomie der europäischen Laubhölzer und Sträucher.* Académiai Kiadó, Budapest.

Greguss, P. (1969). *Tertiary angiosperm woods in Hungary.* Akadémiai Kiadó, Budapest.

Hantke, R. (1954). Die fossile Flora der obermiozänen Oehninger-Fundstelle Schrotzburg (Schienerberg, Süd-Baden). *Denkschriften der Schweizerischen Naturforschenden Gesellschaft* **80**(2), 27–118.

Harms, H. (1930). Hamamelidaceae. In *Die natürlichen Pflanzenfamilien.* ed. 2, **18a** (ed. A. Engler and K. Prantl), pp. 303–343. Englemann, Leipzig.

Harms, H. (1932). Unsere Freiland-Hamamelidaceen. *Mitteilungen der Deutschen Dendrologischen Gesellschaft* **44,** 1–24.

Hegnauer, R. (1966). *Chemotaxonomie der Pflanzen,* **4.** Birkhäuser-Verlag, Basel.

Holm, T. (1930). Leaf-variation in *Liquidambar styraciflua* L. *Rhodora* **32,** 95–100.

Houlbert, C. (1910). Les formes diverses de la vie dans les Faluns de la Touraine. Les bois des Faluns der Touraine. *La Feuille des Jeunes Naturalistes,* IV Série **40** (No 473), 69–76.

Huang, T.-C. (1972). *Pollen flora of Taiwan.* National Taiwan University Botany Press, Taipei.

Hunter, A. G. and Goggans, J. F. (1969). Variation in fiber length of sweetgum in Alabama. *Tappi* **52,** 1952–4.

Ikuse, M. (1956). *Pollen grains of Japan.* Hirokawa, Tokyo.

Jay, M. (1968). Distribution des flavonoides chez les Hamamélidacées et familles affines. *Taxon* **17**(2), 136–47.

Kapp, R. O. (1969). *Pollen and spores.* W. C. Brown, Dubuque.

Kesercioğlu, T. (1973). Türkiye Bitkileri Üzerinder Sitotaksonomik, Anatomik ve Morfolojik Araştirmalar. 1. *Liquidambar orientalis* Mill. *Ege Üniversitesi Fen Fakültesi Ilmi Raporlar Serisi* **173,** 1–14 (In Turkish; English summary).

Knobloch, E. and Kvaček, Z. (1976). Miozäne Blätterfloren vom Westrand der Böhmischen Masse. *Rozpravy Ústředního Ústavu Geologického* **42,** 1–131.

Knowlton, F. H. (1917). Fossil floras of the Vermejo and Raton Formations of Colorado and New Mexico. *United States Geological Survey Professional Paper* **101,** 223–445.

Koorders, S. H. and Valeton, T. (1895). Bijdrage No. 2 tot de kennis der boomsoorten van Java. *Mededeelingen uit 's Lands Plantentuin* **14,** 000–212.

Kuprianova, L. A. (1960). Palynological data contributing to the history of *Liquidambar. Pollen et Spores,* **2,** 71–88.

Lesquereux, L. (1874). Contributions to the fossil flora of the Western Territories, part I, The Cretaceous flora. *Report of the United States Geological and Geographical Survey of the Territories* **6,** 1–136.

Litke, R. (1967). Kutikularanalytischer Nachweis für einen Wechsel von Warmgemässigten zu warmem Klima im Jungtertiär. *Abhandlungen des Zentralen Geologischen Instituts* **10,** 123–7.

Mai, D. H. (1968). Zwei ausgestorbene Gattungen im Tertiär Europas und ihre Florengeschichtliche Bedeutung. *Palaeontographica* **123**(*B*), 184–99.

Mai, D. H. and Walther, H. (1978). Die Floren der Haselbacher Serie im

Weisselster-Becken (Bezirk Leipzig, DDR). *Abhandlungen des Staatlichen Museums für Mineralogie und Geologie zu Dresden* **28,** 1–200.

Manchester, S. R. (1986). Vegetative and reproductive morphology of an extinct plane tree (Platanaceae) from the Eocene of western North America. *Botanical Gazette* **147**(2), 200–26.

Makarova, Z. I. (1957). K Istorii roda *Liquidambar* L. *Botanicheskii Zhurnal SSSR* **42,** 1182–95 (In Russian).

Meikle, R. D. (1977). *Flora of Cyprus* **1.** Royal Botanic Gardens, Kew.

Merrill, E. D. (1927). New Chinese ligneous plants. *Journal of the Arnold Arboretum* **8,** 3–19.

Metcalf, F. P. (1931). Botanical notes on Fukien and southeast China I. *Lingnan Science Journal* **10**(4), 413–14.

Metcalfe, C. R. and Chalk, L. (1950). *Anatomy of the dicotyledons.* Clarendon Press, Oxford.

Moll, J. W. and Janssonius, H. H. (1914). *Mikrographie des Holzes der auf Java vorkommenden Baumarten* **III.** E. J. Brill, Leiden.

Muller, J. (1981). Fossil pollen records of extant angiosperms. *Botanical Review* **47,** 1–142.

Nagy, F. and Petrescu, I. (1969). Prezenţa resturilor de *Quercoxylon* şi *Liquidambaroxylon* în Terţiarul Transilvănean. *Contributiuni Botanice din Cluj* **1969,** 275–9 (In Romanian; French summary).

Newberry, J. S. (1898). The later extinct floras of North America. *United States Geological Survey Monograph* **35,** 1–295.

Peşmen, H. (1972). *Liquidambar* L. In *Flora of Turkey and the East Aegean Islands* **4,** (ed. P. H. Davis), pp. 264–5.

Plouvier, V. (1961). Acides quinique et shikimique, bergénine et hétérosides chez quelques Hamamelidacées. *Comptes Rendus de l'Academie des Sciences, Paris* **252,** 599–601.

Plouvier, V. (1964). Arbutoside et aspéruloside de quelques Rubiacées, présence de monotroposide chez *Liquidambar. Comptes Rendus de l'Académie des Sciences Paris* **258,** 735–7.

Prakash, U. and Barghoorn, E. S. (1961). Miocene fossil woods from the Columbia Basalts of central Washington. *Journal of the Arnold Arboretum* **42,** 165–203.

Rao, K. R. and Purkayastha, S. K. (1972). *Indian woods. Their identification, properties and uses,* **3.** Forest Research Institute, Dehra Dun.

Reinsch, A. (1890). Über die anatomischen Verhältnisse der Hamamelidaceae mit Rücksicht auf ihre systematische Gruppierung. *Botanische Jahrbücher für Systematik, Pflanzengeschichte und Pflanzengeographie* **11,** 347–95.

Samorodova-Bianki, G. (1957). De genere *Liquidambar* L. *Notulae Systematicae ex Herbario Instituti Botanici Academiae Scientiarum URSS* **18,** 77–89.

Sharma, G. K. and Tyree, J. (1973). Geographic leaf cuticular and gross morphological variations in *Liquidambar styraciflua* L. and their possible relationship to environmental pollution. *Botanical Gazette* **134,** 179–84.

Smith, R. F. (1967). The leaf dimorphism of *Liquidambar styraciflua* L. *American Midland Naturalist* **77,** 42–50.

Takhtajan, A. (1969). *Flowering plants. Origin and dispersal.* Oliver and Boyd, Edinburgh.

Tanai, T. (1976). The revision of the Pliocene Mogi flora, described by Nathorst (1883) and Florin (1920). *Journal of the Faculty of Science, Hokkaido University,* Series IV **17,** 277–346.

Tang, Y. (1943). Systematic anatomy of the woods of the Hamamelidaceae. *Bulletin of the Fan Memorial Institute of Biology* New Series **1,** 8–63.

Tattje, D. H. E., Bos, R. and Bruins, A. P. (1980). Constituents of essential oil from leaves of *Liquidambar styraciflua* L. *Planta Medica* **38,** 79–85.

Thomas, J. L. (1961). *Liquidambar. Arnoldia* **21**(10), 59–64.

Tippo, O. (1938). Comparative anatomy of the Moraceae and their presumed allies. *Botanical Gazette* **100,** 1–99.

Turner, W. B. (1971). *Fungal metabolites.* Academic Press, London.

Unger, F. (1845). *Synopsis plantarum fossilium.* L. Voss, Lipsiae.

Van der Burgh, J. (1964). Hölzer der niederrheinischen Braunkohlenformation, I. Hölzer der Braunkohlengrube 'Anna' zu Haanrade (Neiderländisch Limburg). *Acta Botanica Neerlandica* **13,** 250–301.

Van der Burgh, J. (1973). Hölzer der niederrheinischen Braunkohlenformation, 2. Hölzer der Braunkohlengruben 'Maria Theresia' zu Herzogenrath, 'Zukunft West' zu Eschweiler und 'Victor' (Zülpich Mitte) zu Zülpich. Nebst einer systematisch-anatomischen Bearbeitung der Gattung *Pinus* L. *Review of Palaeobotany and Palynology* **15,** 73–275.

Vink, W. (1957). Hamamelidaceae, In *Flora Malesiana* Series I. **5**(1) (ed. C. G. G. J. van Steenis), pp. 363–79. Nijhoff, Groningen.

Walther, H. (1970). Die Gattung Hedera L. im Tertiär von Salzhausen. *Abhandlungen des Staatlichen Museums für Mineralogie und Geologie zu Dresden* **16,** 211–34.

Watari, S. (1952). Dicotyledonous woods from the Miocene along the Japan-Sea side of Honsyu. *Journal of the Faculty of Science, Tokyo University,* Section III Botany **6,** 97–134.

Winstead, J. E. (1973). Fiber tracheid length and wood specific gravity of seedlings as ecotypic characters in *Liquidambar styraciflua* L. *Ecology* **53,** 165–72.

Wisniewski, M. and Bogle, A. L. (1982). The ontogeny of the inflorescence and flower of *Liquidambar styraciflua* L. (Hamamelidaceae). *American Journal of Botany* **69,** 1612–24.

Wolfe, J. A. (1973). Fossil forms of Amentiferae. *Brittonia* **25**(4), 334–55.

Zavada, M. S. and Dilcher, D. L. (1986). Comparative pollen morphology and its relationship to phylogeny of pollen in the Hamamelidae. *Annals of Missouri Botanical Garden* **73,** 348–81.

14. A review of the family Buxaceae Dumortier

CHARLES E. JARVIS

Department of Botany, British Museum (Natural History), London SW7 5BD, UK

Abstract

The Buxaceae is a small but widespread family of mostly evergreen woody plants in tropical, temperate, and subtropical areas. It contains three to five genera and 50–110 species. *Buxus* (30–90 species, $x = 14$), with opposite leaves and four to six stamens, is the most widespread genus. *Notobuxus* (eight species, $x = 10$, Africa), with opposite leaves and sessile anthers, *Sarcococca* (11 species, $x = 12$, 14, Asia), with alternate leaves and four stamens, and the herbaceous genus *Pachysandra* (four species, $x = 12$, 13, North America/eastern Asia) are the other core genera. *Styloceras* (four species, South America) has been variously placed in the Buxaceae or referred to its own family. However, the monotypic *Simmondsia* (Jojoba, $x = 13$, southwestern North America) is now usually given its own family. Fossil pollen believed allied to the Buxaceae is known from the Late Cretaceous, and leaf remains from the Oligocene have been attributed to *Buxus*. Although frequently included within the Euphorbiales, the family has also been referred to the orders Celastrales, Terebinthales, Hamamelidales, Pittosporales, and Buxales.

Introduction

The family Buxaceae Dumortier is a small but widespread family of evergreen woody plants or rhizomatous herbs in tropical, temperate, and subtropical areas. The family (see Baillon 1859; Pax 1890; van Tieghem 1897; Mathou 1939; Channell and Wood 1987) contains the

Evolution, Systematics, and Fossil History of the Hamamelidae Volume 1: Introduction and 'Lower' Hamamelidae. (eds P. R. Crane and S. Blackmore), Systematics Association Special Volume No. 40A, pp. 273–8. Clarendon Press, Oxford, 1989.

genera *Buxus* L. (sometimes including *Notobuxus* Oliver), *Sarcococca* Lindley, *Pachysandra* Michaux, and sometimes *Styloceras* Jussieu. *Simmondsia* Nuttall, often included in the Buxaceae, is now referred to its own family by many authors.

Genera

Buxus, with 30 (Mabberley 1987) to 90 species (Köhler 1985), occurs in the temperate northern hemisphere, Asia (see Hatusima 1942), Africa, Central America and Venezuela, and the Antilles. It includes evergreen, woody plants, with opposite leaves, which are monoecious; the inflorescence heads comprising a terminal female flower surrounded by male flowers. Flowers are apetalous. The male flowers have four sepals and four to six exserted stamens with dorsifixed anthers; the female flowers have four to six sepals, a trilocular ovary with two pendulous, anatropous ovules per locule, and three styles. The fruit is an explosive loculicidal capsule. Chromosome number: $x = 14$ (Raven 1975). Studies on leaf morphology (Kvaček *et al.* 1982; Köhler 1984), anatomy (Carlquist 1982), embryology (Wiger, 1935; Dang-Van-Liem 1959) and pollen morphology (Köhler 1981) have also been made. *Buxus*, *Notobuxus*, *Sarcococca*, and *Pachysandra* all contain numbers of steroidal alkaloids, which Gibbs (1974) regards as a diagnostic character of the family. Razafindrambao (1973) records 105 alkaloids which have been isolated from species of *Buxus*.

Notobuxus, with eight species (Verdcourt 1962; Radcliffe-Smith 1981), occurs in tropical and southern Africa, and Madagascar. It differs from *Buxus* mainly in having sessile anthers, no rudimentary ovary in the male flowers (Phillips 1943), and a chromosome number of $x = 10$ (Raven 1975). Pollen characters, too, are interpreted by Köhler and Brückner (1982) as supporting generic separation from *Buxus*.

Sarcococca, with 11 species (Sealy 1986), occurs widely in southeast Asia from Afghanistan, south and east through India, Sri Lanka, China, and Thailand to Sumatra and Java. It is a genus of evergreen shrubs or small trees with alternate leaves, and inflorescences composed of axillary compact clusters of flowers, or erect spikes. The male flowers have four stamens; the female flowers have green bracteoles, and a syncarpous ovary with two to three styles. The fruit is a berry with a fleshy endocarp. Chromosome number: $n = 12, 14$ (Raven 1975, Singh *et al.* 1982).

Pachysandra has one North American and three eastern Asian species (Robbins 1968) of shrubs or perennial herbs with alternate leaves, and spicate inflorescences of unisexual, apetalous flowers. There are four stamens, with long filaments; the ovary is superior with (two or) three carpels, each with two locules containing an anatropous ovule. The fruit is an indehiscent capsule or berry. Chromosome number: $n = 12, 13, 24, 27$ (Raven 1975; Kurosawa 1981).

Although sometimes included within the Buxaceae, *Styloceras*, with four species in western, tropical South America (Gentry and Foster 1981), is frequently referred to its own family. The genus comprises glabrous trees with alternate leaves, the male flowers with 6–30 stamens. The female flowers have five sepals and a (two to) four-celled ovary and solitary ovules. The fruit is a fleshy drupe.

Simmondsia, (Jojoba), with one species from southwestern North America, was included in the Buxaceae until van Tieghem (1897) argued for its separation as a monotypic family in the Centrospermae. Pollen characters (Köhler and Brückner 1984; Nowicke and Skvarla 1984), the absence of alkaloids, S-type sieve-element plastids (Behnke 1982), the presence of an unusual liquid wax in the seed, and other features have supported van Tieghem's view. Chromosome number: $x = 13$ (Raven 1975).

Fossil record

Leaf remains from the Oligocene have been attributed to *Buxus* (Kvaček *et al.* 1982), and fossil pollen (*Erdtmanipollis*), believed to be allied to *Pachysandra* and *Sarcococca*, occurs from the Campanian onwards (Gray and Sohma 1964; Muller 1981). *Buxus*-type pollen has been found in the Lower Miocene (Krutsch 1966; van der Hammen *et al.* 1971).

Systematic position

The systematic position of the family is still unclear (see reviews of Takhtajan 1980; Webster 1987). Segregated from the Euphorbiaceae on ovule characters by Baillon (1859), the Buxaceae is sometimes still treated as a member of the Euphorbiales (e.g. by Cronquist, 1981). However, it has been included in the Celastrales (e.g. Dahlgren 1980), the Pittosporales (Thorne 1983), the Terebinthales (Chadefaud and Emberger 1960), and also the Hamamelidales (Takhtajan 1980). Corner (1976) supports the latter relationship on evidence from seed anatomy, and argues for a close relationship between the Daphniphyllaceae and the Buxaceae, as do Barabé *et al.* (1982). The presence of steroidal alkaloids in both *Buxus* and *Didymeles* (see Hardman 1987) suggests a close affinity between the Buxaceae and Didymelaceae. Webster (1987), reviewing much of the evidence, supports the placement of the Buxaceae in a separate order Buxales adjacent to the Euphorbiales and Pandales (as suggested by Novàk 1961).

References

Baillon, H. (1859). *Monographie des Buxacées et des Stylocérées.* Masson, Paris.

Barabé, D., Bergeron, Y. and Vincent, G. A. (1982). La position des Daphniphyllaceae, Buxaceae, Simmondsiaceae et Cecropiaceae dans le sous-classe Hamamelididae. Étude numérique. *Comptes Rendus des Séances de l'Académie des Sciences, Paris, Série III* **294,** 891–6.

Behnke, H. D. (1982). Sieve-element plastids, exine sculpturing and the systematic affinities of the Buxaceae. *Plant Systematics and Evolution* **139,** 257–66.

Carlquist, S. (1982). Wood anatomy of Buxaceae: correlations with ecology and phylogeny. *Flora* **172,** 463–91.

Chadefaud, M. and Emberger, L. (1960). *Traité de botanique.* Masson, Paris.

Channell, R. B. and Wood, C. E. (1987). The Buxaceae in the southeastern United States. *Journal of the Arnold Arboretum* **68,** 241–57.

Corner, E. J. H. (1976). *The seeds of dicotyledons.* Cambridge University Press, Cambridge.

Cronquist, A. (1981). *An integrated system of classification of flowering plants.* Columbia University Press, New York.

Dahlgren, R. (1980). A revised system of classification of the angiosperms. *Botanical Journal of the Linnean Society* **80,** 91–124.

Dang-Van-Liem (1959). Embryogénie des Buxacées. Développement de l'embryon chez le *Buxus sempervirens* L. *Comptes Rendus Hebdomaires des Séances de l'Académie des Sciences, Paris* **248,** 1844–7.

Gentry, A. H. and Foster, R. (1981). A new Peruvian *Styloceras* (Buxaceae): discovery of a phytogeographical missing link. *Annals of the Missouri Botanical Garden* **68,** 122–4.

Gibbs, R. D. (1974). *Chemotaxonomy of flowering plants.* McGill-Queen's University Press, Montreal.

Gray, J. and Sohma, K. (1964). Fossil *Pachysandra* from western America with a comparative study of pollen in *Pachysandra* and *Sarcococca. American Journal of Science* **262,** 1159–97.

Hammen, T. van der, Wymstra, T. A. and Zagwijn, W. H. (1971). The floral record of the late Cenozoic in Europe. In Turekian, K. K. (ed.), *The late Cenozoic glacial ages.* Yale University Press, New Haven.

Hardman, R. (1987). Recent developments in our knowledge of steroids. *Planta Medica* **53,** 233–8.

Hatusima, S. (1942). A revision of the Asiatic *Buxus. Journal of the Department of Agriculture, Kyusyu Imperial University* **6,** 261–342.

Köhler, E. (1981). Pollen morphology of the West Indian–Central American species of the genus *Buxus* L. (Buxaceae) with reference to taxonomy. *Pollen et Spores* **23,** 37–91.

Köhler, E. (1984). Zur Blattnervatur der neotropischen *Buxus*-Arten und ihre Bedeutung für die Systematik (Buxaceae). *Flora* **175,** 345–74.

Köhler, E. (1985). Vorstellungen zur Evolution und Chorogenese der neotropischen *Buxus*-Arten. *Feddes Repertorium* **96,** 663–75.

Köhler, E. and Brückner, P. (1982). Die Pollenmorphologie der afrikanischen

Buxus-Arten und *Notobuxus*-Arten (Buxaceae) ihre systematische Bedeutung. *Grana* **21,** 71–82.

Köhler, E. and Brückner, P. (1984). Zur Pollenmorphologie und systematischen Stellung der Gattung *Simmondsia* Nutt. *Wissenschaftliche Zeitschrift Friedrich-Schiller-Universität Mathematisch-Naturwissenschaftliche Reihe* **32,** 945–55.

Köhler, E. and Brückner, P. (1989). The genus *Buxus* (Buxaceae): aspects of its differentiation in space and time. *Plant Systematics and Evolution* **162,** 267–83.

Krutzsch, W. (1966). Die sporenstratigraphische Gleiderung des älteren Tertiär im Nördlichen Mitteleuropa (Paläozän-Mitteloligozän). *Abhandlungen Zentrallen Geologische Institüt* **8,** 112–49.

Kurosawa, S. (1981). Notes on chromosome numbers of Spermatophytes (3). *Journal of Japanese Botany* **56,** 245–51.

Kvaček, Z., Bůžek, Č. and Holý, F. (1982). Review of *Buxus* fossils and a new large-leaved species from the Miocene of Central Europe. *Review of Palaeobotany and Palynology* **37,** 361–92.

Mabberley, D. (1987). *The plant-book.* Cambridge University Press.

Mathou, T. (1939). *Recherches sur la famille des Buxacées; étude anatomique, microchimique et systématique.* Douladoure, Toulouse.

Muller, J. (1981). Fossil pollen records of extant angiosperms. *Botanical Review* **47,** 1–146.

Novàk, F. A. (1961). *Vyšší Rostliny. Tracheophyta.* Ceskoslovenska Academie, Praha.

Nowicke, J. W. and Skvarla, J. J. (1984). Pollen morphology and the relationships of *Simmondsia chinensis* to the order Euphorbiales. *American Journal of Botany* **71,** 210–5.

Pax, F. (1897). Buxaceae. In *Die naturlichen Pflanzenfamilien* **3**(5), 130–35. Engelmann, Leipzig.

Phillips, E. P. (1943). Some changes in nomenclature. IV: 3. *Journal of South African Botany* **9,** 137–40.

Radcliffe-Smith, A. (1981). A remarkable new species of *Notobuxus* (Buxaceae) from Tanzania. *Kew Bulletin* **36,** 39–41.

Raven, P. H. (1975). The bases of angiosperm phylogeny: cytology. *Annals of the Missouri Botanical Garden* **62,** 724–64.

Razafindrambao, R. S. (1973). Étude d'une plante médicinale malgache, *Buxus madagascarica* Baill. et ses variétés. *Travaux et documents de l'O.R.S.T.O.M. Number 25.* O.R.S.T.O.M., Paris.

Robbins, H. C. (1968). The genus *Pachysandra* (Buxaceae). *Sida* **3,** 211–48.

Sealy, J. R. (1986). A revision of the genus *Sarcococca* (Buxaceae). *Botanical Journal of the Linnean Society* **92,** 117–59.

Singh, G. Bir, S. S. and Gill, B. S. (1982). In IOPB Chromosome number reports LXXVII (ed. A. Löve). *Taxon* **31,** 761–77.

Takhtajan, A. L. (1980). Outline of the classification of flowering plants. *Botanical Review* **46,** 225–359.

Thorne, R. F. (1983). Proposed new realignments in the angiosperms. *Nordic Journal of Botany* **3,** 85–117.

Tieghem, P. van (1897). Sur les Buxacées. *Annales des Sciences Naturelles*, Série 10 **5,** 289–338.

Verdcourt, B. (1962). In *Flora of tropical East Africa* (eds. C. E. Hubbard and E. Milne-Redhead) Buxaceae, pp. 1–3.

Webster, G. L. (1987). The saga of the spurges: a review of classification and relationships in the Euphorbiales. *Botanical Journal of the Linnean Society* **94,** 3–46.

Wiger, J. (1935). *Embryological studies on the families Buxaceae, Meliaceae, Simarubaceae and Burseraceae.* Thesis, University of Lund. Ohlsson, Lund.

15. The Didymelales: a systematic review

DAVID A. SUTTON

Department of Botany, British Museum (Natural History), London SW7 5BD, UK

Abstract

The Didymelales consists of the monogeneric family Didymelaceae with two species endemic to the Malagasy Republic (Madagascar). Species of *Didymeles* are glabrous, evergreen, dioecious trees with alternate, exstipulate leaves and small, apetalous flowers. Their chemistry is atypical for the Hamamelidae (*sensu* Cronquist) as they have both iridoids and steroidal alkaloids. An unusual anatomical character is the abundance of coarse, short fibres in non-xylem tissues of the bark, stems, and leaves. The fruit is a large drupe with a single seed. The Didymelaceae is usually considered allied to the Leitneriaceae though some authors have concluded that floral similarities result from parallel reduction. Other authors have placed the family close to the Buxaceae, Euphorbiaceae, or Thymelaeaceae. The Didymelaceae has been included in the Euphorbiales or Pittosporales but most recent reviews have placed the family in a separate order of the Hamamelidae, close to the Daphniphyllaceae and Leitneriaceae. The fossil record for the Didymelaceae is restricted to dispersed pollen from Paleocene and more recent deposits. This paper reviews critical characters used to determine the problematic placement of *Didymeles* and provides a survey of the literature.

Taxonomy

Didymelales (Takhtajan 1966)

Evolution, Systematics, and Fossil History of the Hamamelidae Volume 1: Introduction and 'Lower' Hamamelidae. (eds P. R. Crane and S. Blackmore), Systematics Association Special Volume No. 40A, pp. 279–84. Clarendon Press, Oxford, 1989.

Didymelaceae (Leandri 1937)

Glabrous, evergreen, dioecious trees; wood with vessels, perforation plates scalariform with 6–25 cross-bars, imperforate tracheary elements with small, bordered pits, wood-rays heterogeneous, uniseriate and pluriseriate, the pluriseriate rays narrow, mostly with elongate ends, wood-parenchyma replaced by lignified cells; non-xylem tissues of bark, stem, and leaves with abundant coarse, short fibres. *Leaves* alternate, simple, petiolate, exstipulate; lamina entire with pinnate venation, stomata apparently encyclocytic. *Inflorescence* axillary, racemose, or paniculate. *Flowers* inconspicuous, apetalous, pedicellate, bracteate. *Staminate flowers* in open panicles; perianth apparently absent; stamens 2, the filaments short, connate, the anthers basifixed, tetrasporangiate, extrorse, opening by 2 longitudinal slits; pollen 3-zonocolporate with 2-orate, operculate colpi, oblate to spheroidal, reticulate, tectate-columellate with footlayer. *Pistillate flowers* in racemes, paired, solitary or in threes, hypogynous, usually with 1–4 scale-like sepals; gynoecium unilocular, apparently monocarpellate; stigma sessile or with short style, papillate; ovule solitary, more or less apical, hemitropous, epitropous, bitegmic, the integuments prolonged at apex forming collar. *Fruit* an indehiscent, fleshy, ovoid, sulcate drupe with persistent stigma; seed solitary, exalbuminous; embryo straight with 2 thick cotyledons.
Type genus: *Didymeles* Thouars, *Pl. Îles Afriq. Austral.*, 23, tab. 1 (1804). Type species: *Didymeles integrifolia* J. St. Hil., *Expos. Fam. Nat.*, **2,** 388 (1805). Type: MADAGASCAR: *Noronha s.n.* (syn.? (cf. Leandri 1937) P).

Introduction

A monogeneric family with two species endemic to eastern and northern Malagasy Republic (Madagascar). First collected before 1787 by Fernand de Noronha, *Didymeles* is represented by scant herbarium material. The original generic description lacked a specific epithet. That most widely used for the type species (*D. madagascarensis* Willd.) is generally misquoted and apparently antedated by *D. integrifolia* J. St. Hil. (Saint-Hilaire 1805, cf. Willdenow 1806). Interpretation of the flowers is controversial, especially whether scale-like structures represent perianth or bracts. Though Drake del Castillo (1897) illustrated a bicarpellate, apocarpous flower, Leandri (1937) interpreted this as two monocarpellate flowers. Pollen of *Didymeles* is distinctive and thus dispersed pollen from Paleocene and Eocene deposits is probably reliably attributed to this taxon (Muller 1981).

Ordinal placement

After much hesitation, Baillon (1873) placed *Didymeles* near *Xanthoxylum* (Rutaceae) but later (Baillon 1876) placed it with *Leitneria* in a tribe

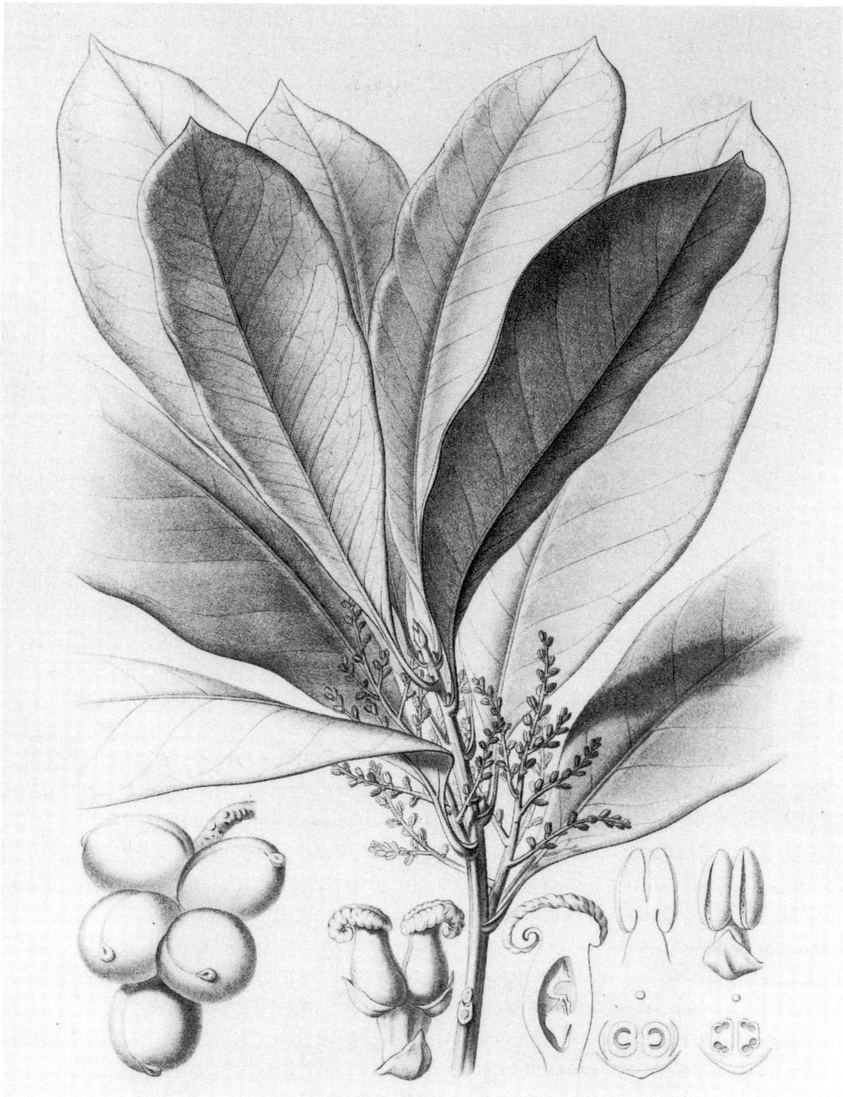

Fig. 15.1 *Didymeles integrifolia* J. St. Hil. Flowering shoot of male plant, with details of male flowers, female flowers and fruits. Reproduced from Drake del Castillo, E. (1897: tab. 308A).

Leitnerieae of the 'Castanéacées' (Fagaceae). Bentham (1880) adopted the latter suggestion but de Candolle (1873) only noted that *Didymeles* doubtfully approached Myricaceae. Van Tieghem and Lecomte (1886), finding that *Didymeles* differed from *Leitneria* in the abundant fibres and absence of secretory canals, suggested that it might be related to the 'Ternstroemiacées' (Theaceae). Hallier (1912) first postulated affinity with Hamamelidaceae but many authors (cf. Webster 1987) placed *Didymeles* near the Euphorbiaceae. Leandri (1937) compared *Didymeles* with these families and concluded that it represented a monotypic family allied to Leitneriaceae. Melchior (1964) placed Didymelaceae with Leitneriaceae in the Leitneriales, next to Balanopales. Takhtajan (1966) considered that Didymelaceae formed the monotypic order Didymelales (subclass Hamamelididae: superorder Hamamelidanae), but Stebbins (1974) included it in the Hamamelidales (superorder Hamamelidae).

Wolfe (1974) found the leaf-venation of *Didymeles* aberrant for the Hamamelidales *sensu stricto* but similar to the Thymelaeaceae. *Didymeles* also differs from the Hamamelidales in the structure of the testa (Melikian 1973) and pollen morphology (Köhler 1980). The latter author suggested that the Didymelaceae belonged to an alliance including Buxaceae, Euphorbiaceae, and Thymelaeaceae. Cronquist (1981) recognized Didymelales in the Hamamelidae, considering that placement with either Euphorbiaceae or Thymelaeaceae required a pseudomonomerous (not monomerous) ovary and different xylem anatomy. Carlquist (1982) found some similarities in the wood of Didymelaceae and Buxaceae. Using this evidence and the resemblance of the encyclocytic stomata to those of Buxaceae, Thorne (1983) placed the Didymelaceae with Balanopaceae, Buxaceae, and Daphniphyllaceae as a suborder (Buxinae) of the Pittosporales (superorder Roseflorae). However, Carlquist (1982) argued against the stomatal similarity and stated that *Didymeles* probably lacked subsidiary cells. Dahlgren (1983), referring to Köhler's work, provisionally placed Didymelaceae in the Buxales (superorder Rosiflorae) with Buxaceae and Daphniphyllaceae. Webster (1987) considered that the Didymelaceae belonged with Daphniphyllaceae and showed general hamamelidalean affinity. In phenetic analyses, Barabé *et al.* (1987) found that Didymelaceae was consistently grouped with Leitneriaceae and only more distantly linked to Balanopaceae, Buxaceae and Daphniphyllaceae.

Though little is known of its chemistry, *Didymeles* differs from most Hamamelidae in having iridoids (cf. Giannasi 1986) and highly unusual steroidal alkaloids (Ahond *et al.* 1980). Twelve of these aminopregnane and conane alkaloids have been isolated from *Didymeles* and a similar number (cf. Hardman 1987) from the Buxaceae, suggesting close affinity.

Summary

Further collections of *Didymeles* are required for investigation into the floral structure, pollen, embryology, cytology, and chemistry. Available evidence suggests that Didymelaceae are only distantly related to the Hamamelidaceae and Leitneriaceae. Distinctive steroidal alkaloids indicate close relationship to the Buxaceae, supported to some extent by wood anatomy, leaf anatomy, and pollen morphology, but whether these families are included in the Hamamelidae/Hamamelidiflorae or Rosidae/Rosiflorae remains an open question.

References

Ahond, A., Debray, M.-M., Picot, F., Poupat, C., Sánchez, V., and Potier, P. (1980). Alcaloïdes stéroïdiques de *Didymeles* cf. *madagascariensis*. *Planta Medica* **39,** 204.

Baillon, H. E. (1873). *Histoire des plantes. Ochnacées, Rutacées* **4,** 357–520, i-iii. L. Hachette & Cie, Paris.

Baillon, H. E. (1876). *Histoire des plantes. Castanéacées–Rhizophoracées* **6,** 217–304. L. Hachette & Cie, Paris.

Barabé, D., Bergeron, Y., and Vincent, G. A. (1987). La répartition des caractères dans la classification des Hamamelididae (Angiospermae). *Canadian Journal of Botany* **65,** 1756–67.

Bentham, G. (1880). Ordo CLV. Leitneriaceae. In *Genera Plantarum ad Exemplaria Imprimis in Herbariis Kewensibus Servata Definita,* (eds G. Bentham and J. D. Hooker), **3**(1), 396–7. Lovell Reeve & Co., London.

Carlquist, S. (1982). Wood anatomy of Buxaceae: Correlations with ecology and phylogeny. *Flora* **172,** 463–91.

Cronquist, A. (1981). *An integrated system of classification of flowering plants,* 178–9. Columbia University Press, New York.

Dahlgren, R. M. T. (1983). General aspects of angiosperm evolution and macrosystematics. *Nordic Journal of Botany* **3,** 119–49.

De Candolle, A. L. P. P. (1873). Dicotyledonearum genera prodromo vetustiora, sed variis causis omissa, ordine alphabetico breviter memorata. In *Prodromus* (ed. A. P. de Candolle) **17,** 289–99. G. Masson, Paris.

Drake del Castillo, E. (1897). Histoire naturelle des plantes. Tome VI.- Atlas.- IV, 1re Partie. In Grandidier, A., *Histoire physique, naturelle et politique de Madagascar* **36(45),** tab. 308A. L'Imprimerie Nationale, Paris.

Du Petit-Thouars, L. M. A. A. (1804). *Plantes des îles de l'Afrique australe . . .*, 23, tab. 1. Auteur, Madame Huzard, Levrault, Schoell et Compagnie, Madame Devaux, Giguet et Compagnie, Paris.

Giannasi, D. E. (1986). Phytochemical aspects of phylogeny in Hamamelidae. *Annals of Missouri Botanical Garden* **73,** 417–37.

Hallier, H. (1912). L'origine et le système phylétique des Angiospermes exposés à l'aide de leur arbre généalogique. *Archives Néerlandaises des Sciences Exactes et Naturelles* Sér. 3 b **1**, 146–234.

Hardman, R. (1987). Recent developments in our knowledge of steroids. *Planta Medica* **53,** 233–8.

Köhler, E. (1980). Zur Pollenmorphologie und systematischen Stellung der Didymelaceae Leandri. *Feddes Repertorium* **91,** 581–91, tabs XLVII–XLVIII.

Leandri, J. (1937). Sur l'aire et la position systématique du genre malgache *Didymeles* Thouars. *Annales des Sciences Naturelles, Botanique* sér 10 **19,** 309–18.

Melchior, H. (1964). 4. Reihe Leitneriales. In *A. Engler's Syllabus der Pflanzenfamilien,* (ed. H. Melchior) ed. 12 **2,** 44–5. Gebrüder Borntraeger, Berlin.

Melikian, A. P. (1973). Seed-coat types of Hamamelidaceae and allied families in relation to their systematics. *Botanicheskiĭ Zhurnal SSSR* **58,** 350–9.

Muller, J. (1981). Fossil pollen records of extant angiosperms. *Botanical Review* **47,** 1–142.

Saint-Hilaire, J. H. H. (1805). *Exposition des Familles Naturelles et de la germination des plantes* **2,** 388. Paris.

Stebbins, G. L. (1974). *Flowering plants: Evolution above the species level,* pp. xviii, 399. Belknap Press of Harvard University Press, Cambridge, Massachusetts.

Takhtajan, A. L. (1966). *Systema et phylogenia Magnoliophytorum,* pp. 610. Nauka, Moscow.

Thorne, R. F. (1983). Proposed new realignments in the angiosperms. *Nordic Journal of Botany* **3,** 85–117.

Van Tieghem, P. and Lecomte, H. (1886). Structure et affinités du *Leitneria. Bulletin, Société Botanique de France* **33,** 181–4.

Webster, G. L. (1987). The saga of the spurges: a review of classification and relationships in the Euphorbiales. *Botanical Journal of the Linnean Society* **94,** 3–46.

Willdenow, C. L. (1806). *Caroli a Linné Species Plantarum . . .,* **4(2),** 648. G. C. Nauk, Berolini.

Wolfe, J. (1974). Fossil forms of Amentiferae. *Brittonia* **25,** 334–55.

16. The Daphniphyllales: a systematic review

DAVID A. SUTTON

Department of Botany, British Museum (Natural History), London SW7 5BD, UK

Abstract

The Daphniphyllales comprises the monogeneric family Daphniphyllaceae, with less than 40 species, all restricted to eastern Asia. Species of *Daphniphyllum* are evergreen, more or less dioecious trees of shrubs, with inconspicuous, apetalous flowers. Their chemistry is unusual for the Hamamelidae as they have unique alkaloids of the daphniphylline group, iridoid compounds, clusters of calcium oxalate crystals in parenchymatous tissues, and some accumulate aluminium. In fruit *Daphniphyllum* is also exceptional, the drupe containing a single seed with copious oily, proteinaceous endosperm and a minute embryo. In spite of its minor economic importance and few species, *Daphniphyllum* has been relatively well studied, with three monographs within the space of a century. The genus has had a varied history and was for long considered referable to the Euphorbiales. The fate of the Daphniphyllaceae has often followed that of the Buxaceae, both being transferred to the Pittosporales, though the two families have quite different sieve-tube plastids. A consensus of current taxonomic opinion favours a position in the Hamamelidae. This chapter reviews the controversial placement of *Daphniphyllum* with an examination of critical characters and a summary of the main literature.

Taxonomy

Daphniphyllales (Hurasawa 1954).

Evolution, Systemtics, and Fossil History of the Hamamelidae Volume 1: Introduction and 'Lower' Hamamelidae. (eds P. R. Crane and S. Blackmore), Systematics Association Special Volume No. 40A, pp. 285–91. Clarendon Press, Oxford, 1989.

Daphniphyllaceae (Mueller 1869)

Evergreen trees or shrubs, more or less dioecious; wood with small, solitary vessels, the scalariform perforation plates with 20–30 cross-bars, the imperforate tracheary elements with bordered pits, the wood-rays heterogeneous, 1–2 cells wide, the axial parenchyma diffuse. *Leaves* alternate or rarely opposite, sometimes fasciculate and apparently verticillate, simple, petiolate, exstipulate; lamina elliptic, oblong, obovate or ovate, entire, somewhat coriaceous, venation pinnate, stomata paracytic. *Inflorescence* axillary or rarely subterminal, racemose. *Flowers* inconspicuous, regular, hypogynous, apetalous, pedicellate, bracteate; calyx absent or with 2–6 lobes, the lobes almost united to free, imbricate. *Staminate flowers* rarely with androgonophore; stamens 5–14, the filaments free or apically coherent, the anthers basifixed, tetrasporangiate, dithecal, latrorse, opening by 2 longitudinal slits; pollen 3-(4-)colpate, colporoidate or rarely colporate, spherical to oblate, somewhat psilate to verrucate, microperforate, tectate–columellate with thin footlayer, the endexine thin except around apertures; staminodes frequent; pistillodes rarely present. *Pistillate flowers* often with staminodes; gynoecium of 2(–4) carpels more or less united to form 2-(4-)locular, imperfectly septate ovary; style 1-2–(4-) lobed, the lobes almost free, divaricate, recurved, the stigmatic areas dry, papillate; ovules (1–)2 per loculus, more or less apical, pendulous, anatropous with ventral raphe, bitegmic, crassinucellate, the endosperm development cellular. *Fruit* an indehiscent, fleshy, usually ovoid drupe, often with persistent style and staminodes; seeds 1(–2); endosperm copious, oily proteinaceous; embryo minute, straight with 2 cotyledons; germination epigeous. $2n=32$, $x=8$.

Type genus: *Daphniphyllum* Blume *Bijdr. Fl. Nederl.*, **17,** 1152 (1826). Type species: *D. glaucescens* Blume *Bijdr. Fl. Nederl.*, **17,** 1153 (1826). Type: INDONESIA: Java; Mt Salak, (vi) *Blume 1908* (holo. L, iso. L NY, para. L US).

Introduction

A small, monogeneric family from eastern Asia and Malaysia, ranging from New Guinea to South Korea and Japan in the east, and from Sri Lanka to Himalayan India in the west. Estimates of the number of species vary from nine to about forty. In spite of minor economic importance and few species, *Daphniphyllum* has received considerable attention. Mueller (1869) prepared the first monographic study and recognized 13 species. Rosenthal (1916) recognized 24 species in her treatment and contributed accounts to Engler's *Das Pflanzenreich* and *Die Natürlichen Pflanzenfamilien* (Rosenthal 1919, 1931). The most recent monographic study was by Huang (1965, 1966), who adopted a broad species concept and recognized only 9 species but listed over 50 further specific epithets in synonymy. There have been recent contributions to the Chinese Daphniphyllaceae, mostly using narrower species concepts (Cheng and Ming 1980, T.-C. Huang 1977, C. C. Huang 1980, Wang 1981). There is no apparent fossil record for the family (Tiffney 1986).

Ordinal placement

The taxonomic position of *Daphniphyllum* has been widely disputed. Though Blume (1826) and some contemporaries (Meisner 1836, Endlicher 1840) tentatively attributed the genus to the Rhamnaceae, many subsequent authors placed *Daphniphyllum* either with genera of the subfamily Phyllanthoideae of the Euphorbiaceae (Baillon 1858; Bentham 1880) or as a separate family (Mueller 1866, 1869; Scholz 1964; Stebbins 1974) of the Euphorbiales (or Tricoccae). The main objections to this placement (cf. Webster 1987) are the vessels with scalariform end plates, indehiscent, drupaceous fruit, absence of an obturator or caruncle, but especially the tiny embryo in copious, cellular (Rao 1970) endosperm.

Prior to a revision of the Chinese and Japanese species of *Daphniphyllum* (Croizat and Metcalf 1941), Croizat (1941) gave a detailed reappraisal of the phylogenetic position of the genus and concluded that it was related to the Eucommiaceae and Cercidiphyllaceae, both regarded as pittosporalean. Thorne (1983) also considered that *Daphniphyllum* belonged to the Pittosporales (superorder Rosiflorae), but in a suborder Buxineae with the Balanopaceae, Buxaceae, and Didymelaceae. The fate of the Daphniphyllaceae has often followed that of the Buxaceae (cf. Webster 1987, p. 10) though Behnke (1982) showed that the two families have different sieve-tube plastids.

Hallier (1904) proposed a radically different treatment from contemporary authors and placed *Daphniphyllum* near the Hamamelidaceae. Support is found in the primitive wood structure (Carlquist 1982; Metcalfe and Chalk 1983) with similarities to the Hamamelidaceae and especially the Balanopaceae. Pollen morphology (Bhatnagar and Garg 1977) and embryology (Mohana Rao 1974) indicate an affinity with Hamamelidaceae but not with Euphorbiaceae or Pittosporaceae, though *Daphniphyllum* parallels some Euphorbiaceae in the precocious degeneration of antipodal cells (Sato 1972). Hurasawa (1954), working primarily with Japanese species (Hurasawa 1942*a*,*b*), considered that the family should be placed in a separate order allied to the Euphorbiales. Cronquist (1981, 1983) agreed on the ordinal rank, but as an isolated element of the Hamamelidae comparable to the Didymelales, Eucommiales and Leitneriales. Supporting evidence was given as the primitive xylem anatomy, reduced unisexual flowers with latrorse stamens, almost free styles with stigmatic inner surfaces, and the tiny embryo with abundant endosperm.

Hamamelidae are characterized by a wide range of phenolic compounds with tannins, proanthocyanidins, ellagic acid, and myricetin common to most taxa (Gianassi 1986) but the Daphniphyllaceae differs in lacking ellagic acid, myricetin, and probably proanthocyanidins. Flavones are common to *Daphniphyllum* (apigenin and luteolin) and *Leitneria*. The iridoids asperuloside and daphniphylloside have been found in *Daphniphyllum*. While unusual for Hamamelidae, iridoids occur in Hamamelidaceae (*Liquidamabar*), and Eucommiaceae. The Daphniphyllaceae also yield unique alkaloids (cf. Rizk 1987) including daphnicadine, daphnicaline, daphnicamine, daphnigraciline, daphnimacine, daphniphylline, neodaphniphylline, and yuzurimine. These are diterpene alkaloids, differing from the isoquinoline alkaloids of the Magnoliidae.

Phenetic analyses by Barabé *et al.* (1982, 1987) showed that the Daphniphyllaceae were distant from the main families of the Hamamelidae, but allied to 'fringe' elements including the Barbeyaceae, Eucommiaceae, Leitneriaceae, and particularly the Balanopaceae. In a cladistic analysis, Barabé (1984) grouped the Daphniphyllaceae and Buxaceae on the basis of venation, exstipulate leaves, and pollen structure, though the states recorded for the latter character are questionable. Zavada and Dilcher (1986) used cluster analysis on pollen characters of the Hamamelidae and found that *Daphniphyllum* pollen formed a group with that of *Leitneria*, *Fagus*, and possibly *Barbeya*.

Summary

There is significant evidence to refute the placement of the Daphniphyllaceae in the Euphorbiales. Though showing general hamamelid affinity, the family differs from more typical elements in several important respects and deserves similar status to the Balanopales and Leitneriales, with the former perhaps more closely related. In spite of some gross differences, there are many similarities in detail in the placement of the Daphniphyllaceae in treatments by Thorne, Cronquist, and in the phenetic or cladistic analyses. The Daphniphyllales apparently represent one of the products of an early radiation of the dicotyledons involving elements of the hamamelids and rosids. A critical reappraisal of this larger group is required for more detailed assessment of the position of the Daphniphyllales.

References

Baillon, H. E. (1858). *Étude générale du groupe des Euphorbiacées.* Victor Masson, Paris.

Barabé, D. (1984). Application du cladisme à la systématique des Angiospermes: cas des Hamamélidales. *Candollea* **39,** 51–70.

Barabé, D., Bergeron, Y. and Vincent, G. A. (1982). La position des Daphniphyllaceae, Buxaceae, Simmondsiaceae et Cecropiaceae dans la sous-classe des Hamamelididae. Étude numérique. *Comptes Rendus Hebdomadaires des Séances de l'Académie des Sciences. Paris,* série III **294,** 891–6.

Barabé, D., Bergeron, Y. and Vincent, G. A. (1987). La répartition des caractères dans la classification des Hamamelididae (Angiospermae). *Canadian Journal of Botany* **65,** 1756–67.

Behnke, H.-D. (1982). Sieve-element plastids, exine sculpturing and the systematics of the Buxaceae. *Plant Systematics and Evolution* **139,** 257–66.

Bentham, G. (1880). Ordo CLI. Euphorbiaceae. In Bentham, G. and Hooker, J. D. (eds.), *Genera Plantarum . . .*,**3,** 239–340. Lovell Reeve & Co., London.

Bhatnagar, A. K. and Garg, M. (1977). Affinities of *Daphniphyllum*—Palynological approach. *Phytomorphology* **27,** 92–7.

Blume, C. L. (1826). *Bijdragen tot de flora van Nederlandisch Indië,* **17,** pp. 1152–3. Lands Drukkerij, Batavia.

Carlquist, S. (1982). Wood anatomy of Daphniphyllaceae: Ecological and phylogenetic considerations, review of Pittosporalean families. *Brittonia* **34,** 252–66.

Cheng, M. and Ming, T.-L. (1980). Daphniphyllaceae. In *Flora Reipublicae Popularis Sinicae,* (eds M. Cheng and T.-L. Ming) **45**(1), pp. 1–11. Science Press, Beijing.

Croizat, L. (1941). On the systematic position of *Daphniphyllum* and its allies. *Lingnan Science Journal,* **20,** 79–103.

Croizat, L. and Metcalf, F. P. (1941). The Chinese and Japanese species of *Daphniphyllum. Lingnan Science Journal,* **20,** 104–27, plates 1–3.

Cronquist, A. (1981). *An integrated system of classification of flowering plants,* pp. 178–9. Columbia University Press, New York.

Cronquist, A. (1983). Some realignments in the dicotyledons. *Nordic Journal of Botany* **3,** 75–83.

Endlicher, S. L. (1840). *Genera plantarum secundum ordines naturales disposita,* part **14,** 1104. Fr. Beck, Wien.

Giannasi, D. E. (1986). Phytochemical aspects of phylogeny in Hamamelidae. *Annals of Missouri Botanical Garden* **73,** 417–37.

Hallier, H. (1904). Ueber die Gattung *Daphniphyllum,* ein Uebergangsglied von den Magnoliaceen und Hamamelidaceen zu den Kätzchenblüthlern. *Botanical Magazine, Tokyo* **18,** 55–69.

Huang, C.-C. (1980). *Daphniphyllum* Bl. In *Flora Reipublicae Popularis Sinicae* (eds M. Cheng and T.-L. Ming) **45**(1), p. 139. Science Press, Beijing.

Huang, T.-C. (1965). Monograph of *Daphniphyllum* (I). *Taiwana* **11,** 57–98.

Huang, T.-C. (1966). Monograph of *Daphniphyllum* (II). *Taiwania* **12,** 137–234.

Huang, T.-C. (1977). Daphniphyllaceae. In *Flora of Taiwan,* (eds H.L. Li *et al.*) **3,** pp. 501–5. Epoch Publishing Co., Taipei, Taiwan.

Hurusawa, I. (1942*a*). Notes on the Japanese species of *Daphniphyllum* (I). *Journal of Japanese Botany* **18,** 155–63.

Hurusawa, I. (1942*b*). Notes on the Japanese species of *Daphniphyllum* (II). *Journal of Japanese Botany* **18,** 261–76.

Hurusawa, I. (1954). Eine nochmalige Dursicht des herkömmlichen Systems der Euphorbiaceen im weiteren Sinne. *Journal of the Faculty of Science, Tokyo University,* Sect. 3., Bot. **6,** 209–342, tabs I-IV.

Meisner, C. F. (1837). *Plantarum Vascularium Genera* **1(3),** 71. Weidmann, Lipsiae.

Metcalfe, C. R. and Chalk, L. (1983). *Anatomy of the dicotyledons,* (2nd edn), Vol. 2. Clarendon Press, Oxford.

Mohana Rao, P. R. (1974). Seed anatomy in some Hamamelidaceae and phylogeny. *Phytomorphology* **24,** 113–39.

Mueller, J. (1866). Ordo CLXXVII. Euphorbiaceae (1), In de Candolle, A. P. (ed.), *Prodromus,* **15**(2/2), 189–1286. Victoris Masson et fils, Paris.

Mueller, J. (1869). Ordo CLXXIX. Daphniphyllaceae, In *Prodromus,* (ed. A. P. de Candolle) **16**(1), 1–6. Victoris Masson et fils, Paris.

Rao, P. N. (1970). Euphorbiaceae. In *Proceedings of the Symposium on Comparative Embryology of Angiosperms. Bulletin of the Indian National Science Academy* **41,** 136–41.

Rizk, A-F. M. (1987). The chemical constituents and economic plants of the Euphorbiaceae. *Botanical Journal of the Linnean Society* **94,** 293–326.

Rosenthal, K. (1916). *Monographie der Gattung Daphniphyllum.* Inaugural Dissertation, pp. 1–32. Breslau.

Rosenthal, K. (1919). Daphniphyllaceae. In *Das Pflanzenreich,* (ed. H. G. A. Engler) IV. **147a** (Heft 68), pp. 1–15. Wilhelm Englemann, Leipzig.

Rosenthal, K. (1931). Daphniphyllaceae. In *Die Natürlichen Pflanzenfamilien,* (eds H. G. A. Engler and H. Harms) ed. 2 **19c,** pp. 233–5. Wilhelm Engelmann, Leipzig.

Sato, Y. (1972). Development of the embryo sac of *Daphniphyllum macropodum* var. *humile* (Maxim) Rosenth. *Science Reports of the Tôhoku University, Sendai,* Fourth Series, Biology **36,** 129–133.

Scholz, H. (1964). 26. Reihe Geraniales (Gruinales). In *A. Engler's Syllabus der Pflanzenfamilien,* (ed. H. Melchior) ed. 12 **2,** 246–262. Gebrüder Borntraeger, Berlin.

Stebbins, G. L. (1974). *Flowering plants: Evolution above the species level.* Belknap Press of Harvard University Press, Cambridge, Massachusetts.

Thorne, R. F. (1983). Proposed new realignments in the angiosperms. *Nordic Journal of Botany* **3,** 85–117.

Tiffney, B. H. (1986). Fruit and seed dispersal and the evolution of the Hamamelidae. *Annals of Missouri Botanical Garden* **73,** 394–416.

Wang, J.-X. (1981). A revision of Daphniphyllaceae in China. *Acta Phytotaxonomica Sinica* **19,** 75–84.

Webster, G. L. (1987). The saga of the spurges: a review of classification and relationships in the Euphorbiales. *Botanical Journal of the Linnean Society* **94,** 3–46.

Zavada, M. S. and Dilcher, D. L. (1986). Comparative pollen morphology and its relationship to phylogeny of pollen in the Hamamelidae. *Annals of Missouri Botanical Garden* **73,** 348–81.

INDEX

Systematics Association Publications

---

1. Bibliography of key works for the identification of the British fauna and flora *3rd edition* (1967)
   Edited by G. J. Kerrich, R. D. Meikle and N. Tebble
2. Function and taxonomic importance (1959)
   Edited by A. J. Cain
3. The species concept in palaeontology (1956)
   Edited by P. C. Sylvester-Bradley
4. Taxonomy and geography (1962)
   Edited by D. Nichols
5. Speciation in the sea (1963)
   Edited by J. P. Harding and N. Tebble
6. Phenetic and phylogenetic classification (1964)
   Edited by V. H. Heywood and J. McNeill
7. Aspects of Tethyan biogeography (1967)
   Edited by C. G. Adams and D. V. Ager
8. The soil ecosystem (1969)
   Edited by H. Sheals
9. Organisms and continents through time (1973)†
   Edited by N. F. Hughes

Published by the Association (out of print)

Systematics Association Special Volumes

1. The new systematics (1940)
   Edited by Julian Huxley (Reprinted 1971)
2. Chemotaxonomy and serotaxonomy (1968)*
   Edited by J. G. Hawkes
3. Data processing in biology and geology (1971)*
   Edited by J. L. Cutbill
   Out of print
4. Scanning electron microscopy (1971)*
   Edited by V. H. Heywood
   Out of print

5. Taxonomy and ecology (1973)*
   Edited by V. H. Heywood
6. The changing flora and fauna of Britain (1974)*
   Edited by D. L. Hawksworth
   Out of print
7. Biological identification with computers (1975)*
   Edited by R. J. Pankhurst
8. Lichenology: progress and problems (1976)*
   Edited by D. H. Brown, D. L. Hawksworth, and R. H. Bailey
9. Key works to the fauna and flora of the British Isles and northwestern Europe, *4th Edition* (1978)*
   Edited by G. J. Kerrich, D. L. Hawksworth, and R. W. Sims
10. Modern approaches to the taxonomy of red and brown algae (1978)*
    Edited by D. E. G. Irvine and J. H. Price
11. Biology and systematics of colonial organisms (1979)*
    Edited by G. Larwood and B. R. Rosen
12. The origin of major invertebrate groups (1979)*
    Edited by M. R. House
13. Advances in bryozoology (1979)*
    Edited by G. P. Larwood and M. B. Abbot
14. Bryophyte systematics (1979)*
    Edited by G. C. S. Clarke and J. G. Duckett
15. The terrestrial environment and the origin of land vertebrates (1980)*
    Edited by A. L. Panchen
16. Chemosystematics: principles and practice (1980)*
    Edited by F. A. Bisby, J. G. Vaughan, and C. A. Wright
17. The shore environment: methods and ecosystems (2 Volumes) (1980)*
    Edited by J. H. Price, D. E. G. Irvine and W. F. Farnham
18. The Ammonoidea (1981)*
    Edited by M. R. House and J. R. Senior
19. Biosystematics of social insects (1981)*
    Edited by P. E. Howse and J.-L. Clément
20. Genome evolution (1982)*
    Edited by G. A. Dover and R. B. Flavell
21. Problems of phylogenetic reconstruction (1982)*
    Edited by K. A. Joysey and A. E. Friday
22. Concepts in nematode systematics (1983)*
    Edited by A. R. Stone, H. M. Platt, and L. F. Khalil
23. Evolution, time and space: the emergence of the biosphere (1983)*
    Edited by R. W. Sims, J. H. Price, and P. E. S. Whalley
24. Protein polymorphism: adaptive and taxonomic significance (1983)*
    Edited by G. S. Oxford and D. Rollinson
25. Current concepts in plant taxonomy (1983)*
    Edited by V. H. Heywood and D. M. Moore
26. Databases in systematics (1984)*
    Edited by R. Allkin and F. A. Bisby
27. Systematics of the green algae (1984)*
    Edited by D. E. G. Irvine and D. M. John

28. The origins and relationships of lower invertebrates (1985)‡
Edited by S. Conway Morris, J. D. George, R. Gibson, and H. M. Platt
29. Infraspecific classification of wild and cultivated plants (1986)‡
Edited by B. T. Styles
30. Biomineralization in lower plants and animals (1986)‡
Edited by B. S. C. Leadbeater and R. Riding
31. Systematic and taxonomic approaches in palaeobotany (1986)‡
Edited by R. A. Spicer and B. A. Thomas
32. Coevolution and systematics (1986)‡
Edited by A. R. Stone and D. L. Hawksworth
33. Key works in the fauna and flora of the British Isles and north-western Europe, *5th edition* (1987)‡
Edited by R. W. Sims, P. Freeman, and D. L. Hawksworth
34. Extinction and survival in the fossil record (1987)‡
Edited by G. P. Larwood
35A. The phylogeny and classification of the tetrapods. Volume 1: Amphibians, reptiles, birds (1988)‡
Edited by M. J. Benton
35B. The phylogeny and classification of the tetrapods. Volume 2: Mammals (1988)‡
Edited by M. J. Benton
36. Prospects in systematics (1988)‡
Edited by D. L. Hawksworth
37. Biosystematics of haematophagous insects (1988)‡
Edited by M. W. Service
38. The chromophyte algae: problems and perspectives (1989)‡
Edited by J. C. Green, B. S. C. Leadbeater, and W. Diver
39. Electrophoretic studies of agricultural pests (1989)‡
Edited by H. D. Loxdale and J. den Hollander
40A. Evolution, systematics, and fossil history of the Hamamelidae. Volume 1: Introduction and 'lower' Hamamelidae (1989)‡
Edited by P. R. Crane and S. Blackmore
40B. Evolution, systematics, and fossil history of the Hamamelidae. Volume 2: 'Higher' Hamamelidae (1989)‡
Edited by P. R. Crane and S. Blackmore

*Published by Academic Press for the Systematics Association
†Published by the Palaeontological Association in conjunction with the Systematics Association
‡Published by the Oxford University Press for the Systematics Association